SAP S/4HANA® Sourcing and Procurement

Fabienne Bourdelle

SAP S/4HANA® Sourcing and Procurement Certification Guide

Application Associate Exam

Editor Will Iobst
Acquisitions Editor Emily Nicholls
Copyeditor Rachel Paul
Cover Design Graham Geary
Photo Credit Shutterstock.com/159353360/© Evgeny Karandaev
Layout Design Vera Brauner
Production Hannah Lane
Typesetting SatzPro, Krefeld (Germany)
Printed and bound in the United States of America, on paper from sustainable sources

ISBN 978-1-4932-1989-6

1st edition 2021

Library of Congress Cataloging-in-Publication Data
Names: Bourdelle, Fabienne, author.
Title: SAP S/4Hana sourcing and procurement certification guide : application associate exam / Fabienne Bourdelle.
Description: 1st edition. | Boston : Rheinwerk Publishing, [2020] | Includes index.
Identifiers: LCCN 2020043925 (print) | LCCN 2020043926 (ebook) | ISBN 9781493219896 (paperback) | ISBN 9781493219902 (ebook)
Subjects: LCSH: Industrial procurement--Examinations--Study guides. | Purchasing--Examinations--Study guides. | SAP HANA (Electronic resource)--Examinations--Study guides.
Classification: LCC HD39.5 .B68 2020 (print) | LCC HD39.5 (ebook) | DDC 658.7/202855376--dc23
LC record available at https://lccn.loc.gov/2020043925
LC ebook record available at https://lccn.loc.gov/2020043926

Contents at a Glance

Dear Reader,

Sometimes, it's all we can do to prepare.

You know the feeling—something coming up on the horizon, looming over our day-to-day lives: a cross-country move, a go-live at a client, or a tough book publication date.

Naturally, we focus on all the possibilities—What if our belongings get damaged en route? What if the go-live gets delayed? What if an editor gets swamped?—and look for someone who can help us prepare for anything. Someone who's done this before, with the experience to assuage our worries and the wisdom to set us on the path for success.

If you're prepping for an SAP S/4HANA Sourcing and Procurement certification, that *someone* is author Fabienne Bourdelle. Her expertise is distilled in these pages, ready to guide you through your exam review. Worry not, you're in great hands!

What did you think about *SAP S/4HANA Sourcing and Procurement Certification Guide*? Your comments and suggestions are the most useful tools to help us make our books the best they can be. Please feel free to contact me and share any praise or criticism you may have.

Thank you for purchasing a book from SAP PRESS!

Will Jobst
Editor, SAP PRESS

willj@rheinwerk-publishing.com
www.sap-press.com
Rheinwerk Publishing · Boston, MA

Contents

Preface

The SAP PRESS certification guides are designed to provide those preparing for an SAP-certified exam with the overall picture, insight, and practice necessary to pass the exam. Written in a practical, easy-to-understand language, these guides provide targeted content that focuses on what you need to know to successfully take your exam.

Target Audience

This book is specially written for those preparing for the SAP Certified Application Associate – SAP S/4HANA Sourcing and Procurement exam. It focuses on the core SAP S/4HANA materials management capabilities: consumption-based planning, purchasing, inventory management and inventory valuation, invoice verification, special procurement processes, and more.

With this book you will gain a thorough understanding of the exam structure and what to expect when taking the exam. You will receive a refresher on the key concepts covered in the exam, and you will be able to test your skills using sample practice questions and answers. The book is closely aligned with the course content and exam structure, so that all information is relevant and applicable to what you need to know in preparation. We explain SAP products and functions using practical examples and in simple language, so that you can prepare for the exam and improve your skills in your daily work. The book is structured and designed to highlight what you really need to know.

Structure of the Book

Each chapter dealing with an exam topic is organized in a similar way so that you can familiarize yourself with the structure and easily find the information you need. Here is an example of a typical structure:

- **List of the techniques you must master**
  Each chapter begins with a clear list of learning objectives.
- **Introduction to the topic**
  This section gives you an overview of the present subject and contains objectives for the exam topic covered.

- **Real-world scenario**
  This part describes situations where these skills would be beneficial for you or your company.
- **Objectives**
  This section is intended to explain the purpose of the current topic.
- **Key concept refresher**
  This section covers the most important concepts of the chapter. It reviews or remembers the tasks you need to understand or perform correctly in order to answer the questions of the certification exam.
- **Important terminology**
  After having refreshed the content, we offer a section for reviewing important terminology.
- **Practice questions**
  The chapter contains several practical questions related to the topic of the chapter. The questions are structured in a similar way to the actual questions for the certification exam.
- **Practice question answers and explanations**
  After the questions you will find their solutions. As part of the answer we discuss why an answer is considered correct or incorrect.
- **Takeaway**
  This section provides an overall understanding of the areas that you should now understand.
- **Summary**
  Finally, we conclude with a summary of the chapter.

Now that you have an idea of how the book is structured, the following overview will introduce the individual topics of each chapter:

- **Chapter 1: SAP S/4HANA Essentials**
  The focus of this chapter is on the architecture of SAP HANA, the scope for and deployment options of SAP S/4HANA, and the basic functions of the SAP Fiori user Interface.
- **Chapter 2: Procurement in SAP S/4HANA**
  This chapter reviews the procurement processes in SAP S/4HANA, the organizational structure, and the main master data for procurement.
- **Chapter 3: Procurement of Stock Material**
  This chapter reviews the key SAP S/4HANA features related to stock material procurement.
- **Chapter 4: Procurement for Direct Consumption**
  This chapter explains the key SAP S/4HANA features related to procurement for direct consumption.

- **Chapter 5: Sources of Supply**
  This chapter explains the possible sources of supply in SAP S/4HANA-run procurement processes.
- **Chapter 6: Purchasing Optimization**
  This chapter shows how steps in the procurement process can be optimized by, for example, source determination.
- **Chapter 7: Special Functions in Purchasing**
  This chapter deals with special purchasing functions such as purchase document release and output and confirmation management.
- **Chapter 8: Consumption-Based Planning**
  This chapter explains the materials planning process, the MRP data in the material master record, and the most important differences between consumption-based and demand-driven planning. The manual reorder point planning procedure and the source determination are covered in detail in this chapter.
- **Chapter 9: Inventory Management**
  This chapter reviews the functions of inventory management and the stages of physical inventory.
- **Chapter 10: Valuation and Account Determination**
  This chapter reviews the logic of automatic account determination and gives an overview of split valuation.
- **Chapter 11: Invoice Verification**
  This chapter deals with invoice and credit memo entry as well as variance handling, invoice release, and invoice reduction.
- **Chapter 12: Configuration Cross Topics**
  This chapter gives a brief overview of selected general settings and introduces to the control options provided for purchasing and inventory management in SAP S/4HANA.

Practice Questions

We would like to give you some background information on the test questions before you encounter the first questions in the chapters. Just like in the exam, each question has a basic structure:

- **The actual question**
  Read the question carefully and pay attention to all the words used in the question, as they can have an impact on the answer.
- **The question hint**
  This is not a formal term, but we call it a hint because it tells you how many answers are correct when more than one is correct, as in the actual certification exam.

- **The answer**
  The answers you can select depend on the question type. The following question types are possible:
  - *Multiple answer*: There is more than one correct answer possible.
  - *Multiple choice*: Only one correct answer is possible
  - *True/false*: Only one answer is correct. This type of question is not used in the exam but is used in the book to test your understanding.

Summary

This certification guide will teach you how to approach the content and key concepts highlighted for each exam topic. You will also have the chance practice with sample questions in each chapter. After you answer the practice questions, you can review the answers, which explain why they are correct or incorrect. The practice questions give you an insight into the type of questions you can expect, how the questions look like and how the answers relate to the questions. It is just as important to understand the structure of the questions and see how the questions and answers interact as it is to understand the content. This book gives you the tools and understanding you need to be successful. With these skills, you are on the path to becoming an SAP Certified Application Associate in SAP S/4HANA Sourcing and Procurement.

Introduction: The Path to Certification

Techniques You'll Master

- Understand the different certification offerings for sourcing and procurement in SAP S/4HANA.
- Find the courses required for the certification.
- Learn techniques for taking the certification exams.
- Identify further relevant SAP Training and Adoption offerings in SAP S/4HANA.
- Explore additional resources for sourcing and procurement in SAP S/4HANA.
- Expand your knowledge and keep your skills up to date.

This chapter presents the overall picture of the examination and certification as an SAP S/4HANA Sourcing and Procurement application associate. We provide information about the different levels and offerings for SAP S/4HANA Sourcing and Procurement certifications, highlight resources for study, and offer advice for taking certification exams.

With SAP S/4HANA, SAP has introduced a new type of enterprise resource planning solution to help customers become a digital enterprise. SAP S/4HANA is the most important product in the SAP portfolio, with sourcing and procurement still at the heart of the processes. As more and more customers implement SAP S/4HANA, the number of projects worldwide is also growing rapidly, resulting in the need for more implementation and support consultants who understand core procurement functions.

In this chapter, we will discuss the details of the SAP S/4HANA Sourcing and Procurement application associate certification and the SAP training materials required to obtain the certification. We will discuss how to access the SAP Training and Adoption materials and then go through some tips for preparing and successfully passing the certification exam. Finally, we will look at some additional resources that could be useful in preparing for the exam and for your continued journey with SAP S/4HANA.

Who This Book Is For

There is almost no SAP logistics implementation project in the world that does not require project staff who are familiar with the procurement and sourcing processes and configuration. This was the case for SAP ERP with materials management and applies to SAP S/4HANA. This makes sourcing and procurement consultants with good skills and solid experience a valuable and demanded resource for both SAP partner companies and SAP customers.

The book covers a broad and deep scope of configuration and business processes for sourcing and procurement in the SAP S/4HANA system. Therefore, this book is an excellent starting point for readers who are just starting with sourcing and procurement. For example, you might be a new employee at an SAP implementation partner, an internal support consultant whose company is upgrading to SAP S/4HANA, or a recent graduate looking to start a career in the SAP environment.

The book is also a good opportunity for experienced consultants to revisit the basics. Often, as experience increases, one specializes in a part of materials management. This book will help you to revisit the overall picture of materials management and at the same time familiarize yourself with the special features of SAP S/4HANA.

Those who work as users in SAP materials management often want to go beyond this role to gain a deeper and broader understanding of the implementation or to work as key users.

Developers who want to focus on sourcing and procurement or gain a good understanding of common business processes configured with SAP S/4HANA will also find this book useful, even if they are not interested in the certification aspect.

The book is designed around the latest SAP S/4HANA 1909 release and certification. However, it covers core processes that are now quite mature, so the knowledge found here should be helpful for both younger and future releases of the product, as well as for the corresponding certifications.

Certifications for Which This Book Is Designed

This book can be used to study for the following tests:

- **SAP Certified Application Associate—SAP S/4HANA Sourcing and Procurement (C_TS452_1909)**
  This book is primarily intended as preparation for the latest certification C_TS452_1909 SAP Certified Application Associate—SAP S/4HANA Sourcing and Procurement. It tests your process and configuration knowledge in the area of materials management and is based on the latest SAP S/4HANA release 1909 and the corresponding SAP Training and Adoption courses.
- **SAP Certified Application Associate—SAP S/4HANA Sourcing and Procurement (w/o Inventory Management) (C_TS451_1809)**
  Since SAP offers two certification versions in parallel, you can also opt for certification C_TS451_1809, which is based on SAP S/4HANA release 1809 and the corresponding courses. Note that inventory management is not part of this certification. This book, which focuses on the core processes and configuration topics, is also perfectly suited for preparing for this earlier test.
- **SAP Certified Application Associate—SAP S/4HANA Sourcing and Procurement—Upskilling for ERP Experts (C_TS450_1909)**
  In addition to the "standard" certification C_TS452_1909, SAP also offers a special exam for ERP materials management experts. This exam is based on the ERP courses and on the delta course "SAP S/4HANA Sourcing & Procurement—Functions and Innovations."

  As an ERP specialist, you can choose whether to take the standard or the upskilling exam. Both test your basic knowledge of materials management processes and their configuration. In SAP S/4HANA, these do not differ significantly from SAP ERP, and we discuss them in detail in this book. The innovations are integrated in the standard exam, while they are tested separately and more specifically in the upskilling exam.

  In this book, we discuss the innovations as much as possible together with the associated processes. Stand-alone topics are specifically discussed in Chapter 7.
- **SAP Certified Application Associate—SAP S/4HANA Sourcing and Procurement—Upskilling for ERP Experts (C_TS450_1809)**
  This test is the upskilling test based on SAP S/4HANA release 1809.

Note
You will find information on available certifications at *https://training.sap.com/certification/validity*.

Format of the Certification Exams

The mentioned certifications have the following format:

- Length of the exam: Up to three hours.
- Exam questions: Eighty questions of the following types:
  - Multiple choice, for which you must select one correct answer out of four available options
  - Multiple select type 1, for which you must select two correct answers out of four available options
  - Multiple select type 2, for which you must select three correct answers out of five available options

  There are no other question types than these. Note that you must get all answers correct for each question for the question to be considered correctly answered.

Note
In the practice questions of this book you will also find true or false questions. This type of questions is for practice but are not part of the certification exam.

  To pass the C_TS452_1909 certification, for example, you must answer 65% of the questions correctly. This means that you must answer 52 out of 80 questions completely correctly. The percentage for the pass rate varies depending on the certification and may change in future versions of an exam, as it depends on the rated difficulty of the question items in the certification. You can find the cut score and further details on each certification at *https://training.sap.com/certification/validity*, as shown in Figure 1. Choose the desired certification (e.g., C_TS452_1909, as in Figure 2) to obtain all relevant information about the exam, such as cut score, possible exam locations, which SAP courses you need for preparation, and so forth.

- Exam location: Can be taken either at a certification center or online through the SAP Certification Hub. Note that the SAP Certification Hub is a cloud offering on a subscription basis. You subscribe annually and can make six certification attempts during this period. That means if you always pass on the first try, you can potentially earn six different certifications. For each individual certification, you will get up to three attempts, which means you can fail a maximum of two times per subscription.

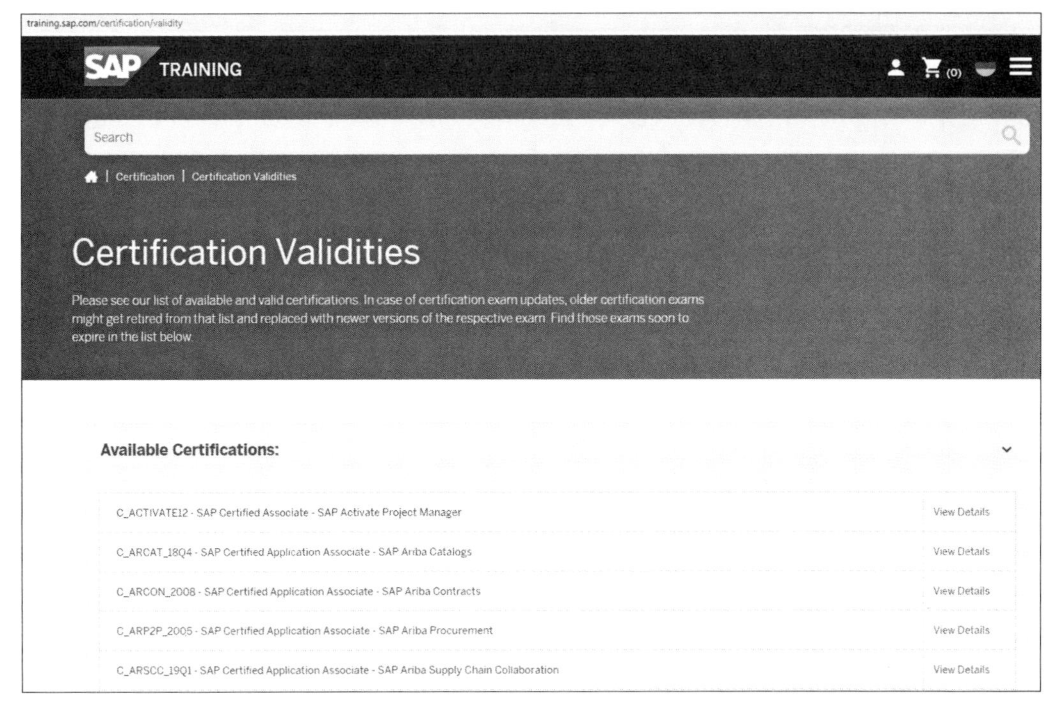

Figure 1 Access to the List of Available Certifications

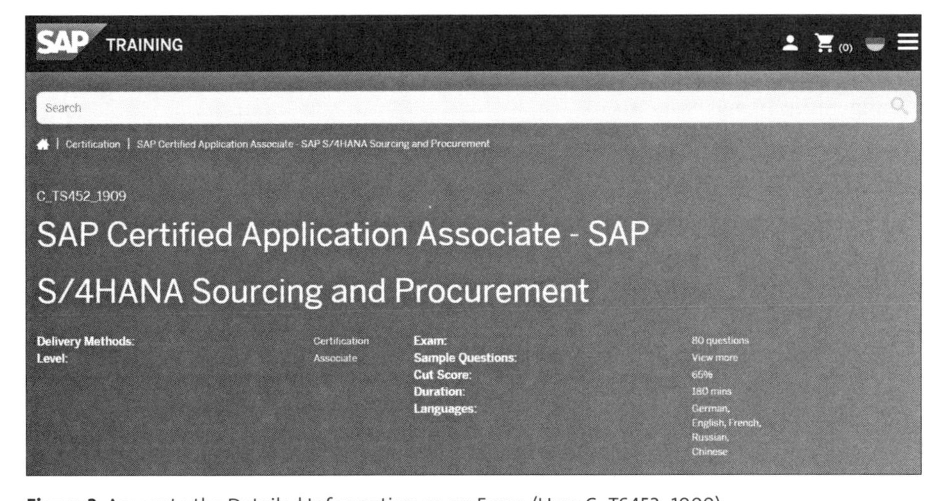

Figure 2 Access to the Detailed Information on an Exam (Here C_TS452_1909)

Scope of the Certification Exams

The book covers the same exact areas and scope as the exams and the academy courses on which the certifications are based. Table 1 lists the topic and weight distribution for the C_TS452_1909 exam, and Table 2 lists the corresponding topics and weight distribution for the C_TS452_1809 exam.

Topic Area	Topic Weight on Exam
Inventory management and physical inventory	8–12%
Valuation and account assignment	8–12%
Configuration of purchasing	8–12%
Configuration of master data and enterprise structure	8–12%
Invoice verification	8–12%
Sources of supply	8–12%
Document release procedure	<8%
Purchasing optimization	<8%
Source determination	<8%
Procurement analytics	<8%
Consumption-based planning	<8%
Enterprise structure and master data	<8%
Specific procurement processes	<8%
Basic procurement processes (including self-service procurement)	<8%
SAP S/4HANA user experience	<8%

Table 1 Topic Areas for C_TS452_1909 (Standard 1909)

Topic Area	Topic Weight on Exam
Configuration of purchasing	8–12%
Basic procurement processes (including self-service procurement)	8–12%
Enterprise structure and master data	8–12%
Consumption-based planning	8–12%
Valuation and account assignment	8–12%
Configuration of master data and enterprise structure	8–12%
Invoice verification	8–12%
Source determination	8–12%
Sources of supply	<8%
Document release procedure	<8%
Purchasing optimization	<8%
Procurement analytics	<8%
Specific procurement processes	<8%
SAP S/4HANA user experience	<8%

Table 2 Topic Areas for C_TS452_1809 (Standard 1809)

The main difference between the two "standard" certifications C_TS451_1809 and C_TS452_1909 is that inventory management, valuation, and account assignment topics were added with the 1909 version. Also, the number of questions for the following topics has decreased:

- Basic procurement processes
- Enterprise structure and master data
- Consumption-based planning
- Source determination

For upskilling certifications, in addition to the process and configuration topics covered in SAP ERP courses, there are specific questions about SAP S/4HANA innovations: SAP S/4HANA Essentials, Logistics Processes in SAP S/4HANA, and SAP S/4HANA Innovations in Sourcing and Procurement, as shown in Figure 3.

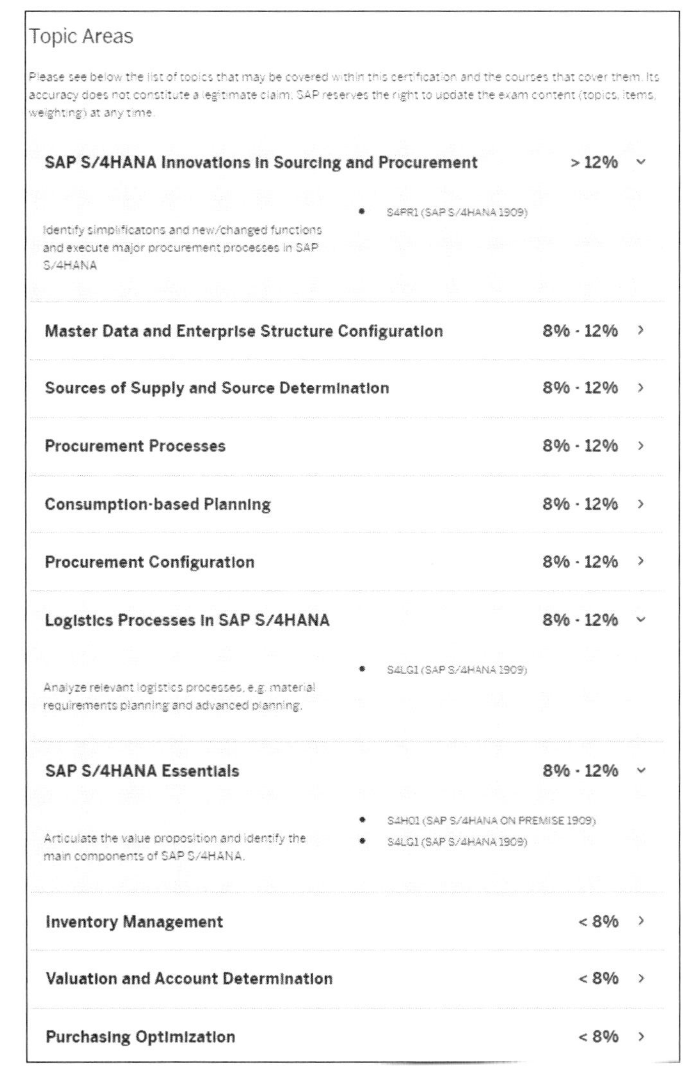

Figure 3 Topic Areas for C_TS451_1909 (Upskilling 1909)

SAP Training and Adoption Courses

You can select from different options to gain the knowledge required for the certification exams.

Standard Certification C_TS452_1909 and C_TS451_1809

The following options are available:

- **Individual courses**
  - S4500
  - S4510
  - S4515
  - S4520
  - S4525
  - S4550
- **Academy courses**
  - TS450
  - TS451 (for C_TS451_1809)
  - TS452 (for C_TS452_1909)

Upskilling Certification C_TS451_1809 and C_TS451_1909

The following options are available:

- **Individual courses**
  - S4H01
  - S4LG1
  - S4PR1
  - SCM500
  - SCM510
  - SCM520
  - SCM525
  - SCM550
- **Academy courses**
  - TSCM50
  - TSCM52

Note

The individual courses are also available as e-learning; just add an "e" to the end of the course names: S4520e, S4550e, and so on.

For all these courses, the latest materials as of the date of this writing are in collection 14 and refer to SAP S/4HANA 1909.

Note

Collections are course version numberings that generally map a course to a specific product release.

Always go for the newest collection of materials available. You can find the newest material referenced in the relevant learning journey. The scope of the content for these courses isn't expected to change dramatically, but each version of the course gets a little better, as is true for the product as well.

In order to stay up to date on which learning materials are available for sourcing and procurement in SAP S/4HANA, you should check the corresponding learning journey. Here you will see the offers categorized into sections. The learning journey is interactive—that is, if you click on the course link, you will get to the latest version of the respective course. If you have an active SAP Learning Hub subscription, you will be taken directly to the course materials (choose **eBook** or **eLearning**). Assessment questions are also available for each course. Finally, there is a link to SAP Live Access, which provides a PDF document with all the relevant exercises and details for booking a Live Access training system for practice.

Ways to Learn

You can choose to learn and consume the standard SAP Training and Adoption courses in different ways:

- Classroom training is the most obvious way to learn, especially if you are new to the subject. The inherited and undeniable advantages are an environment that provides immediate live feedback from an expert trainer, networking with other participants with whom you can share your goals, and a system at your disposal to perform exercises during the training. At an SAP training center, you will receive your printed version of the relevant course material (one for each individual course or each week of the academy).
- SAP Live Class is a virtual online training with video technology. It is like a classroom experience in that you still have a live trainer you can ask questions, system access to perform exercises, and learners you can interact with.

- SAP Learning Hub is a cloud-based offering with an annual subscription, including the following elements:
  - Access to learning content such as learning journeys, e-books, recorded demo simulations, videos, and so forth.
  - Access to the SAP Learning Rooms, which are expert-led social learning forums. Here you can ask questions, interact, and find materials and links. The SAP S/4HANA Enterprise Management Learning Room is the relevant learning room for sourcing and procurement topics.
  - Access to the SAP Live Access environment, which enables SAP Learning Hub subscribers to work with fully configured SAP software systems to carry out class exercises, cross-train, and experiment. When you use SAP Live Access, your system environment is specifically configured for the course you are studying. There's no system setup on your side—just launch and learn with a system designed specifically to accompany your training course content and exercises. You can purchase access in blocks (contingents) of 20 hours. Each block gives you up to 30 days to carry out exercises or explore the fully configured system.

Tips for Taking the Certification Exam

Now let's explore some tips that will help you better prepare for the exam. Studying the course material is the key to passing the certification, but we will also analyze the types of questions you will answer and finally list some tips and tricks for use before and during the exam.

Some General Considerations

The main objective of this book is to help you pass one of the certifications for SAP S/4HANA Sourcing and Procurement. This is not the same as preparing for a consulting project. Although the requirements for the exam and consulting overlap to a large extent in practice and SAP publishes materials with content and exercises that are relevant for consultants, you should remember that certification at the application consultant level tests knowledge of the specific materials on which it is based. In other words, although your own experience will contribute greatly to your success in certification and experience is probably the most important indicator of real-world skills, it is not enough for certification. For certification, knowledge and understanding of the materials and scope of the course are the most important.

In any case, we want to invite you to study the teaching materials. The exam is certainly not a memorization exercise, but it is based on the content of the books.

The Questions Types

As already mentioned, the exam consists of multiple-choice and multiple-selection questions. For all questions, the answers must be derived directly or easily from the content of the course book. In this section, we will look at an example from each category based on the sample questions on the certification exam page.

The multiple-choice question type is the most straightforward and is usually a very specific question with a single very specific answer. In the sample question in Figure 4, you are asked what is required to enter two types of confirmation for a purchase order item. To answer this, you need to know the meaning and possible configuration of a confirmation control key.

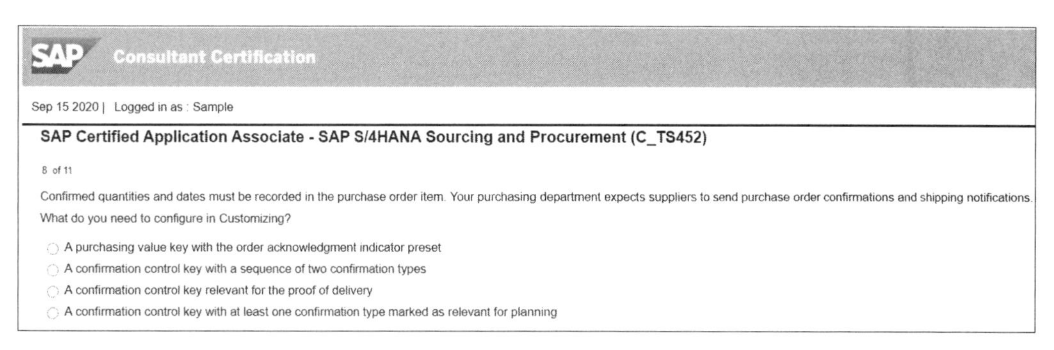

Figure 4 Multiple-Choice Example

SAP avoids adding scenario information to a question. Most of the time, every element of a question is important (in our example, entering quantities and dates, confirmation, and shipping notification expected), so read the question carefully and even twice to make sure you have it. You have enough time for this.

Figure 5 shows a sample multiple-choice question where you must select two of four options as correct. You must select both correct answers to have the answer marked as correct. SAP does not award points for half-correct answers. The example question here is correct if you select both proposed consequences of posting an invoice with reference to a purchase order: accounting documents are created, and the order history is updated.

SAP Consultant Certification

Sep 15 2020 | Logged in as : Sample

SAP Certified Application Associate - SAP S/4HANA Sourcing and Procurement (C_TS452)

1 of 11

What happens when you post a logistics invoice for a purchase order that you have received in the system?

Note: There are 2 correct answers to this question.

- The payment is posted against open liabilities.
- The accounting documents are created.
- The purchase order history is updated.
- Open liabilities are evaluated.

Figure 5 Multiple Selection Example 1

One level higher on the difficulty ladder is the multiple-selection question, with three out of five correct answers, as shown in Figure 6. An additional answer option adds another parameter to be considered and thus increases the probability that the question will be answered incorrectly. In addition, it usually takes a little longer to simply read through and assimilate everything. If you are sure of an answer, then it is correct—do not reconsider it. If you mark down the ones you are sure of, you are left with an easier question. You know that one answer is correct, so now you have a two-out-of-four question instead of a three-out-of-five question. When you know two answers for sure, you are left with a one-out-of-three question. It seems obvious, but eliminating options is the best way to get them out of your mind. The same applies to wrong-answer options. Composing wrong answers can be a challenge for exam designers, so take advantage of it: If something appears out of place, you should know that it probably is. If the direct knowledge approach does not work, continue with the process of elimination. Knowing what is wrong is as important as knowing what is right for certification.

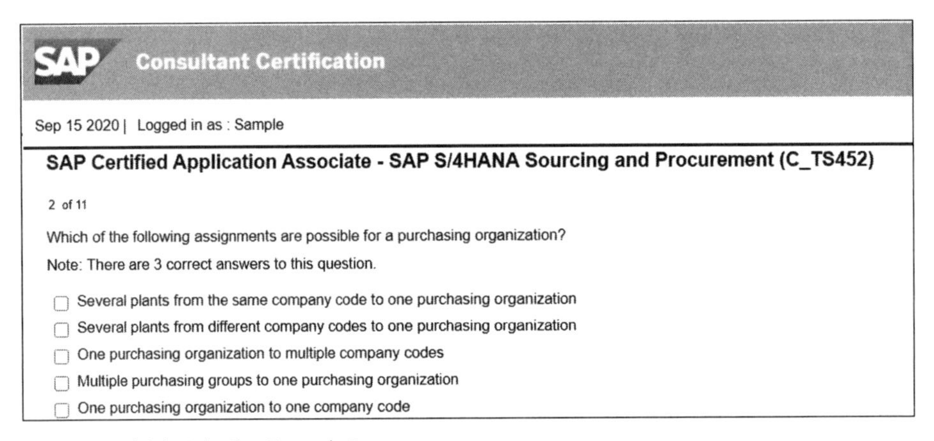

Figure 6 Multiple Selection Example 2

You have three hours to go through all the exam questions. This gives you more than two minutes per question. The questions are designed to be readable in an average of 30 seconds, giving you enough time to work through the entire exam. Stay relaxed and don't worry about the time. Focus on the question at hand, and if you are not sure of the answer, continue. The system keeps track of your unanswered questions. Always answer all questions as best you can before you submit the final exam. There is no penalty for incorrect answers, so not submitting an answer is a wasted opportunity.

SAP allows you to repeat the exam up to three times. Both the order of the questions and the order of the answers in a question are randomized. You will not get new questions if you repeat the exam, but the questions will be reshuffled.

Tips and Tricks

The following are some useful tips when preparing for the exam:

- Bulleted lists in course materials are a favorite source for creating questions, so be mindful of these.
- Use the exam questions in the course materials or in this book to evaluate your knowledge of a topic. If you don't understand something, reread the relevant section, check the correct answer details, and, if necessary, post a question in an online forum or community.
- There is a lot to learn and understand, so take your time.
- Do not spend too much time answering so-called certification questions outside of official SAP resources. These are not always written correctly, and you have no guarantee of the correctness of the answers. The SAP Learning and Adoption courses and this book contain many test questions.
- Work on the system as much as possible. In addition to the exercises offered with the SAP courses, you can reproduce the examples, processes, and configurations presented in this book. It sometimes helps enormously to understand the theory.

Here are some recommendations for during the exam:

- Answer all questions and bookmark the ones you are not sure about so you can work on them again.
- Do not worry too much about time. In most cases, you can go through the questions two or three times if you want.
- Read questions and answers very carefully.
- Eliminate the answer options that make no sense or are obviously wrong. Usually one of the wrong answers stands out as wrong, and the fewer options you have, the more chance you have of choosing the right one.

Additional Learning Resources

There is no doubt that it is important to acquire knowledge in order to pass a certification exam: a certification is valuable in the job market, and it confirms at least a certain level of knowledge. But especially in the IT area, innovation never stands still. The following resources are useful to consider not only when preparing for a certification but also in general, to learn more and keep your knowledge up to date:

- **Stay-current materials**
  Stay-current materials are created and provided by SAP Training and Adoption in direct cooperation with the development departments and are available on the SAP Learning Hub.

- **SAP Best Practices Explorer**
  The SAP S/4HANA system deployment and the SAP Activate Implementation Methodology are closely integrated with the SAP Best Practices. SAP Best Practices consist of preconfigured and documented business process scenarios you can review in the SAP Best Practices Explorer at *https://rapid.sap.com/bp*.
- **SAP Help Portal**
  You can use the SAP Help Portal at *https://help.sap.com* as a single point of entry to find help content, product documentation, simplification lists, learning journeys, and more.
- **SAP Fiori apps reference library**
  In the SAP Fiori app library, at *https://fioriappslibrary.hana.ondemand.com*, you can find key information for each SAP app, including all the technical data you need for installation and configuration.

Summary

This chapter described the SAP S/4HANA certification options available for sourcing and procurement in SAP S/4HANA. You should now know how to discern the exam structure, subject weighting, and minimum score required for each exam. You should also know about which SAP Training and Adoption courses you can review or attend for your certification exam, and which SAP programs and resources complement and enhance your knowledge and skills.

In the chapters that follow, we will dive directly into the exam topics, starting with an introduction and overview of SAP S/4HANA.

Chapter 1

SAP S/4HANA Essentials

Techniques You'll Master

- Explain the SAP strategy for digital transformation.
- Review basic SAP HANA technology.
- Choose the relevant options for SAP S/4HANA.
- Explain the SAP Fiori design.
- Use and configure the SAP Fiori launchpad.
- Describe the basic SAP Fiori application types.

In this chapter, we will examine the architecture of the SAP HANA database, describe the possible deployment of SAP S/4HANA, and discuss the basic functions of the SAP Fiori user interface.

Real-World Scenario

As a consultant, you must understand the underlying architecture on which the SAP S/4HANA system is built. SAP S/4HANA is the core of SAP's strategy, and you must feel comfortable explaining this technology to customers in simple terms. SAP HANA is a faster database, but there is some more information you should be able to share without scaring off non-technical staff.

The SAP S/4HANA system is a separate product line from the old SAP ERP system. SAP ERP is no longer the standard enterprise resource planning (ERP) platform from SAP. This was a huge change and a challenge for SAP. But it has proven to be successful, and more and more customers are being convinced of the advantages of the new system. But what is so special about it? How difficult is the implementation? How can customers switch to SAP S/4HANA? You must be able to understand and explain the reasons for this change by SAP and what it means for new and existing customers.

Finally, SAP Fiori is the new end-user interface in SAP S/4HANA, so you must be able to use and configure it with confidence, explain its components, and describe the benefits it offers.

Objectives of This Portion of the Test

The purpose of this part of the certification exam is to test your general knowledge of the SAP HANA, SAP S/4HANA, and SAP Fiori applications. The certification exam expects you to have a good understanding of the following topics:

- Overview of SAP HANA in-memory architecture
- Positioning of SAP S/4HANA in the SAP product portfolio
- SAP S/4HANA content
- Deployment options for SAP S/4HANA
- SAP Fiori launchpad features
- SAP Fiori applications types

Key Concept Refresher

The digital transformation drives innovation in the economy, and in this section, we will describe SAP's intelligent enterprise vision and strategy and show how SAP HANA and SAP S/4HANA are at the core of this strategy. We will go through the basic SAP HANA architecture and look at the technology that enables SAP HANA to be the innovation platform of the future for SAP.

We will also discuss the SAP S/4HANA solution, deployment options, and system landscape. Finally, we will look at the SAP Fiori user experience.

Intelligent Enterprise Framework

The effects of the digital transformation are visible in all corporate functions, such as marketing, production and supply chain, human resources, administration, and so on. It is driving companies to rethink and optimize their business processes so that they can benefit from increasing automation in their daily activities. SAP is strongly committed to supporting all the different aspects of the digital enterprise and has developed a strategic framework that encompasses business processes as well as technologies and platforms.

SAP's strategy to support the "intelligent enterprise" is based on three key product packages:

- **Intelligent suite**
  The intelligent suite consists of all SAP applications that map and support the end-to-end business processes of a company. SAP S/4HANA is part of it and represents the so-called digital core. In addition to the digital core, which is made of SAP S/4HANA or SAP S/4HANA Cloud, the intelligent suite contains specific applications for the following lines of business:
  - For customer experience, SAP offers SAP C/4HANA.
  - For manufacturing and supply chain, SAP Integrated Business Planning and SAP Digital Manufacturing are available.
  - For employee engagement, SAP proposes SAP SuccessFactors.
  - SAP Fieldglass, SAP Ariba, and SAP Concur are available for network and spend management.
- **Digital platform**
  The digital platform relates to the SAP Cloud Platform, an open platform-as-a-service for the creation, delivery, and integration of business applications and extensions to existing solutions. SAP's approach is to shift development focus to the SAP Cloud Platform to ensure maximum flexibility for both SAP and third-party developers while maintaining a streamlined and stable core.
- **Intelligent technologies**
  The third component of the intelligent enterprise is a technology framework that combines several modern technological tools that can be used to improve applications through automation. Intelligent technologies are embedded in core processes and applied to processes that integrate both SAP and third-party data and applications. Customers can use their data to identify patterns, predict

outcomes, and provide recommendations for action. These intelligent technologies include analytics, machine learning (ML), artificial intelligence (AI), the Internet of Things (IoT), and blockchains.

Previous classic ERP software solutions did not meet the requirements of fully supporting companies in this digital transformation. SAP saw this growing gap as an opportunity, and in 2011, with SAP HANA, it created the foundation for a business software that not only supports the digital transformation but also drives it forward. SAP HANA stands today at the center of the entire SAP product development. The next section will explore what SAP HANA is all about.

SAP HANA and Simplification of the Data Model

First, SAP HANA is a database management system. It performs all the usual database functions for storing and retrieving data for applications based on it, but SAP HANA also has specific features that make it unique for performing these tasks, which we will examine later in this section.

In addition, SAP HANA has embedded advanced analysis capabilities, such as predictive analytics, text analysis, mining, and search. SAP HANA also offers application development services that support a variety of programming languages.

Finally, SAP HANA provides a range of data access, management, and security support services. Obviously, SAP HANA is much more than just a fast database.

We could write a lot about SAP HANA, as many of its features are highly interesting. Here we will focus on the characteristics that make SAP HANA unique as a database from the perspective of the application. SAP HANA owes its originality to three characteristics:

- **In-memory database**
  An in-memory database means that all data from the source system is stored in random access memory (RAM), whereas in a traditional database system, all data is stored on the hard disk. The SAP HANA in-memory database does not waste time loading data from disk to RAM. Thus RAM is much faster than the hard disk, and although the supported data storage capacities are not as large, this limitation is rarely critical due to the other features of SAP HANA that we will introduce in the following discussion.
- **Compression**
  SAP HANA uses intelligent techniques to make compression a practical option while sacrificing minimum speed. You can assume that the data in an SAP HANA database is compressed by a factor of 10, on average. One of the ways that SAP HANA avoids sacrificing speed is to insert only new data instead of editing existing entries. Decompressing and recompressing when entries change is a costly process. By adding only entries and appending the old ones, SAP HANA does not go through the compression/decompression cycle. With versioning,

the system always reads the latest entry for a given record. This brings us to another important technical difference that allows for greater compression: columnar data storage.

- **Columnar data storage**
  The column memory allows efficient compression of the data. This makes it more cost-effective for the SAP HANA database to keep the data in RAM. In addition, search processes and calculations are faster. SAP HANA can process both online transaction processing (OLTP) and online analytical processing (OLAP) from a single data model, eliminating the need to move transaction data to a separate system. This means that transaction and analytical applications run on the same tables, the same copy of data. In addition, analytical applications have real-time data available at every level of detail. For data aggregation and parallel processing, column storage is highly efficient.

The performance of SAP HANA allows aggregation from each line item table in seconds, so that no ready-made aggregates are needed. SAP HANA can generate any view of the data from the same source tables at runtime.

Column storage means that indexes are not normally needed. They can still be created but usually offer only minor improvements. Therefore, in addition to eliminating aggregates and indexes from the database, it was also possible to eliminate huge amounts of application code dealing with aggregates and indexes. SAP HANA enables a simplified core data model and consequently simplified application code. It is now much easier to expand the applications and integrate additional functions.

SAP S/4HANA

SAP S/4HANA is the current intelligent ERP solution from SAP and only runs on the SAP HANA database. SAP S/4HANA stands for the SAP Business Suite 4 SAP HANA. Since the business suite only runs on SAP HANA, the database and business suite are packaged as one product. SAP S/4HANA can be deployed on-premise, in the cloud (public or private), or through a hybrid model that supports a variety of deployment scenarios.

Note

SAP naming conventions stipulate that the product is referred to as SAP S/4HANA in discussions about on-premise editions. The cloud edition is referred to as SAP S/4HANA Cloud.

This book and the previously mentioned certifications refer exclusively to on-premise SAP S/4HANA.

Moving to SAP S/4HANA

Customers have three major options when moving to SAP S/4HANA:

- **New implementation**
  This is a greenfield implementation of SAP S/4HANA. Customers with older, highly customized ERP solutions may prefer this option to clean up and redesign their processes.
- **System conversion**
  This is a complete conversion of an existing SAP ERP system to SAP S/4HANA. This scenario is technically based on Software Update Manager (SUM) with Database Migration Option (DMO), in case the customer is not yet on SAP HANA as the underlying database.
- **Landscape transformation**
  This is a consolidation of current regional SAP systems into one global SAP S/4HANA system or a split out of different parts of a system. This option is for customers who want to consolidate their landscape or separate selected units (e.g., a company code) or processes into a single SAP S/4HANA system.

SAP S/4HANA System Landscape

A new installation for SAP S/4HANA typically requires two basic production systems: the SAP S/4HANA backend server and an SAP Gateway server. The SAP S/4HANA backend server is a mandatory component. The SAP Gateway server is the place where the connections and settings to SAP Fiori are made. You can access both with the traditional SAP GUI. Figure 1.1 shows the SAP Logon window, with the two systems used in SAP Training and Adoption for most SAP S/4HANA Sourcing and Procurement courses: T41 as the backend system, and T4N as the gateway server.

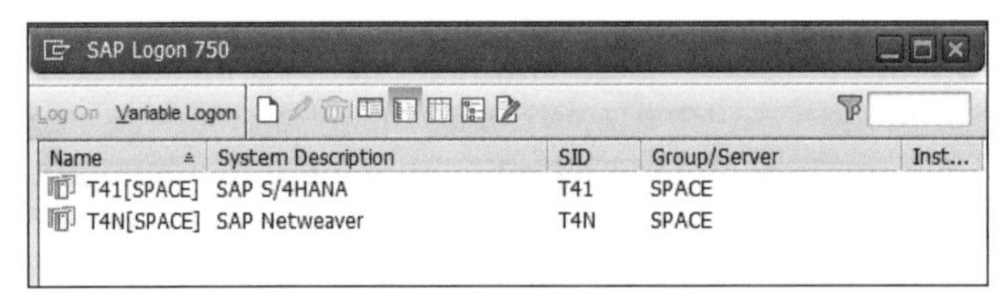

Figure 1.1 SAP Logon Screen

SAP Fiori

SAP Fiori is the common design paradigm for SAP. The SAP Fiori design language affects the appearance of everything in SAP, not just the interface of the new SAP S/4HANA system.

To create apps in SAP Fiori, use the SAPUI5 framework. This framework is in turn based on the open-source framework OpenUI5, with some additional SAP-specific tools. The most important development tools for working with SAPUI5 include HTML5, CCS3, jQuery, and JavaScript.

SAP Fiori Design Principles

The five design principles of SAP Fiori form the core of every SAP Fiori application to achieve these goals:

- **Role-based**
  A role-based user experience means that end users get all the information and functionality they need for their daily work—but nothing more. By comparison, the classic SAP user interface often provides a single complex transaction for many user roles.
  SAP Fiori breaks these large transactions down into several discrete applications tailored to user roles. All apps are interconnected so that all tasks of the transaction can still be performed. However, they are only executed when the user really wants to perform them. The SAP Fiori launchpad then serves as a central entry point for all the user's apps.
- **Responsive**
  The application interface is responsive and adapts to the size and device of the user accessing it.
- **Simple**
  The application is simple: one user, one use case, and up to three screens for each application.
- **Coherent**
  Applications are developed with a coherent structure. All applications speak the same language and can be implemented in multiple landscapes and environments.
- **Instant value**
  Instant value is gained through low barriers to adoption, both on the IT system side and on the user acceptance side.

SAP Fiori for SAP S/4HANA

SAP Fiori for SAP S/4HANA refers specifically to the applications available for the system. SAP development teams are constantly creating new native applications for SAP Fiori, replacing the traditional and usually more complex SAP GUI applications that SAP S/4HANA has inherited. The native applications can also use all the new technologies available with SAP S/4HANA, such as conversational UI, machine learning, and so on.

In addition to the native SAP Fiori applications, you can also launch SAP GUI applications via the SAP Fiori launchpad. These applications are launched in the web browser with a design theme in the SAP Fiori look. The scope of SAP ERP applications is huge, and since it is not necessary to replace everything at once, SAP has some latitude to create new useful and simpler native applications without compromising coverage of the application area.

SAP Fiori applications fit into one of the following three categories:

- **Transactional apps**
  Transactional apps offer task-based access to functions such as changing, creating, and displaying data (e.g., documents, master records) or to entire processes with guided navigation. Figure 1.2 shows an example of a transactional application: Manage Purchase Orders.

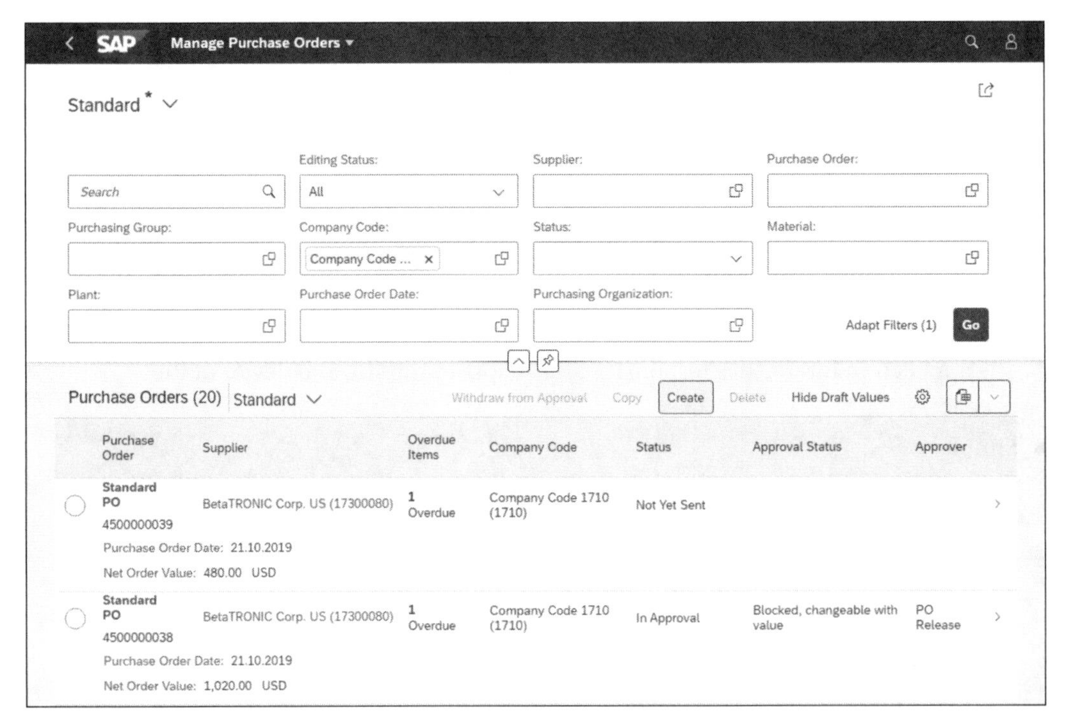

Figure 1.2 Example of a Transactional Application: the Manage Purchase Orders App

- **Analytical apps**
  Analytical apps provide insight into the actions. They are detailed reports used to drill down to finer details, for example, and they give you a visual overview of complex issues for monitoring or tracking. Figure 1.3 shows an example of an analytical app: Monitor Purchase Contract Items.
- **Factsheets**
  These apps give you the opportunity to search and explore your data. They provide a 360-degree view of essential information about an object (e.g., master data or documents) and contextual navigation between related objects. Figure 1.4 shows an example of this with the Business Partner Core Search app.

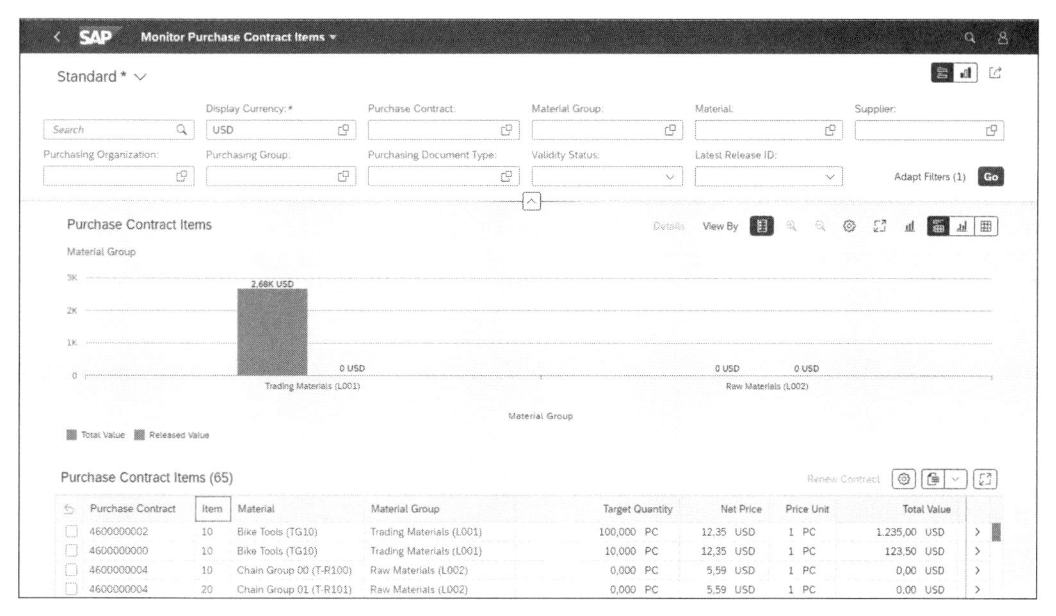

Figure 1.3 Example of an Analytical Application: the Monitor Purchase Contract Items

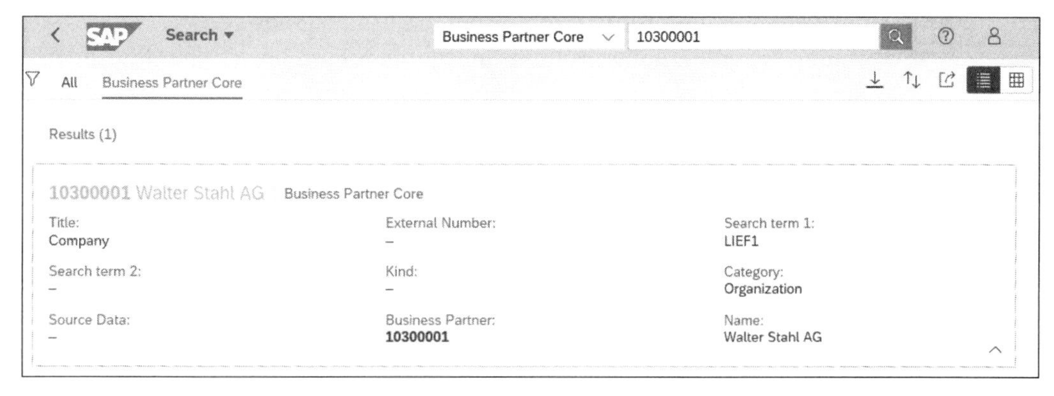

Figure 1.4 Example of a Factsheet Application: the Business Partner Core Search App

The SAP Fiori Launchpad

The SAP Fiori launchpad is the access point for users to the system (see Figure 1.5). It is a customizable web-based interface from which all applications that a user has assigned can be launched. The applications must not originate from a specific system. The tiles can launch applications and links for any system that is connected and compatible.

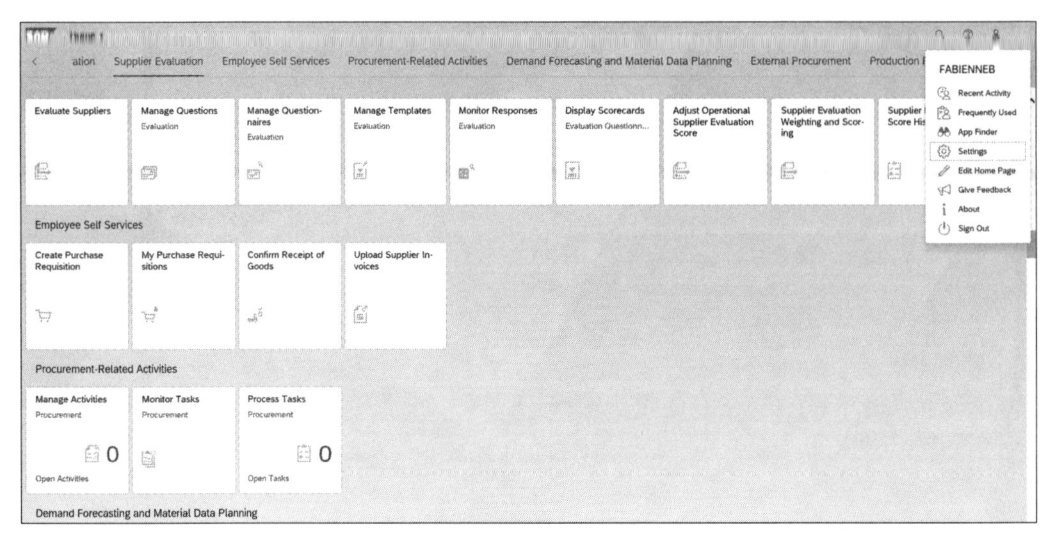

Figure 1.5 SAP Fiori Launchpad with the User Action Menu Open

The tiles are more than just buttons to start an application. They can display important information directly on the tile interface. For example, a tile for starting the application for monitoring overdue purchase orders shows on the surface how many purchase orders are overdue, as in Figure 1.6. Even before you click the tile, you can know how many purchase orders you must process.

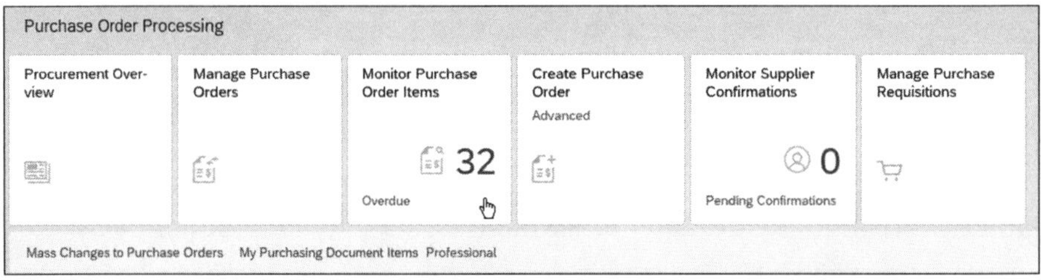

Figure 1.6 Tiles for Purchase Order Processing Apps

Therefore, the SAP Fiori launchpad is a collection of tiles that provides the user with a ready-made cockpit of key information, with the ability to click on any tile to either launch an application or drill down for deeper analysis.

The SAP Fiori launchpad also provides excellent enterprise search capabilities. A user can search for an application or even for a business object, such as an employee or a sales order. When a business partner calls with a delivery reference number, you simply enter it in the launchpad search field and, in a few seconds, you have every document related to the delivery number right in front of you.

Users can select applications from tile catalogs and customize their launchpad to organize their own tiles into groups to improve their productivity and user experience. The user opens the **User Action** menu shown in Figure 1.7 to configure their launchpad.

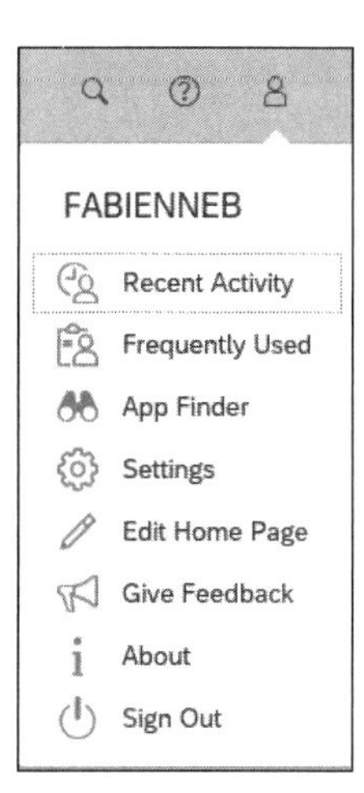

Figure 1.7 The User Action Menu in the SAP Fiori Launchpad

Selecting the **Edit Home Page** option from the **User Action** menu will allow you to move, remove, or add apps, and move, add, remove, or create groups, as shown in Figure 1.8.

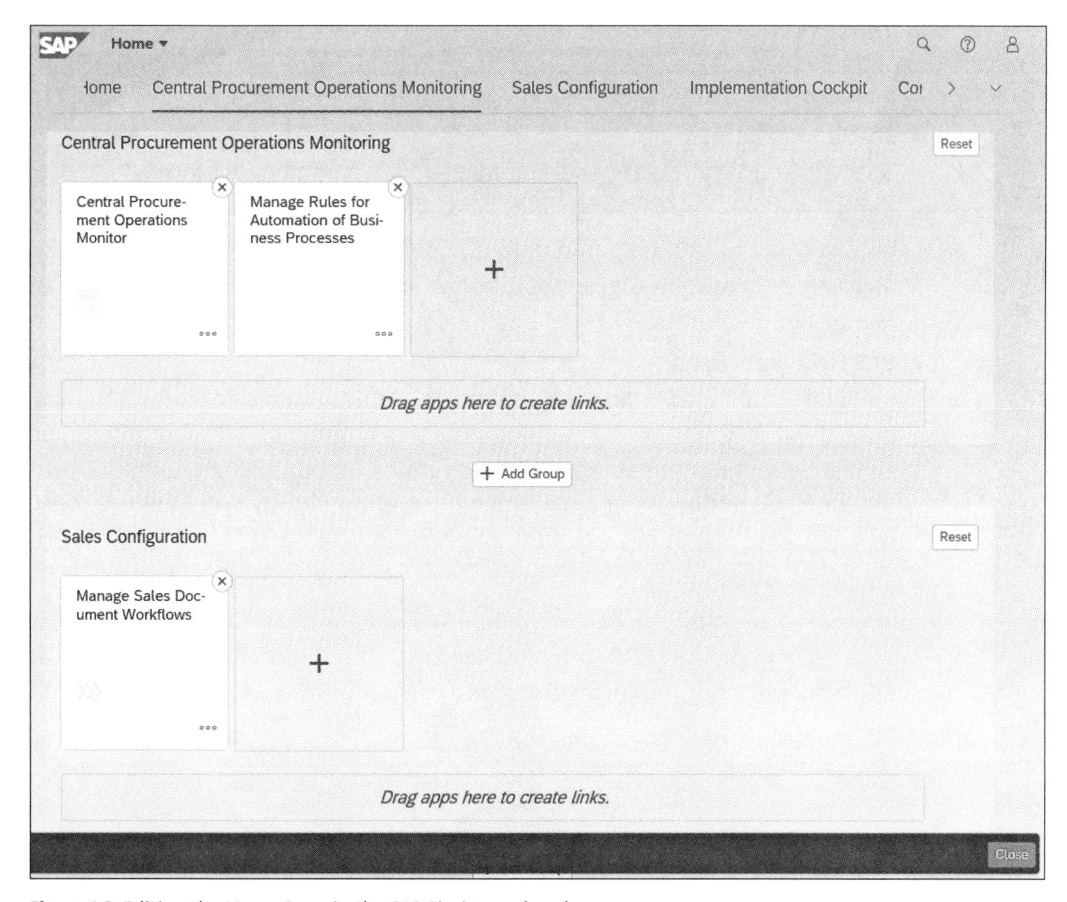

Figure 1.8 Editing the Home Page in the SAP Fiori Launchpad

Important Terminology

In this chapter, the following terminology was used:

- **SAP HANA**
  The SAP HANA database was created in 2010 through the Hasso Plattner Institute, with support from Berlin University and Intel. SAP HANA, which stands for *high-performance analytic appliance*, combines transactional and analytical data, and holds the data in-memory, as opposed to reading it from a disk. This enables businesses to process massive amounts of real-time data in a short time, without disrupting operational processing.
- **In-memory technology**
  Data storage in RAM instead of disks to take advantage of the reduced delay and read/write times. The technology uses disks for historical data and backup purposes.
- **Columnar store**
  The data is stored in columns instead of rows. This allows for higher reading performance and better compression capabilities.
- **SAP S/4HANA**
  In 2015, SAP introduced SAP S/4HANA, the new generation of intelligent ERP systems running only on the SAP HANA database. SAP S/4HANA stands for SAP Business Suite 4 SAP HANA.
- **SAP Cloud Platform**
  The SAP Cloud Platform is a development and deployment platform.
- **SAP Fiori**
  SAP Fiori is the new common design paradigm for all SAP applications. The objective is to make business applications intuitive to use and available on all platforms and devices.
- **SAP Fiori launchpad**
  SAP Fiori launchpad is an entry point for users to access SAP S/4HANA and other applications. It can be accessed via a web browser and displays tiled apps according to the user's role and authorizations. The SAP Fiori launchpad provides many configuration options for the end user, such as theme selection, tiles grouping, and group creation.
- **SAP Gateway server**
  The gateway server is used to connect SAP Fiori to one or more SAP S/4HANA systems or even SAP ERP or other backend systems. It is there that both the SAP Fiori configuration and the maintenance of the SAP Fiori users is performed.

Practice Questions

These questions will help you evaluate your understanding of the topics covered in this chapter. They are similar in nature to those on the certification examination. Although none of these questions will be found in the exam itself, they will allow you to review your knowledge of the subject.

Select the correct answers, and then check the completeness of your answers in the next section. Remember that on the exam, you must select all correct answers—and only correct answers—to receive credit for the question.

1. You must develop a comprehensive extension in the area of procurement. Where does SAP recommend that you do this development?

☐ **A.** SAP Cloud Platform

☐ **B.** SAP S/4HANA backend server

☐ **C.** SAP Fiori launchpad

☐ **D.** SAP S/4HANA gateway server

2. What is SAP Fiori?

☐ **A.** A database

☐ **B.** An application

☐ **C.** A user experience

☐ **D.** A development platform

3. True or false: SAPUI5 is the leading UI technology in the SAP user experience strategy.

☐ **A.** True

☐ **B.** False

4. What are the different types of SAP Fiori applications? (There are three correct answers.)

☐ **A.** Transactional

☐ **B.** Responsive

☐ **C.** Factsheet

☐ **D.** Role-based

☐ **E.** Analytical

5. True or false: Although SAP HANA uses a complete in-memory database, disk storage is still required.

☐ **A.** True

☐ **B.** False

6. Which kind of data storage does SAP HANA provide? (There are two correct answers.)

☐ **A.** Row

☐ **B.** Flat

☐ **C.** Random

☐ **D.** Column

7. Which of the following configurations can a user make on their SAP Fiori launchpad? (There are three correct answers.)

☐ **A.** Create their own tile group.

☐ **B.** Switch a tile from one group to another.

☐ **C.** Change the name of a tile.

☐ **D.** Save a selection variant as a new tile.

8. True or False: SAP GUI has been replaced by SAP Fiori in SAP S/4HANA and is no longer available.

☐ **A.** True

☐ **B.** False

9. True or False: In an environment using SAP HANA, you can perform both OLAP and OLTP processing without duplicating data.

☐ **A.** True

☐ **B.** False

10. Which of the following are SAP Fiori design principles? (There are two correct answers.)

☐ **A.** Simple

☐ **B.** Comprehensive

☐ **C.** Logical

☐ **D.** Coherent

11. Where can a user edit their home page to create new tile groups, for example?

☐ **A.** In the **Tile Group** area of the SAP Fiori

☐ **B.** In the **User Action** menu

☐ **C.** By choosing the **Home** button

☐ **D.** In the SAP Gateway server

Practice Question Answers and Explanations

1. Correct answer: **A**
   In principle, you could develop extensions in your backend system, as was the case in SAP ERP times. However, SAP recommends shifting these developments to the SAP Cloud Platform. This ensures that the core remains as close as possible to the standard and the customer still enjoys development flexibility.
2. Correct answer: **C**
   SAP Fiori is a user experience. SAP HANA is a database, and the SAP Cloud Platform is a development platform. You can create apps in SAP Fiori using the SAPUI5 framework.
3. Correct answer: **A**
   True. SAPUI5 is a development toolkit that provides a consistent user experience. Applications created with SAPUI5 can be used across browsers and devices—they run on smartphones, tablets, and desktop computers. The SAPUI5 framework is used to create apps in SAP Fiori.
4. Correct answers: **A, C, E**
   SAP Fiori applications can be divided into three different types: transactional, factsheet, and analytical applications. Responsive and role-based are two of the five principles of the SAP Fiori design, but they are not application types.
5. Correct answer: **A**
   True. Disks are used for historical data and backup purposes. SAP HANA stores data in persistent disk volumes (either hard disk or solid-state drives) that ensure that changes are permanent and that the database can be reset to the last commit state after a restart.
6. Correct answers: **A, D**
   Column store is the preferred method for SAP HANA because it is better suited for reading data and SAP HANA is particularly optimized for column-based storage. However, raw storage is also supported.
7. Correct answers: **A, B, D**
   Changing the name or design of a tile can only be done in the catalog view of the SAP Fiori launchpad designer, which is an administrator interface for managing SAP Fiori tiles.

8. Correct answer: **B**

   False. For the on-premise edition of SAP S/4HANA, the SAP GUI will continue to be supported and can be used alongside SAP Fiori applications to provide an easy transition for existing SAP customers who are familiar with the classic interface. In addition, there are several transactions that have not yet been converted to SAP Fiori, and SAP GUI is still required for these.

9. Correct answer: **A**

   True. With SAP HANA, OLTP and OLAP can now be combined to offer modern applications that combine transactions and analysis. This is called *embedded analytics*.

10. Correct answers: **A, D**

    The SAP Fiori principles dictate that the application should be simple, role-based, coherent, responsive, and of instant value.

11. Correct answer: **B**

    An end user typically has access to the SAP Fiori launchpad, and via the user action menu, the user controls all changeable settings for their SAP Fiori launchpad. SAP Fiori Designer and SAP Gateway are more for administrators.

Takeaway

This chapter provided an overview of the SAP HANA database and its advantages. It also provided you with a definition of the term *intelligent enterprise*. You should now understand the deployment possibilities of SAP S/4HANA, the successor of SAP ERP, and be able to make a recommendation to a customer, depending on their situation and intentions. Finally, you should understand SAP Fiori and its principles, be able to navigate the SAP Fiori launchpad, and configure it, if necessary.

Summary

This chapter provided a high-level overview of SAP HANA, SAP S/4HANA, and SAP Fiori. This is the only chapter in this book not directly related to sourcing and procurement. It contains some technical information as well as the basics of SAP S/4HANA. However, these are important, and we recommend deepening the topics outlined here with the help of SAP course S4H01, especially if you are preparing for one of the upskilling certifications.

In the next chapter, we delve into the heart of the matter by introducing the procurement processes, the core organizational units for sourcing and procurement, as well as the master data.

Chapter 2
Procurement in SAP S/4HANA

2

Techniques You'll Master

- Understand the different procurement processes that are mapped in SAP S/4HANA.
- Set up an organizational structure for procurement.
- Maintain material master records.
- Maintain business partner master records.

In this chapter, we introduce the different procurement processes and some of their control parameters. We explain the most important master data: material and supplier. We explain the organizational units relevant to the procurement processes and analyze how process steps and master data are embedded in the organizational structure.

Real-World Scenario

As a consultant for sourcing and procurement, you will map customer processes in SAP S/4HANA and translate customer requirements into system configuration settings. You must be able to offer and present the appropriate procurement process. The organizational structure is determined in the initial phases of a customer implementation project. You must have an early understanding of your customer's processes and organization in order to define the appropriate organizational structure in SAP S/4HANA. You must always keep in mind that processes, master data, and organizational structure must be coherent.

Objectives of This Portion of the Test

The purpose of this part of the certification exam is to test your knowledge of the various procurement processes, the relevant master data, and organizational units. For the certification exam, you must have a good understanding of the following topics:

- Describe a standard external procurement process.
- Describe a supplier consignment process.
- Describe a subcontracting process.
- Explain the usage of a stock transport order.
- Explain the function as well as the configuration options and restrictions of the item category and the account assignment category.
- Name and define the organizational units for procurement and know the possible assignments.
- Explain the structure and data of the material master.
- Understand the business partner approach.
- Understand the dependencies between organizational structure, master data, and processes.

Key Concept Refresher

In this section, we will discuss the most important features of the various procurement processes available in the SAP S/4HANA system. You will learn about the organizational units relevant to procurement. We will discuss how material and vendor master records are maintained and configured, and finally, we will explore how processes, master data, and operations are correlated.

Overview of the Procurement Processes

External procurement can cover various aspects: for example, you can purchase materials or services. If you procure materials, you can purchase them for stocking or for direct consumption. External procurement involves a sequence of operations. In this section, we will analyze these activities in detail and describe how they can vary, depending on the type of procurement process. We will describe the most important procurement processes and explain how a user can trigger one or the other process.

Standard External Procurement Process

Standard external procurement typically consists of the following main steps:

- Determination of requirements
- Supply source determination
- Creating and processing purchase orders
- Goods receipt
- Invoice receipt
- Payment

Let's cover these in more detail.

Determination of Requirements

A requirement for external procurement of materials or services can be recorded in SAP S/4HANA using a purchase requisition. The purchase requisition stands at the beginning of the external procurement process, represents the requirement, and must be converted into a purchasing document (e.g., purchase order, contract, etc.).

The requirement for stock material can be determined by automatic material requirements planning (MRP). In this process, purchase requisitions are created for the externally procured materials. Purchase requisitions are checked and can be adjusted if necessary before they are converted into purchase orders. They can also be created manually—for example, for consumables or services and also for stock material.

In the system, the purchase requisition is an optional step of the procurement process: if a requirement is reported by telephone or email, for example, you can dispense with the purchase requisition and start ordering directly.

Supply Source Determination

Source of supply determination in SAP S/4HANA means that, in concrete terms, a suitable vendor and valid conditions are searched for in the system for a specific requirement (e.g., purchase requisition for a quantity of material on a specific date). SAP S/4HANA supports this search or even completely takes it over by referring to data already available in the system.

The following objects represent sources of supply in SAP S/4HANA:

- Purchasing info records
- Contracts
- Scheduling agreements

The system attempts to determine a suitable source of supply during the MRP run. The logic of sourcing during the MRP run will be explained in Chapter 8. If you create or process a purchase requisition manually, the system can also support you in determining a valid source: either the system determines a unique source and adopts it, or it finds several valid possibilities and proposes a list for selection. The logic of source determination outside the MRP run is the subject of Chapter 6.

The result of a successful source determination is the assignment of a source of supply to the purchase requisition. In practice, the source number (i.e., the number of the info record or outline agreement) is stored in the purchase requisition.

Figure 2.1 shows a purchase requisition that has been assigned to a source of supply. In this case, the source of supply is a purchasing info record.

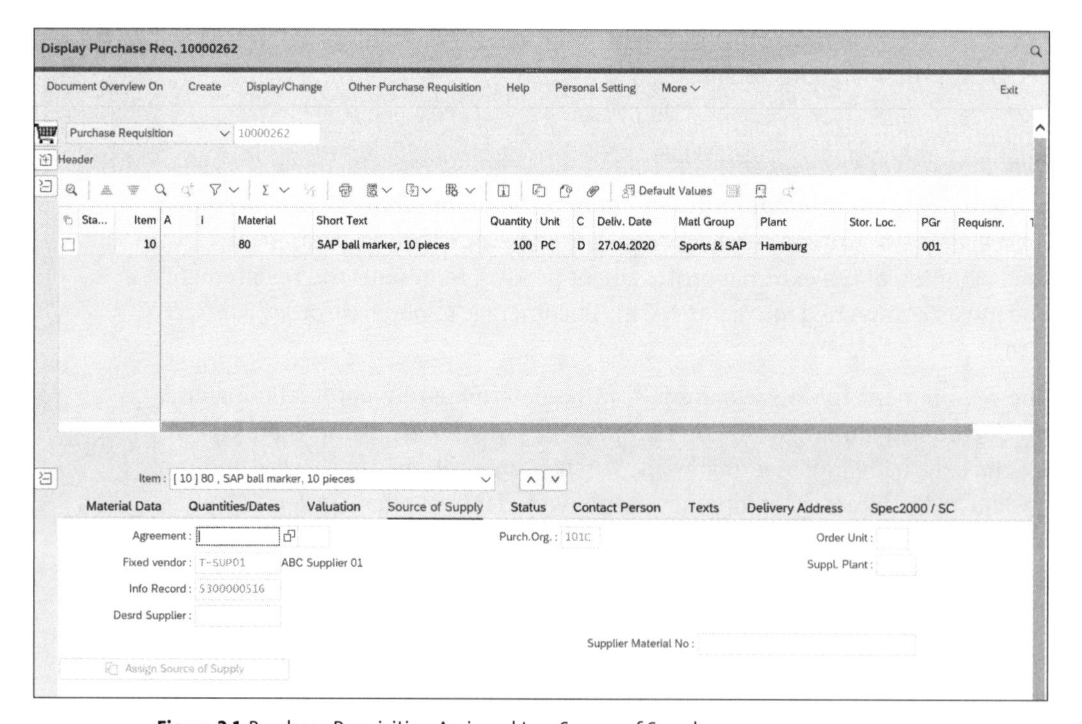

Figure 2.1 Purchase Requisition Assigned to a Source of Supply

Note

If the system does not contain a valid source of supply, you can start an inquiry.

Even if no source of supply could be determined, you can still convert a requisition into a purchase order or create a purchase order manually: in this case, you specify the vendor and the purchase price manually.

Creation and Processing of Purchase Orders

The purchase order is the central object for the entire procurement process. The purchase order consists of a document header and one or more items:

- **Document header**
  The document header contains information that relates to the entire purchase order, such as order type, vendor, purchasing organization, purchasing group, company code, currency, document date, and terms of payment.
- **Items**
  The items in a purchase order describe the materials or services ordered. Necessary data for an item is the material/service, order quantity and unit of measure, delivery date, price, and plant for which the material/service is ordered. You can maintain additional information for each item (e.g., tolerance for over delivery or item-specific text).

  You also use the *item category* to define the procurement process for each purchase order item. For example, you use the item category field to specify whether the item is a standard or a special procurement process, such as vendor consignment, subcontracting, or stock transfer. *You must leave the item category blank for the standard procurement process.*

  Another important indicator in the purchase order item is the *account assignment category*. With the account assignment category, you decide whether you buy for stocking or for direct consumption. You leave the account assignment category blank for a stock item. For a purchase order item for a cost center, for example, you enter the corresponding account assignment category (K in the standard system). We will discuss the account assignment category and procurement for consumption in Chapter 4.

Note

You can adjust existing account assignment categories and create new account assignment categories in Customizing. Item categories, however, are predefined by SAP, and their control parameters cannot be adjusted in Customizing. You can only change their description.

You can use a single purchase order to procure materials or services for different plants. All plants in a purchase order must be assigned to the purchasing organization entered in the purchase order header.

You can create a purchase order manually or have the system generate it automatically. When creating a purchase order, you can use an existing purchase order or a request for quotation as a reference to reduce the time and effort involved in entering it.

You can also create a purchase order item with reference to a contract item, in which case you generate a contract release order. The contract item is stored in the purchase order item and the contract history is updated.

When you create a purchase order with reference to a purchase requisition, we say that you are converting a purchase requisition into a purchase order. There are several ways of converting requisitions into purchase orders: manually, via assignment lists, or even automatically.

Creating purchase orders automatically from purchase requisitions is one way to optimize the procurement process. We discuss the conversion of purchase requisitions in Chapter 6.

Figure 2.2 shows you a purchase order with its header (collapsed in the figure), the item list in the middle, and the item details in the lower part. The account assignment category is shown in ❶ and the item category is shown in ❷. It is a screenshot of Transaction ME21N, also available as the Create Purchase Order—Advanced app on the SAP Fiori launchpad.

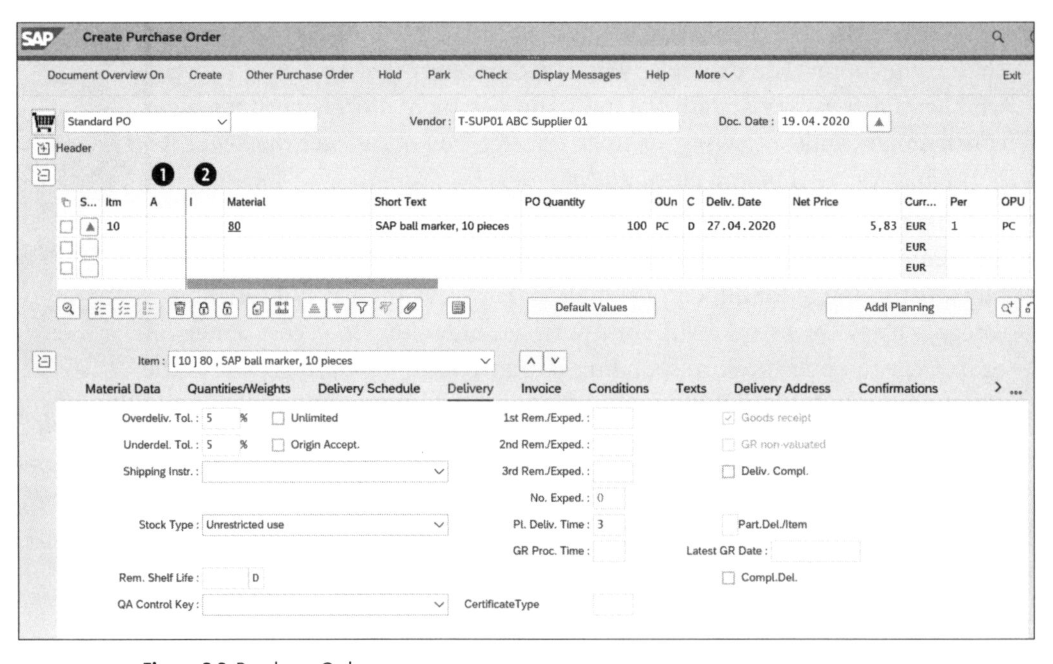

Figure 2.2 Purchase Order

Monitoring of Purchase Orders

Monitoring purchase orders means monitoring the processing status of purchase orders in the system. You can determine, for example, whether a delivery or invoice has already been received for a purchase order item; these process steps are recorded in the purchase order history. You can also remind vendors of outstanding deliveries or outstanding order acknowledgments.

Goods Receipt

When you receive goods in relation to a purchase order, it is important that you enter this goods receipt with reference to the purchase order. When you enter the goods receipt, the system proposes all open items from the purchase order. This facilitates both the recording and the verification of the goods receipt.

When you post a goods receipt with reference to a purchase order, the system automatically updates the purchase order history of the relevant purchase order items. The purchase order history records the progress of a procurement process, enabling the buyer to identify outstanding deliveries and, if necessary, to remind or dun the vendor.

Together with the goods receipt posting, the quantities and values are updated. The system documents the goods receipt posting in a *material document*. The material document is a source of information for subsequent processes, such as further expected goods receipts and invoices. The material document is added to the purchase order history.

The goods receipt is always valuated for stock material order items and is optionally valuated for order items with account assignment. In the case of a goods receipt into the warehouse, the stock value is increased together with the stock quantity. This means that the stock account is updated, and this posting is documented in the form of an *accounting document*. In the case of purchase order items with account assignment and valuated goods receipt, the cost account is updated, and in this case, an accounting document is also created.

Note

Depending on the account assignment category, you can either post a non-valuated goods receipt for a standard purchase order item assigned to an account, or you can completely do without the goods receipt posting. These special features will be discussed in more detail in Chapter 4.

Invoice Receipt

During invoice verification, the user enters some invoice data, such as total gross amount, invoice date, and so forth, and specifies the reference to the corresponding purchase order. This reference enables the system to determine and propose the expected invoice quantities and the expected invoice amount from the purchase order and from the purchase order history. If the data proposed by the system differs from the data from the vendor invoice, the user can overwrite it. Depending on how large the differences are and how the system is configured, warnings are issued and/or the invoice can be posted but blocked for payment.

Figure 2.3 shows a purchase order history with goods receipt and invoice receipt

Document Overview On | Create | Display/Change | Other Purchase Order | Messages | Help | Personal Setting | More | Exit

Standard PO 4500000046 Vendor: T-SUP01 ABC Supplier 01 Doc. Date: 19.04.2020

Header

S...	Itm	A	I	Material	Short Text	PO Quantity	OUn	C	Deliv. Date	Net Price	Curr...	Per	OPU
	10			80	SAP ball marker, 10 pieces	100	PC	D	27.04.2020	5,83	EUR	1	PC

Addl Planning

Item: [10] 80 , SAP ball marker, 10 pieces

Material Data | Quantities/Weights | Delivery Schedule | Delivery | Invoice | Conditions | Purchase Order History | Texts

Sh. Text	MvT	Material Document	Item	Posting Date	Σ	Quantity	Delivery cost quantity	OUn	Σ	Amt.in Loc.Cur.	L.cur	Σ	Qty in OPUn
WE	101	5000000200	1	20.04.2020		100	0	PC		583,00	EUR		100
Tr./Ev. Goods receipt					•	**100**		**PC**	•	**583,00**	**EUR**	•	**100**
RE-L		5105600192	1	20.04.2020		100	0	PC		583,00	EUR		100
Tr./Ev. Invoice receipt					•	**100**		**PC**	•	**583,00**	**EUR**	•	**100**

Figure 2.3 Purchase Order History

Note

You can also assign an invoice to a purchase order using the number of the delivery note or bill of lading, provided these numbers were entered at goods receipt.

Payment

You must run the payment program to pay vendor liabilities. The accounting department is responsible for regularly running this payment program. We will not go into more detail in this book, as payment is not part of the sourcing and procurement certification.

Supplier Consignment

In addition to the normal procurement process already described, various other special procurement processes are also possible—for example, supplier consignment.

You control the procurement process that you wish to carry out with a purchase order item via a special indicator called the item category. In the case of a supplier consignment process, choose item category K. Here, K refers to the German word *Konsignation*.

Consignment means that a vendor provides you with material that is stored on your premises but that is still the property of the vendor. A liability to the vendor does not arise until you withdraw material from consignment stock. The quantity consumed or withdrawn is settled with the vendor according to agreed periods.

Before you procure a material from a vendor on consignment, you and the vendor must agree on a price for the material. It may even be that you procure a certain material from a vendor both normally and on consignment. The price for the consignment may differ from the normal price because, in the case of a consignment, you do not have to pay immediately after delivery but later when the consignment stock is withdrawn.

Note

A material that you procure on consignment must have a material master record, since it is managed in stock.

In the system, the price information must be recorded in a special consignment information record. The creation of a consignment information record is a prerequisite for posting material to the vendor consignment stock.

The vendor consignment process consists of the following steps:

1. You request the material from your vendor with a purchase order for consignment.
2. You post the goods receipt with reference to the consignment purchase order when the material is delivered. The quantity of the material is managed in a special stock (i.e., consignment stock).
3. When a withdrawal is posted from consignment stock, a liability to the vendor is created.
4. SAP S/4HANA provides a special function for settling consignment liabilities. The app is started at intervals agreed upon with the vendor and creates a credit memo with a corresponding message for the vendor.

Note

You can start the procurement process one stage before the purchase order with a purchase requisition. You can manually create a consignment purchase requisition (indicated by item category K). If consignment purchase requisitions (instead of normal purchase requisitions) are to be generated automatically by the MRP run, then you must set a corresponding special procurement key for consignment in the material master record at the plant level. SAP S/4HANA contains the special procurement key 10.

Subcontracting

Subcontracting is the process in which a material that you procure from a vendor (subcontractor) requires certain components. You can provide the components directly to the subcontractor, or you can order the components from a third-party supplier who delivers them directly to the subcontractor.

You order the material via a purchase order item with the item category Subcontracting (L). L refers to the German word *Lohnbearbeitung*, which means subcontracting. The item category Subcontracting (L) allows you to create sub-items for the components to be provided to the subcontractor. You can either enter the individual components manually or generate them using a bill of material (BOM) explosion if a BOM exists for the ordered material. No material master record is required for the material to be ordered. However, components that are to be provided must have a material master record.

Components that are provided to the subcontractor are part of your valuated stock. They are managed under the special stock category *stock of material provided to vendor.*

The subcontracting process consists of the following main steps:

1. You request the material from your vendor with a purchase order for subcontracting.
2. You provide the components to the subcontractor. You can create an outbound delivery in the system for this provision, but you can also enter a transfer posting directly. Although the components provided are no longer physically in your warehouse, they are managed in your stock, since they still belong to you. The provided components are displayed in the stock evaluations under the special stock type *stock of material provided to vendor.*
3. You post the goods receipt with reference to the subcontracting purchase order when the material is delivered. Here, not only is the receipt of the finished products posted, but so is the consumption of the components from the special stock *material provided to vendor.*
4. Finally, you enter and post the invoice that the subcontractor issues for the service provided.

Note

You can start the procurement process one stage before the purchase order with a purchase requisition. You can manually create a subcontract purchase requisition (indicated by item category L). If subcontract purchase requisitions (instead of normal purchase requisitions) are to be created automatically by the MRP run, you must set a corresponding special procurement key for subcontracting in the material master record at the plant level. Standard SAP S/4HANA contains the special procurement key 30 for this.

Figure 2.4 shows a subcontracting purchase order with item category L and the components to be provided.

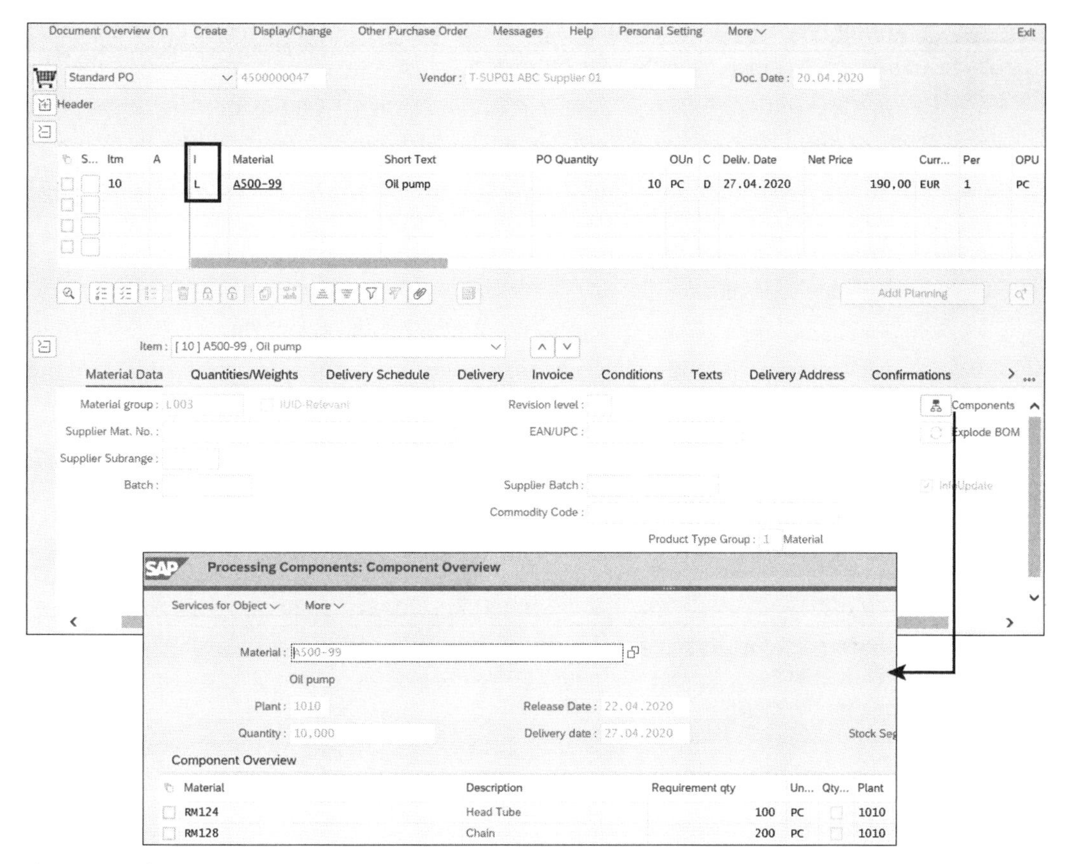

Figure 2.4 Subcontracting

Stock Transfer with Stock Transport Order

There are different procedures for carrying out a stock transfer between two plants within a company code: You can carry out a stock transfer posting using the one-step procedure or the two-step procedure, but you can also process a stock transfer using a stock transport order.

In a procurement with stock transport order, materials are procured and delivered within a company. The plant that requires the materials orders the material from another plant that can supply the materials.

The stock transfer process with a stock transport order consists of the following main processes:

1. You create a stock transport order for the receiving plant. A stock transport order is identified by item category U. Here, U refers to the German word *Umlagerung*, which means *stock transfer*. It is also necessary to use a special purchase order type, especially because some required fields differ from those for other procurement processes—for example, instead of a vendor (vendor number), a plant (plant ID) is entered. In the standard SAP S/4HANA system, the

order type UB stands for stock transport orders without delivery and UD stands for stock transport orders with delivery.

2. In the supplying plant, you enter a goods issue with reference to the stock transport order. Optionally, you can create an outbound delivery for the stock transport order in the supplying plant and post the goods issue with reference to this outbound delivery. After this posting, the quantity is managed in the stock in transit of the receiving plant. The stock in transit is part of the valuated stock of the receiving plant.
3. You can enter a goods receipt in the receiving plant, and the goods are posted with reference to the stock transport order. The quantity is transferred from the stock in transit to the storage location stock of the receiving plant.

Note

You can start the procurement process one stage before the purchase order with a purchase requisition. You can manually create a stock transport requisition (identified with item category U). If stock transport requisitions (instead of normal purchase requisitions) are to be created automatically by the MRP run, you must set an appropriate special procurement key for stock transfer in the material master record at the plant level. A relevant special procurement key must be configured individually in Customizing, as it contains the direction of stock transfer and the plants affected.

Organization Structure

Organizational units in SAP S/4HANA translate real company units into system units. By assigning the organizational units to each other, an entire company can be mapped in SAP S/4HANA. The master data is created along the organizational structure. Business transactions are carried out with reference to the specific organizational units, and reports are analyzed at various organizational levels and combinations of these. In this section, we examine the organizational units relevant for sourcing and procurement, along with their relationships.

Client

The *client* is a legally and organizationally self-contained unit within an SAP system. It contains its own master records and an independent set of tables. Organizational, master, and transactional data are maintained on the client level. For example, the client could represent a corporate group.

The client is the highest hierarchical level in the SAP S/4HANA system. Specifications that you make at the client level, or data that you enter at this level, apply to all organizational units below it. Therefore, you do not need to enter the settings and data at the client level more than once in the system. This guarantees consistent data and avoids redundancy.

Access authorization is assigned on a client-specific basis. A user must have a user master record for each client. Therefore, users log on to a specific client. All user entries are stored, processed, and evaluated client-specifically.

A client is uniquely defined in the system by a three-character numeric key.

Company Code

A *company code* is the most fundamental organizational unit for financial accounting in SAP S/4HANA. It represents the legal unit for which a complete, self-contained chart of accounts can be created. A company code represents an independent unit that creates its own financial statements. It represents a company within the group (of the client). You can set up several company codes in one client.

A company code is defined in the system using a four-character alphanumeric key that is unique to the client.

Plant

A *plant* is the most fundamental organizational unit for logistics in an SAP S/4HANA system. It subdivides the company for production, procurement, maintenance, and material planning. A plant produces materials or provides goods or services. It is an operating unit or subsidiary of a company with the following characteristics:

- A plant belongs to exactly one company code; a company code can contain several plants.
- A plant/division combination can be assigned to only one business area.
- A plant can be assigned to several combinations of sales organizations and distribution channels.
- A plant can have several shipping points; a shipping point can also be assigned to several plants.
- A plant has an address and a language and belongs to a country and a region.
- A plant has its own material master data.
- Plant-specific data can be maintained for various views of the material master record, such as MRP, purchasing, storage, work scheduling, production resources/tools, forecasting, quality management, sales and distribution, and costing.
- Several purchasing organizations can be assigned to a plant.
- Production planning and production are carried out on a plant-specific basis, although certain applications can work across plants.
- Product costing can be carried out across plants.

Storage Location

A *storage location* is the organizational unit where you manage the material stocks on a quantity basis. Physical stocks are always assigned to a storage location in the system. If you do not use warehouse management, then the storage location is the smallest organizational unit on which you can manage your stock quantities, so you carry out your physical inventory at the storage location level.

A storage location also has the following characteristics:

- A storage location always belongs to one plant.
- Unlike a plant, a storage location can have several addresses (e.g., one for the delivery of packaged goods and another for bulk material).

  The storage location addresses can differ from the plant address. Both the storage location addresses and the plant address can be used as a delivery address in the procurement process (in the purchase order item, for example).
- The storage location key must be unique within a plant, but you can use the same storage location key in several or all plants.

  This allows you to define the storage location keys according to the function of the storage location and to set up a uniform storage location structure for all plants. For example, storage location 100A is always the raw materials warehouse, 100B is always the finished goods warehouse, and so on.

Purchasing Organization

A *purchasing organization* is an organizational unit within logistics that subdivides a company according to the requirements of purchasing.

A purchasing organization is responsible for the following tasks:

- A purchasing organization procures materials or services. In other words, each purchase order is created by a purchasing organization; in SAP S/4HANA, this means that each purchase order header contains one purchasing organization.
- A purchasing organization negotiates conditions of purchase with vendors. In SAP S/4HANA, the conditions are always stored at the purchasing organization level.

Note

Conditions can also be stored for a combination of purchasing organization *and* assigned plant, depending on your price agreements and whether Customizing allows it.

A purchasing organization is responsible for the purchasing matters of plants. The area of responsibility of a purchasing organization is determined by its assignment to plants. A purchasing organization can supply several plants; a plant can be covered by several purchasing organizations. An n:m relationship can arise between plants and purchasing organizations.

Let's look at an example to illustrate this. A group of companies with several company codes and plants (including production plants) in different countries, where you are currently implementing SAP S/4HANA Sourcing and Procurement, has the following requirements: Each plant should be responsible for agreeing prices and procuring auxiliary and operating materials. Raw materials, on the other hand, should be negotiated and purchased centrally worldwide. For certain services such as transportation, contracts should be concluded per country or even per legal unit (i.e., per company code). You should make a proposal as to how many purchasing organizations should be created and how they should be assigned.

You could suggest the following:

- One purchasing organization per plant for local purchasing
- One central purchasing organization responsible for all plants for raw materials purchasing
- One purchasing organization per company code for purchasing transportation services

Each purchasing organization is first created and then assigned to one or more plants. If the purchasing organization is company code-specific, you can assign it to the appropriate company code. However, you must still assign the purchasing organization to all the individual plants in the assigned company code. The assignment to a company code is therefore optional.

Figure 2.5 shows the proposed purchasing organizations. For raw materials (group-wide procurement), we have the purchasing organization C100, which is responsible for all plants in all company codes.

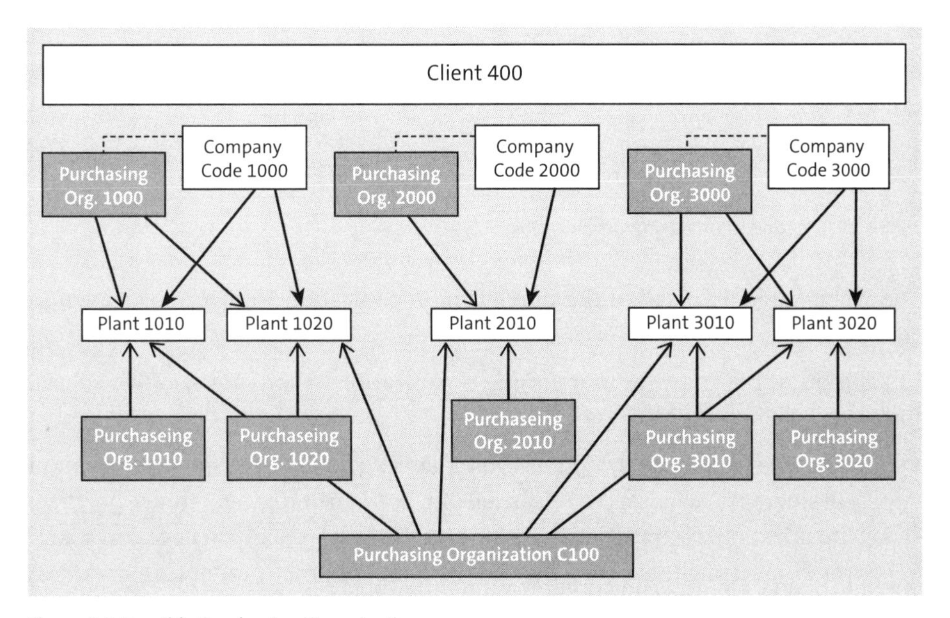

Figure 2.5 Possible Purchasing Organizations

Purchasing organizations 1000, 2000, and 3000 are only responsible for one company code each. (Note that assignment to the company code itself is optional.) Purchasing organizations 1010, 1020, 2010, 3010, and 3020 are plant-specific.

When organizing your purchasing functions and defining purchasing organizations, you should bear the following in mind:

- Each purchasing organization has its own info records and conditions.
- Each purchasing organization has its own business partner master data. In a business partner master record, purchasing data and partner functions are maintained for each purchasing organization.

The following figures show you the Customizing steps for configuring a purchasing organization. The first step is to create it, as shown in Figure 2.6.

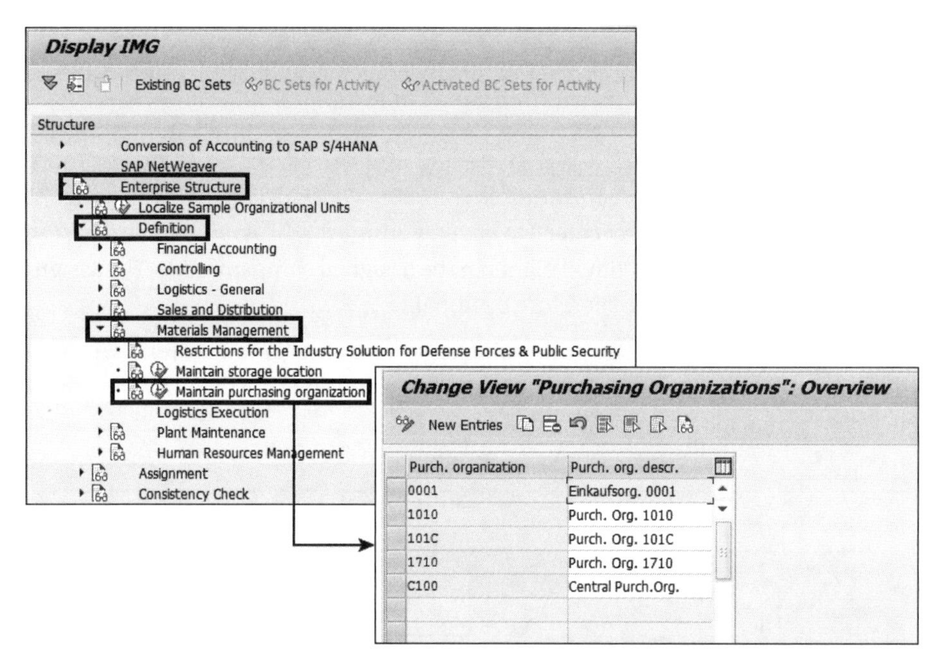

Figure 2.6 Creating a Purchasing Organization

The second step is to assign the purchasing organization to plants, as shown in Figure 2.7.

In addition to the "normal" purchasing organizations, there are also two special forms:

- The *reference purchasing organization* enables cross-purchasing organization procurement transactions to be carried out. You can define a reference purchasing organization as strategic purchasing: at this level, prices are negotiated, and contracts are concluded that the "normal" operational purchasing organizations can access. You must assign "normal" purchasing organizations to reference purchasing organizations in Customizing. The purchasing data in the business partner must also be maintained for the reference purchasing organization.

- The *standard purchasing organization* is required for processes where the system must automatically determine a purchasing organization. These processes include pipeline and consignment procurement, some stock transfer processes, and automatic creation of purchase orders when posting goods receipts. In the source determination process for stock transfers and consignment, SAP S/4HANA then automatically uses this standard purchasing organization. In the case of goods issues of consignment and pipeline materials, the purchasing info records of the standard purchasing organization are determined for the automatic settlement of withdrawals.

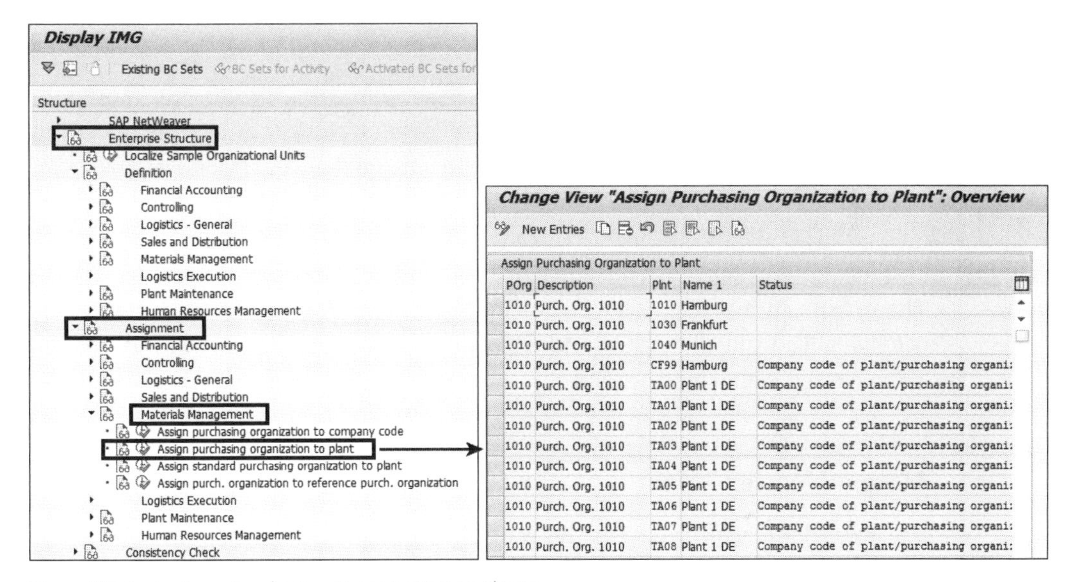

Figure 2.7 Assigning a Purchasing Organization to Plants

Note

In connection with the organizational structure for purchasing, the *purchasing group* is also often mentioned. Strictly speaking, the purchasing group is not an organizational unit but is sometimes mentioned as such! It is also not assigned to any other organizational unit.

The purchasing group corresponds to a buyer or a group of buyers responsible for purchasing certain materials or services. This responsibility is represented in the system by assigning a purchasing group to a material or material group. Specifically, the **Purchasing Group** field in the material master record and in the table is where the material groups are maintained.

The purchasing group is stored as a key with some fields (description, telephone number) in a table that is valid for all clients. Externally, a purchasing group usually represents the contact person for the suppliers.

Valuation Area

The organizational unit where the valuation of material stocks takes place is the *valuation area*.

To define a valuation area, you do not need to create new organizational elements: you simply define a valuation area by specifying the level at which the system valuates material stocks. The valuation area can correspond to the following levels:

- The valuation data for a material is created for each company code. The price control and the price of a material apply per company code. You must therefore valuate the same material in all plants in a company code in the same way.
- The valuation data for a material is created for each plant. The price control and the price of a material are valid for each plant. You can therefore valuate the same material differently in each plant. We recommend valuation at plant level; if you use production and costing, it is mandatory anyway.

Note that this setting applies to the entire client. In a system in which postings have already been made, it is not possible to change the valuation level.

MRP Area

The organizational unit relevant for materials planning in SAP S/4HANA is the *MRP area*. The MRP area corresponds to a plant or part of a plant. The MRP area will be covered in detail in Chapter 8.

Material Master Data

The material master record is the central source for material-specific data. It is used in all areas of logistics. In this chapter, we will concentrate on the structure of the material master data, the maintenance of material master records, and some important fields.

Structure of the Material Master Data

To avoid redundancy, material data is stored in a single database object. A material should only be created once in the system and is assigned a unique number.

All areas, such as purchasing, inventory management, MRP, work scheduling, warehouse, and so forth, all access the stored data together. Since many departments use the material master, the material master contains a lot of data. A specific user department is not interested in all the information stored in the material master, but usually only in the data related to its business activity. For this reason, material master data is structured in so-called *views* that correspond to the user departments: in the material master, you will find purchasing-specific, sales-specific, warehouse-specific data, and so forth.

A material automatically receives a maintenance status, depending on the views maintained. If you want to use a material for a process, the system checks whether the material has the required maintenance status—that is, whether the required views have been maintained.

Depending on the purposes for which you want to use a material, you must have maintained certain views:

- When purchasing stock material, the Purchasing and Accounting views must be maintained (at minimum).
- If you also want to plan this material, you must have maintained the MRP view.
- For the purchasing process of a non-stock material for which you want to create a material master record, you only need the Purchasing view.

In parallel to being structured according to views, the material master is set up along the *organizational structure*. What does that mean in concrete terms? A material master record contains data that is valid for the whole company: the material number, the material description, the assigned material group, the base unit of measure, the weight, and so on. This data is maintained at the client level in the material master.

A material master also contains data that is plant-specific. In MRP, for example, you can define the planning procedure or lot-sizing procedure for each plant. In the area of purchasing, you can, for example, define per plant whether automatic purchase requisitions are allowed or not (using an indicator in the material master). You also decide for each plant (the plant is the valuation level) how a material is to be valuated and at what price. A lot of data relevant for purchasing, MRP, or stock valuation is therefore maintained at the plant level.

Another level where specific material master data can be maintained is the plant storage location level. At this level, you can, for example, maintain a storage bin with the informative character only.

Note

The previously mentioned organizational levels (client, plant, and storage location) are the ones relevant for the external procurement process and inventory management. Other organizational levels can be relevant for other activities. For example, you would maintain sales data depending on the sales organization and distribution channel, and you would specify a warehouse number and storage type for warehouse management data.

Maintaining Material Master Records

The breakdown of material data according to user departments and organizational levels is also reflected in material master record maintenance. When you maintain material master records, you must go through several dialog screens before you start creating, changing, or displaying data.

Figure 2.8 illustrates this: After the initial screen, you see two consecutive dialog boxes. In the **Select View(s)** dialog box ❶, you specify the views you want to process. In the **Organizational Levels** dialog box ❷, you enter the relevant organizational levels. Next, you see the data screens ❸. The data of the material is organized according to views and organizational structure. Here you can see the purchasing data for plant 1010.

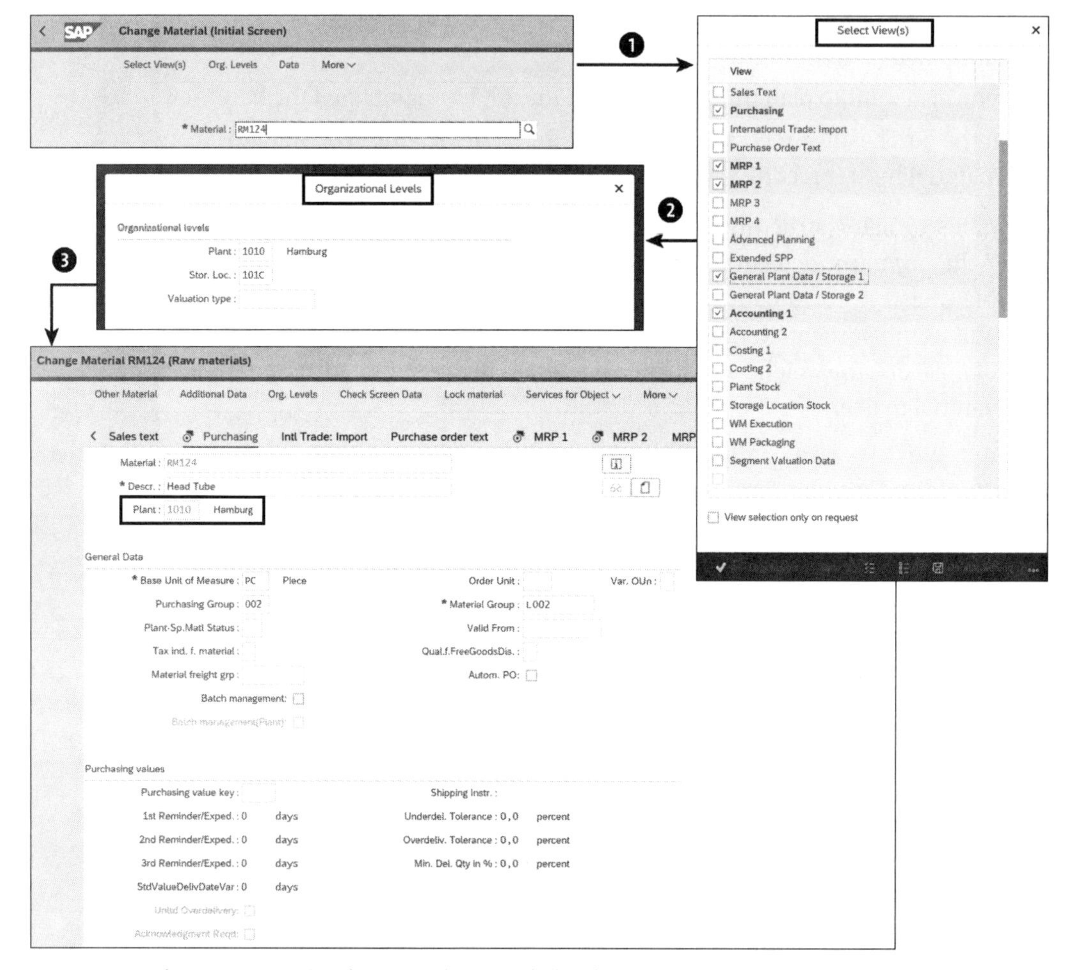

Figure 2.8 Accessing the Material Data with the Change Material App

You can preset both the organizational units and the views to be selected and save this setting. If you have made and saved a default selection, you can even run the dialog boxes in the background.

Example

The following examples illustrate the difference between material change and material extension:

- Example 1: You have already created the Purchasing view of a material for a certain plant, but now you discover that the purchasing group you have stored is no longer correct and you want to store a different purchasing group. To do this, you must choose the Change Material app. You are *changing* the contents of an existing view/organizational unit.

- Example 2: You have already created the Purchasing view and the Accounting view of a material for a certain plant, but now you want to extend the material by the MRP view. To do this, you must choose the Create Material app. You are *creating* a new view.
- Example 3: You have already created the Purchasing view of a material for a certain plant, but now you want to create the Purchasing view for a second plant. To do this, you must choose the Create Material app. You are *creating* data for a new plant.

Important Material Master Fields

In this section, we will describe some important fields that are generally relevant. Other process-related fields are discussed in the respective chapters.

Material Type

Each time you create a new material master record, you must select a material type to which the material is to be assigned. Materials with the same properties are assigned to the same material type. Examples of material types are raw materials, semi-finished products, and finished products.

The material type is a key, each material is assigned to a material type, and the material type performs important control functions. Among other things, the material type influences the following:

- Which number the material should have (internally assigned, user assigned, numeric, alphanumeric, number range, and so on)
- Whether the material should be a product or a service
- Which screens appear in which sequence during material master maintenance
- Which fields in the master record are required fields, optional fields, or hidden fields that should only be displayed (in other words, the material type influences field selection control)
- Which procurement type is allowed for the material (produced in-house, procured externally, or both)
- Which valuation classes can be specified in the material master record—that is, which accounts are posted to for goods movements of this material
- Which views can be created
- For which plants are quantity and value-based inventory management allowed, only quantity-based inventory management allowed, or no inventory management allowed at all

In standard SAP S/4HANA, various material types are available for use. If you need additional material types, you can define them in Customizing according to your requirements.

Industry Sector

Like the material type, the industry sector is a key, can be created and maintained in Customizing, and has a control function. When you create a material master record, you must specify an industry sector. The industry sector determines the following:

- Which screens are displayed and in what sequence
- Which industry-specific fields are displayed on each screen

The industry sector that you assigned when you created a material master record cannot be changed later. But in Customizing, you can define new industry sectors to modify the screen and field selection according to your requirements.

The industry sector is usually unique to a company. Once it has been defined, it can be preset so that a user does not have to make the same entry each time.

Material Status

The material status is a field that you can assign in the material master record if you want to restrict the usability of a material. It is a key that is configured in Customizing. Depending on the Customizing settings, a material status can prevent a material from being purchased (e.g., in case of a phase-out material) and/or moved in the warehouse (e.g., because of quality issues) and so on. Alternatively, a warning can be issued. This restriction can be planned with a valid-from date.

You can set either a cross-plant material status or plant-specific material statuses. Figure 2.9 shows possible entries for the cross-plant material status and Figure 2.10 shows the plant-specific material status with the valid-from date on the Purchasing view.

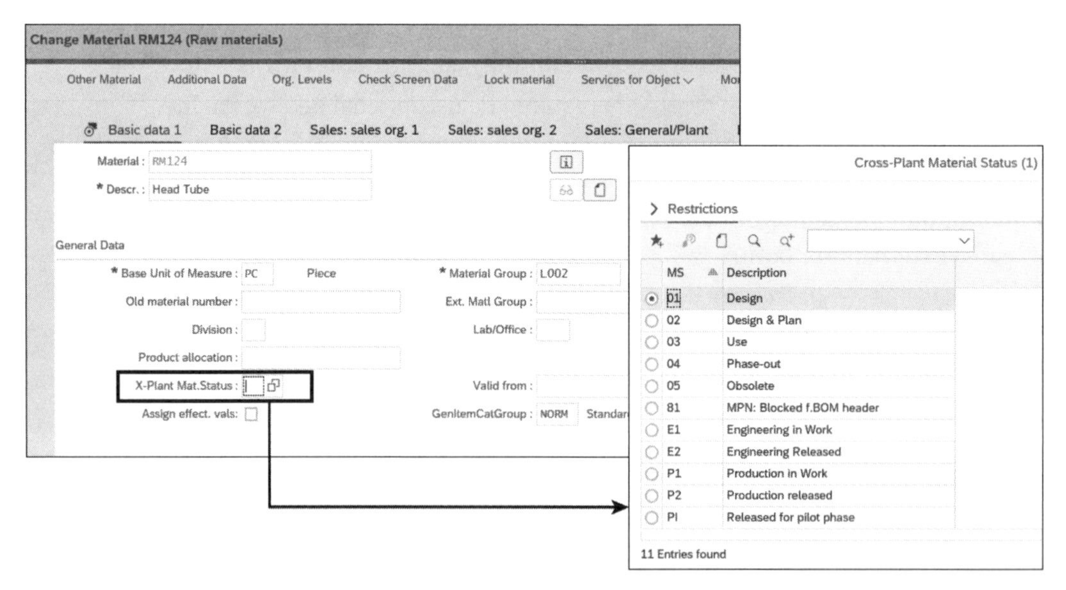

Figure 2.9 Cross-Plant Material Status

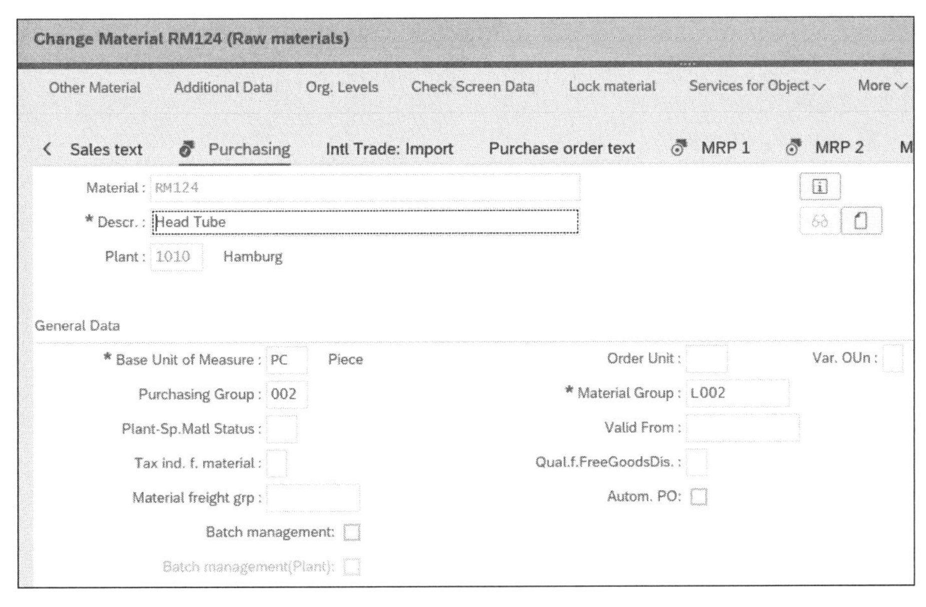

Figure 2.10 Plant-Specific Material Status

Note

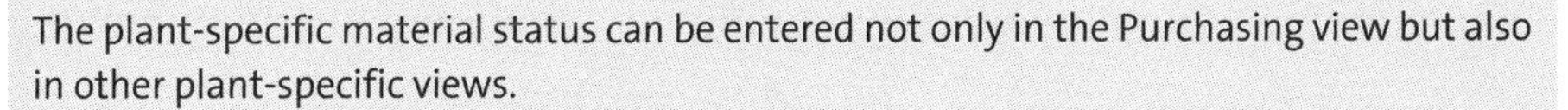

The plant-specific material status can be entered not only in the Purchasing view but also in other plant-specific views.

Business Partner Master Data

In this section, we will first explain the business partner approach. Then we will explain some key terms, review the data structure, and comment on some fields.

The Business Partner Approach

The master data of vendors and customers are managed in SAP S/4HANA via the business partner master data.

The business partner is a central master data object for all natural and legal persons with whom a company has a business relationship. From a technical point of view, the business partner is a bracket for suppliers and customers:

- The basic data (number, name, address, bank data, tax number, and so on) is maintained centrally at the business partner level.
- Depending on the assigned business partner roles, the business partner can be used as a vendor and/or customer, and customer and/or supplier master records are created in the background.

This synchronized update of vendor and customer master records in the background is called customer/vendor integration (CVI), as shown in Figure 2.11.

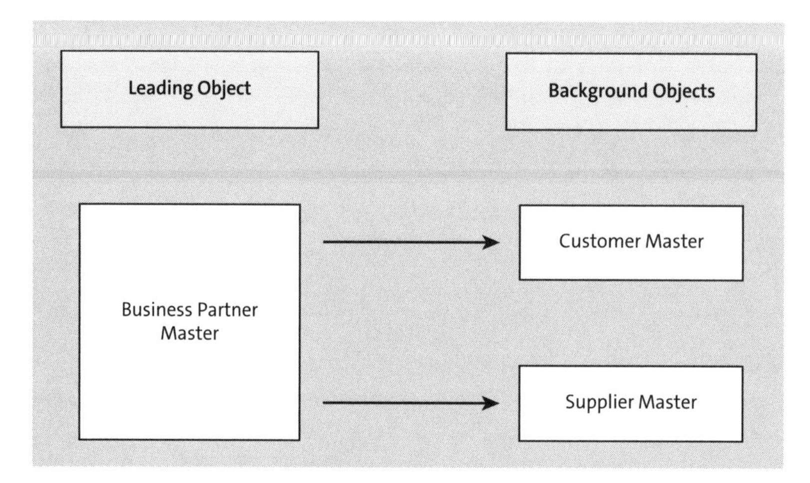

Figure 2.11 Customer/Vendor Integration Synchronization

Together with the business partner approach, some terms have been introduced that can be perplexing at first and therefore need to be explained carefully:

- **Business partner category**
  When you create a business partner, you must select the business partner category. The business partner category is used to classify the business partner as follows:
  - A natural person (such as a private individual or an employee)
  - A group (such as a community of inheritors)
  - An organization (legal entity or part of a legal entity, such as a company or a department of a company)

  The business partner category determines which fields are available for data entry. The assignment of the business partner category is static and cannot be changed once the business partner has been created.
- **Business partner role**
  By selecting the business partner role, you decide whether a business partner—of the category Organization—should appear as a customer and/or supplier. You also decide whether a supplier or customer is only used for financial accounting postings (role FI Vendor or FI Customer) or whether it is relevant for the procurement or sales process (role Supplier or Customer).

  A business partner can assume one or more roles; roles can be deleted or added for a business partner. For example, someone can also be created as a business partner and later assume the role Contact Person.
- **Business partner grouping**
  The grouping determines the numbering of the business partner and the resulting master records in the background (customer and/or supplier). The grouping

is assigned once and cannot be changed. Figure 2.12 shows the important grouping fields on the initial screen of the Maintain Business Partner app.

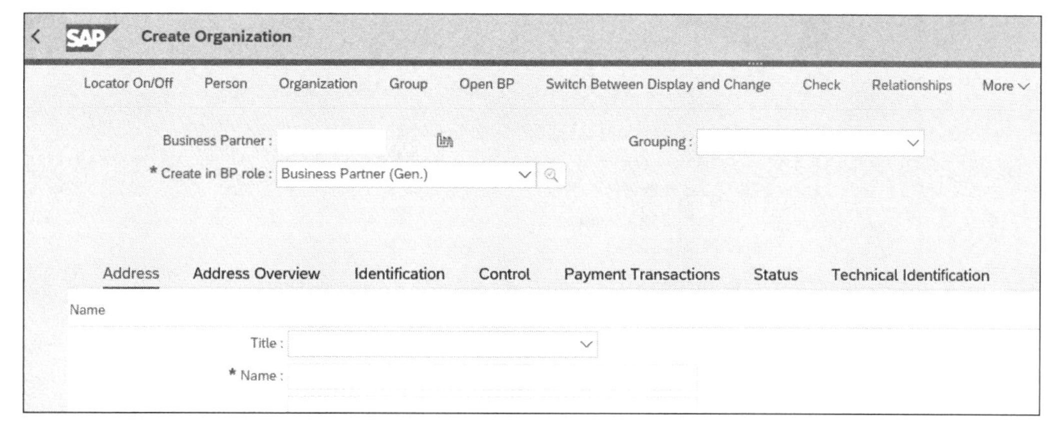

Figure 2.12 Initial Screen of the Maintain Business Partner App

- **Relationship**
  You can use relationships to link your business partners with each other. Figure 2.13 shows a possible relationship between a person and a company; here, Mrs. Miller is the contact person for Company X. Figure 2.14 shows a selection of relationships when maintaining a business partner in SAP S/4HANA.

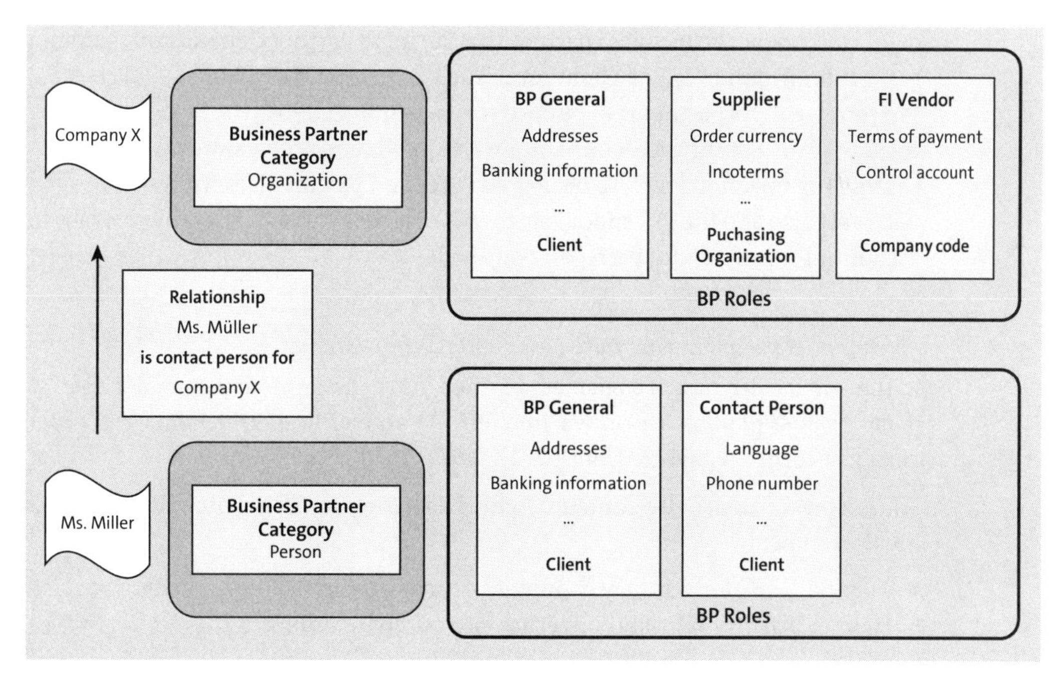

Figure 2.13 Relationships between Business Partners

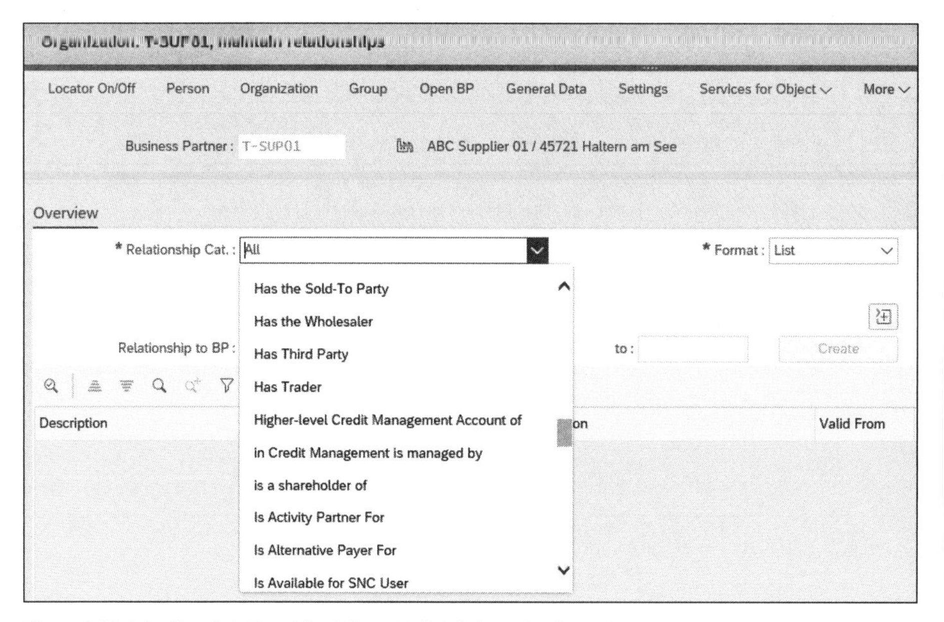

Figure 2.14 Selecting Relationships When Maintaining a Business Partner

Structure of the Supplier Master Data

As in the material master, the supplier data is also organized along the organizational structure: When you create a business partner, you first create the general data. You choose the business partner role Business Partner General and maintain general information at the client level, such as title, name, address, and so on. If this new business partner is to be created as a vendor, choose the business partner role Supplier. You must then maintain the purchasing-relevant data at the purchasing organization level. To be able to enter and pay invoices from this vendor, you must also add the FI Vendor role to this business partner. The data relevant to accounting is entered for each company code.

Purchasing-Relevant Master Data Fields and Control Parameters

In the purchasing data of a business partner, you can maintain data that is used as default values in purchase orders, info records, and outline agreements (e.g., order currency, terms of payment, etc.).

The business partner also contains control data for purchasing processing, including the following:

- The indicator for automatic purchase orders, which allows purchase requisitions to be automatically converted into purchase orders
- The indicator for order confirmation requirement, which allows missing order confirmations to be dunned
- The indicator for automatic goods receipt settlement, which enables the evaluated receipt settlement (ERS) procedure

Partner Functions

During the procurement process, a supplier assumes several functions, including ordering address, goods supplier, invoicing party, and payee. One or more of these roles can also be taken over by other suppliers.

Example

Your purchasing organization buys steel from the Steel, Inc. group. The group subsidiary Southwest Steel receives and processes your purchase orders (ordering address role). The Stuttgart/Germany plant is responsible for delivering goods; return deliveries should also be made to the Stuttgart plant (goods supplier role).

A business partner master record must be created for the Steel, Inc. group, as well as for Southwest Steel and the Stuttgart/Germany subsidiary. In the master record of the Steel Inc. group—in the **Partner Functions** tab—the number of the subsidiary Southwest Steel is entered for the role of ordering address and the number of the Stuttgart/Germany plant for the role of goods supplier.

If you order steel from Steel, Inc., then the system automatically determines the ordering address; if you return goods, then the system also automatically determines the goods supplier address for the return delivery.

Figure 2.15 shows how purchasing-relevant master data fields are stored on different tab pages. If necessary, you maintain different partner functions on the corresponding tab.

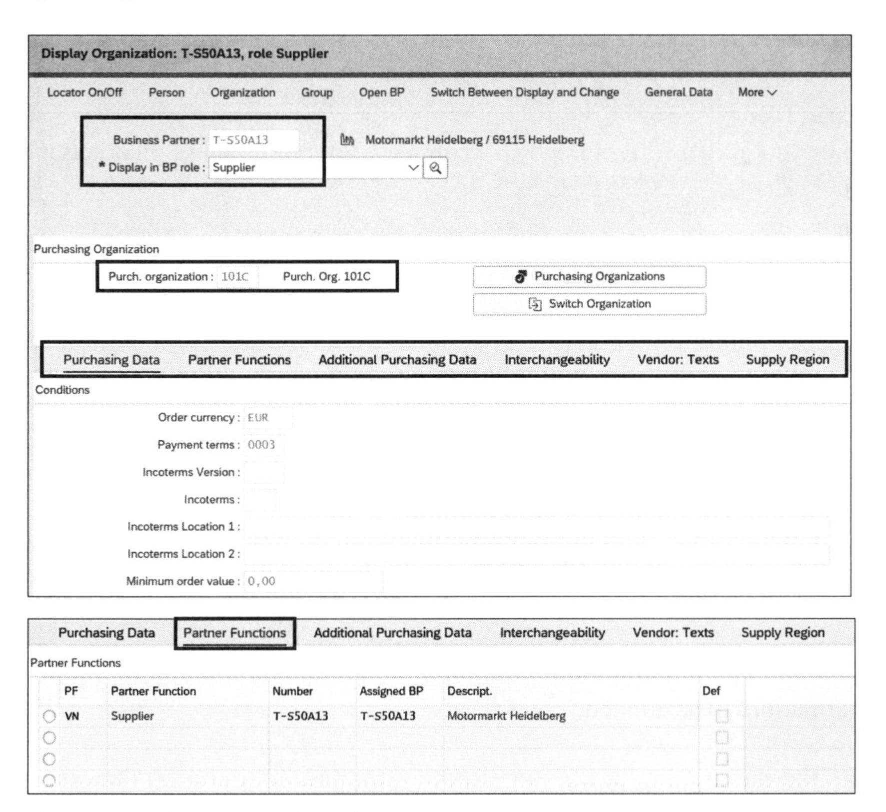

Figure 2.15 Purchasing Data of a Business Partner

Important Terminology

The following terminology was used in this chapter:

- **Purchase requisition**
  A purchase requisition is a request or instruction to the purchase to procure a certain quantity of a material or a service so that it is available at a certain time.
- **Item category**
  The item category is a key that controls the type of procurement.
- **Account assignment category**
  The account assignment category is a key that specifies that and to which account assignment objects a purchase order item is assigned.
- **Purchase order history**
  The purchase order history records the document flow to a purchase order item and enables it to be monitored.
- **Material document**
  The material document is generated during a goods movement posting and contains information about the posting.
- **Accounting document**
  The accounting document is generated when a valuated goods movement or invoice entry is made and contains the account updates.
- **Client**
  The client is the highest organizational unit and can be regarded as the business equivalent of a group or a group of subsidiaries.
- **Company code**
  It represents a legal entity for which a complete self-contained set of accounts can be created.
- **Plant**
  The plant is an operating area or branch within a company.
- **Storage location**
  A storage location is a subdivision of a plant where the stocks are physically stored.
- **Purchasing organization**
  The purchasing organization is an organizational unit that procures articles and negotiates purchase price conditions with vendors.
- **Valuation area**
  The valuation area is the organizational level at which materials are valuated—for example, at the plant or company code level.
- **MRP area**
  The MRP area represents an organizational unit where MRP is carried out independently.

- **Material type**
  The material type is a control key for material master maintenance and management.
- **Business partner**
  A person, organization, group of persons, or group of organizations with which a company has business relations.

 Practice Questions

These questions will help you evaluate your understanding of the topics covered in this chapter. They are similar in nature to those on the certification examination. Although none of these questions will be found in the exam itself, they will allow you to review your knowledge of the subject.

Select the correct answers, and then check the completeness of your answers in the next section. Remember that on the exam, you must select all correct answers—and only correct answers—to receive credit for the question.

1. You create a purchase order for a standard procurement manually. With which field value do you control that you start a standard procurement process?

☐ **A.** You leave the account assignment category blank.

☐ **B.** You leave the item category blank.

☐ **C.** You choose the order type UB.

☐ **D.** You choose the purchasing group Standard.

2. True or false: No material master record is required for the material to be ordered on consignment.

☐ **A.** True

☐ **B.** False

3. Which of the following steps belong to the stock transfer process with a stock transport order? (There are three correct answers.)

☐ **A.** Create a purchase order.

☐ **B.** Post goods issue.

☐ **C.** Post invoice.

☐ **D.** Post goods receipt.

4. True or false: In a stock transfer process, the stock in transit belongs to the valuated stock of the supplying plant.

☐ **A.** True

☐ **B.** False

5. You should set up the organizational structure as part of an SAP S/4HANA implementation. In purchasing, it should be possible to agree on prices and conclude contracts centrally. The local organizations must have access to these central conditions in order to create purchase orders. What do you suggest?

☐ **A.** Several local purchasing organizations responsible for one or more plants and a reference purchasing organization assigned to all these purchasing organizations

☐ **B.** Several local purchasing organizations responsible for one or more plants and a standard purchasing organization assigned to all these purchasing organizations

☐ **C.** Several local purchasing organizations responsible for one or more plants and an additional global purchasing organization assigned to all these plants

6. Which of the following tasks does the material type perform? (There are three correct answers.)

☐ **A.** Control field selection

☐ **B.** Control the MRP procedure

☐ **C.** Define the possible procurement types

☐ **D.** Control the number assignment

7. The user departments of the company where you operate create the material master data one after another: first, the purchasing department creates the Purchasing view, and then the planning department creates the MRP view. With which SAP Fiori app (or transaction) does the planning department maintain its data?

☐ **A.** The Create Material app

☐ **B.** The Change Material app

8. True or false: In SAP S/4HANA, the business partner master record replaces the vendor and customer master records.

☐ **A.** True

☐ **B.** False

9. Which of the following is a characteristic of a plant? (There are two correct answers.)

- ☐ **A.** A plant belongs to exactly one company code.
- ☐ **B.** A plant has its own material data.
- ☐ **C.** Material stocks are managed by quantity at the plant level.
- ☐ **D.** Only one purchasing organization can be assigned to a plant.

10. Which views must be created in advance so that assessed stock material can be ordered and put away? (There are two correct answers.)

- ☐ **A.** Basic Data 1
- ☐ **B.** General Plant Data/Storage 1
- ☐ **C.** Purchasing
- ☐ **D.** Accounting 1

11. True or false: If the valuation area corresponds to the plant level, you can valuate a material differently for each plant.

- ☐ **A.** True
- ☐ **B.** False

12. True or false: The supplier consignment stock belongs to your valuated stock.

- ☐ **A.** True
- ☐ **B.** False

13. True or false: A business partner is always simultaneously a supplier and a customer.

- ☐ **A.** True
- ☐ **B.** False

14. When you create a business partner master record, which field is responsible for the business partner number assignment?

- ☐ **A.** The business partner role
- ☐ **B.** The business partner category
- ☐ **C.** The business partner grouping

15. True or false: Purchasing-specific data in the material master is entered at the purchasing organization level.

- ☐ **A.** True
- ☐ **B.** False

16. You support the implementation of SAP S/4HANA and advise on the definition of the enterprise structure. One requirement is that certain materials must be valuated differently depending on the location, since they are produced at some locations and purchased at others. What is your recommendation?

☐ **A.** The locations should be defined as plants.

☐ **B.** The locations should be defined as storage locations.

☐ **C.** The locations should be defined as MRP areas.

17. True or false: To create a purchase order, you need a supplier master record. It is enough that this supplier is created as a business partner with purchasing data. Accounting data is not required.

☐ **A.** True

☐ **B.** False

18. For what reason could you set a material status in a material master record?

☐ **A.** To control the field selection

☐ **B.** To determine the type of procurement

☐ **C.** To restrict the usability

☐ **D.** To control the account determination

19. Which of the following statements apply to a purchase order? (There are three correct answers.)

☐ **A.** One purchase order can contain items for different plants.

☐ **B.** One purchase order can contain items for different purchasing groups.

☐ **C.** One purchase order contains a single purchasing organization.

☐ **D.** A single purchase order can contain standard items and subcontracting items.

Practice Question Answers and Explanations

1. Correct answer: **B**

   The item category controls which procurement process is triggered. *Blank* stands for the standard procurement process.

   The account assignment category determines whether the order item is intended for the warehouse or for direct consumption.

   The purchase order type controls for example number assignment or field selection. The order type UB is intended for a stock transport order where a plant instead of a supplier must be entered.

The purchasing group represents a group of purchasers or an individual purchaser serving externally as a contact and internally as a basis for evaluation purposes.

2. Correct answer: **B**

   False. A material that is ordered on consignment must have a material master record, since it is managed in stock and only materials with a master record can be managed in stock. Do not confuse this with subcontracting: A material master record is not mandatory for the material to be ordered. However, components that are to be provided must have a material master record.

3. Correct answers: **A, B, D**

   In this book, we only deal with the stock transfer process within a company code, and no invoice is created during this process.

4. Correct answer: **B**

   False. In stock transfer, the stock in transit is valuated and assigned to the receiving plant. If the two plants are assigned to different valuation areas, a cross-plant stock transfer leads to a value update in the stock accounts. Valuation takes place at the time of the goods issue from the issuing plant. An accounting document is generated parallel to the material document for the goods issue. The stock transfer is valuated at the valuation price of the material in the issuing plant.

5. Correct answer: **A**

   Answer A with the creation of a reference purchasing organization corresponds exactly to the requirements: central price agreement or contract conclusion and local use of these conditions for operative purchasing. A reference purchasing organization must be assigned to other purchasing organizations that are allowed access to its conditions and/or contracts.

   The reference purchasing organization must not be confused with the standard purchasing organization; the latter is used by the system when it must determine a suitable purchasing organization automatically.

   An additional global purchasing organization (answer C) would allow prices and contracts to be concluded for all plants, but the local purchasing organizations would not be allowed to access the conditions and/or contracts.

6. Correct answers: **A, C, D**

   The tasks of the material type include controlling field selection during material master record maintenance, controlling number assignment, and defining the permitted procurement types. Using an MRP type, you can define the MRP procedure per material and plant with no restrictions.

7. Correct answer: **A**

   The material has already been created by Purchasing, so while it can be irritating to use the Create Material app again, if you want to add a view or data for a new organizational unit to an existing material—that is, if you want to extend a material—then you must use this app. You can only use the Change Material app if you want to change data in existing views or organizational units.

8. Correct answer: **B**

   False. When you create a business partner in SAP S/4HANA, a customer or supplier master record is created in the background, depending on the business partner's role.

9. Correct answers: **A, B**

   A plant belongs to exactly one company code but can be assigned to several purchasing organizations. A plant has its own material master data—that is, some material master data can be maintained on a plant-specific basis. Material stocks are managed by quantity at the storage location level.

10. Correct answers: **C, D**

    Purchasing and Accounting views must be created before you can order a valuated material for storage. Remember that in the Accounting view, you maintain the valuation data necessary for the valuation of the stock. The storage location data for the material can be created automatically in the background at the first goods receipt.

11. Correct answer: **A**

    True. If the valuation area corresponds to the plant level, you can maintain the valuation data of a material differently for each plant.

12. Correct answer: **B**

    False. Although the consignment stock is physically located on your premises, it is still the property of the vendor. A liability only arises when the stock is withdrawn.

13. Correct answer: **B**

    False. Whether a business partner is a supplier or a customer depends on the business partner role that you assign to this business partner.

14. Correct answer: **C**

    The grouping is responsible for the number assignment.

15. Correct answer: **B**

    False. Most purchasing-specific data in the material master is entered at the plant level. The purchasing organization is not relevant for maintaining the material master. Be careful, though: some purchasing-relevant fields (in the Purchasing view) are valid for the whole client and cannot be maintained differently for each plant. These include the purchasing value key, which contains, for example, dunning levels or over-delivery tolerance limits.

16. Correct answer: **A**

    If you want to valuate materials differently per plant, the valuation level must be the plant and each location must be defined as a plant.

17. Correct answer: **B**
    False. If you do not maintain the accounting data for the business partner (role FI Vendor), you cannot create and pay the invoices from this vendor.
18. Correct answer: **C**
    The material status is intended to restrict the usability of a material. For example, if you no longer want to buy a material because it is a phase-out material, then set a suitable material status. Depending on the configuration of the material status (in Customizing), the system issues either an error or a warning message when you try to create a purchase order for this material.
19. Correct answers: **A, C, D**
    The question refers to the structure of a purchase order: since the purchasing organization belongs in the order header, a purchase order can only contain one purchasing organization. The same applies to the purchasing group. On the other hand, the plant is assigned by item, so that a purchase order can contain items for several plants. The procurement process is controlled via the item category at the item level, so that a purchase order can contain items for different processes.

Takeaway

You should now understand the procurement processes offered in SAP S/4HANA and be able to describe the differences between them. You should be able to describe the most important process steps in a standard procurement process, and be able to describe the special features of subcontracting, consignment processes, and stock transport orders. You should know the most important organizational units for procurement and inventory management, along with their relationships to each other. In addition, you should understand the structure of the material master and recognize the dependencies between the material master and the organizational structure. Finally, you should be able to describe the business process approach.

Summary

This chapter has given you an overview of the procurement processes, the organizational structure, and the most important master data. It is important that you have a good understanding of the terms and relationships explained here so that you can use them well when working through the rest of the book. In the following chapters, we will go into more detail about the topics outlined in this introductory chapter. In the next chapter, we will focus on stock procurement, with the valuation of material stock being an important part of this.

Chapter 3
Procurement of Stock Material

Techniques You'll Master

- Create a stock material in SAP S/4HANA.
- Create a purchase order for stock material.
- Post a goods receipt to the warehouse with reference to a purchase order and explain the consequences of this goods receipt.
- Post a vendor invoice for a warehouse purchase order and explain the consequences of this posting.
- Explain the basics of stock valuation and maintain valuation parameters in the material master record.

In this chapter, we will explain the special features of stock material procurement; what requirements must be met in order to procure stock material, how to create a stock material purchase order, what are the effects of the goods receipt posting, and how to perform the invoice posting.

Real-World Scenario

Procurement of stock material is of importance in many companies. Processes can be simplified, and costs can be reduced through reliable demand planning, optimized and largely automated demand conversion, smooth goods receipt processing, and highly automated invoice verification. SAP S/4HANA can make all this possible. But first, you need to master the basics of warehouse material procurement. The value flow plays a very special role here: How are stock materials valued, how do the procurement prices influence inventory values, and what are the consequences of invoice price deviations? We refresh these important items that you as a consultant or key user must absolutely master.

Objectives of This Portion of the Test

The purpose of this part of the certification exam is to test your knowledge of the procurement of stock material. For the certification exam, you must be able to do the following:

- Explain how to create a stock material.
- Describe a purchase order for stock material.
- Describe goods receipt to warehouse and its impacts.
- Perform the posting of an incoming invoice and explain its effects.
- Explain the fundamentals of material valuation.

Key Concept Refresher

In this section, we will go through the main steps of a warehouse procurement process—purchase order, goods receipt, and invoice entry—and we will examine their features and effects. We observe the impacts in terms of value and explain the principles of material valuation in SAP S/4HANA.

External procurement comprises various aspects: for example, you can purchase materials or services; if you procure materials, you can purchase them for stocking or for direct consumption. If you want to order a material for stock, you must previously have created a material master record.

Material Master Record for Stock Material

A stock material master record must be created with at least the Purchasing and Accounting 1 views. The Purchasing view shown in Figure 3.1 and Figure 3.2 allows for the procurement of the material and contains important data that simplifies or limits the procurement process.

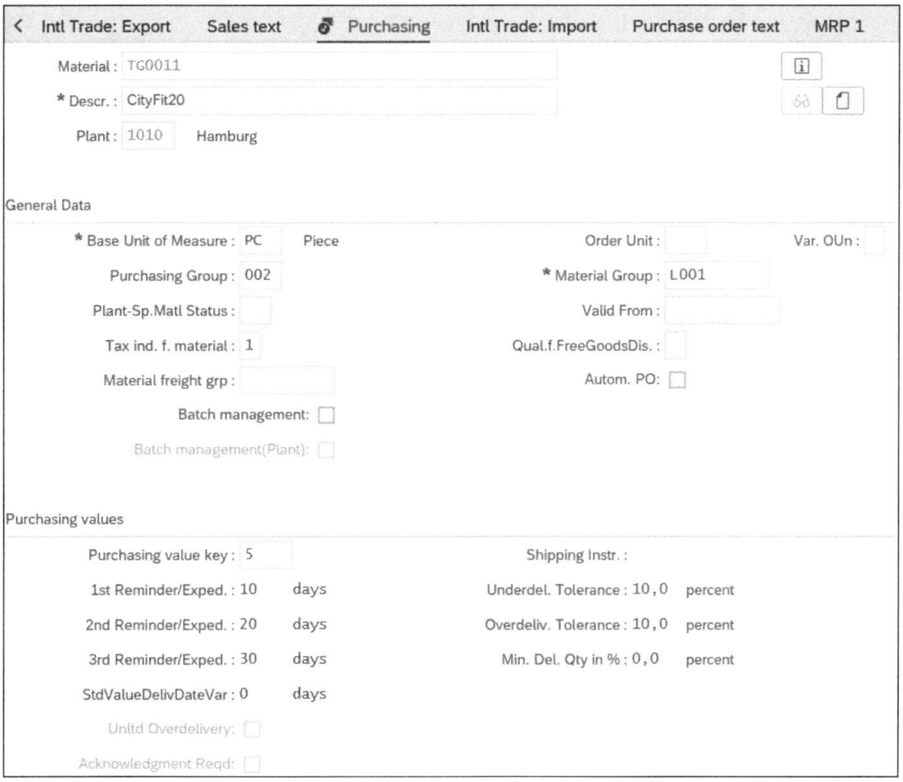

Intl Trade: Export | Sales text | Purchasing | Intl Trade: Import | Purchase order text | MRP 1

Material : TG0011
* Descr. : CityFit20
Plant : 1010 Hamburg

General Data

* Base Unit of Measure : PC Piece
Order Unit :
Var. OUn :
Purchasing Group : 002
* Material Group : L001
Plant-Sp.Matl Status :
Valid From :
Tax ind. f. material : 1
Qual.f.FreeGoodsDis. :
Material freight grp :
Autom. PO:
Batch management:
Batch management(Plant):

Purchasing values

Purchasing value key : 5
Shipping Instr. :
1st Reminder/Exped. : 10 days
Underdel. Tolerance : 10,0 percent
2nd Reminder/Exped. : 20 days
Overdeliv. Tolerance : 10,0 percent
3rd Reminder/Exped. : 30 days
Min. Del. Qty in % : 0,0 percent
StdValueDelivDateVar : 0 days
Unltd Overdelivery:
Acknowledgment Reqd:

Figure 3.1 Purchasing Data in the Material Master Record (I)

Other data / manufacturer data

GR processing time : days
Post to Insp. Stock:
JIT Sched. Indicator :
Critical Part:
Mfr part profile :
Source list:
Manufacturer :
Mfr Part Number :

Figure 3.2 Purchasing Data in the Material Master Record (II)

The material master record generally contains data that is proposed as default values in subsequent process documents. The material master record also contains data that controls or restricts further process steps (control data). The following fields are particularly important:

- The automatic purchase order (**Autom. PO**) indicator is a prerequisite for the automatic conversion of purchase requisitions into purchase orders.
- The **Purchasing Value Key** field is a key that is assigned to a material in the Purchasing view. The key is configured in Customizing and contains the following settings:
  - Reminder intervals for outstanding deliveries
  - Over- and under-delivery tolerances
  - Indicator for order confirmation request
- You use the **Post to Insp. Stock** indicator to plan the goods receipt posting to inspection stock.

Warning

The purchasing value key is valid for the whole client. You cannot define different purchasing value keys for individual plants.

The Accounting view shown in Figure 3.3 is a prerequisite for maintaining the material in stock. In the Accounting view, you specify the valuation procedure, the valuation price, and the valuation class—usually for each plant. The valuation class is a parameter for automatic account determination.

Figure 3.3 Accounting Data in the Material Master Record

If you also want to plan the requirements of your material automatically using MRP, you must also maintain at least the MRP 1 and optionally MRP 2 views.

You do not necessarily have to create the storage location views (Plant Data/Storage 1 and 2). You can use a Customizing setting to control the storage location view of a material, so that it is automatically created in the background with the first goods receipt posting. You can adjust this setting per plant and movement type.

Figure 3.4 shows the path in Customizing for inventory management to allow the automatic creation of the storage location view of a material at the first goods receipt.

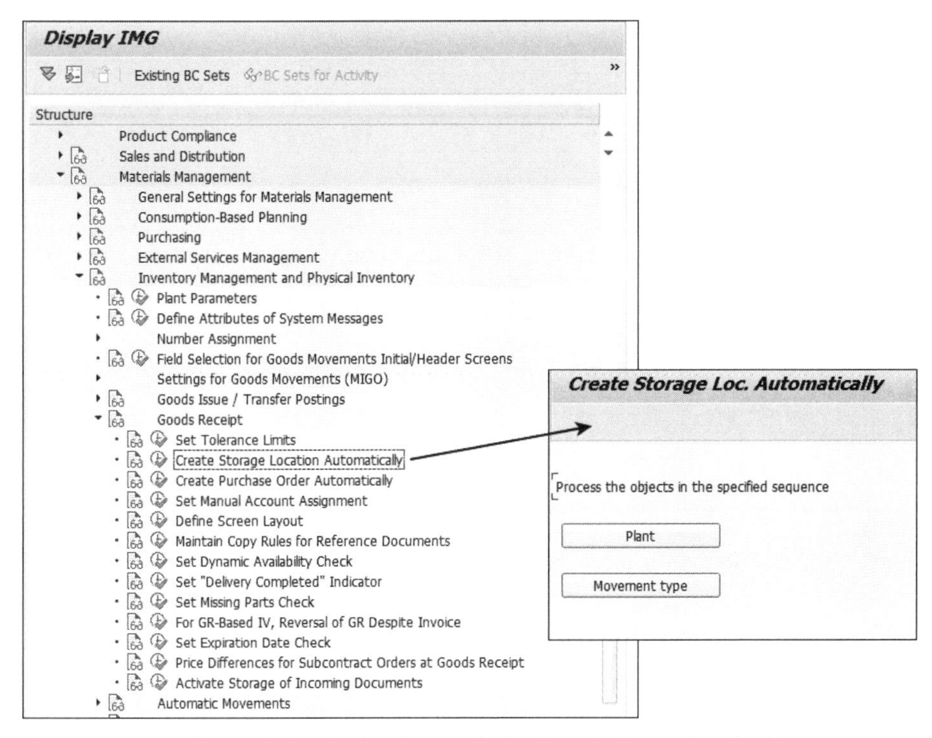

Figure 3.4 Customizing Path for Allowing Automatic Creation of a Storage Location View

For each plant, you can decide whether the storage location view of a material may be automatically created at the first goods receipt, as shown in Figure 3.5.

Change View "Autom. Creation of SLoc per Plant": Overview

Plant	Name 1	Create SLoc. automat.
1010	Hamburg	☑
1020	Berlin	☑
1030	Frankfurt	☑
1040	Munich	☑
1710	Plant 1 US	☐
CF99	Hamburg	☑
TA00	Plant 1 DE	☐

Figure 3.5 Allowing Automatic Creation of a Storage Location View per Plant

You can then define for each movement type whether the storage location view of a material may be created automatically at the first goods receipt. Figure 3.6 shows that this is allowed for movement type 101 (goods receipt for a purchase order or production order) but not for 103—the movement type for goods receipt for purchase order into goods receipt blocked stock.

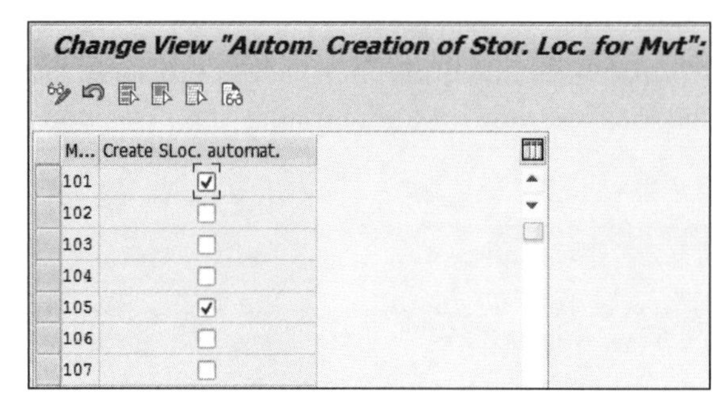

Figure 3.6 Allowing Automatic Creation of a Storage Location View per Movement Type

The terminology in Customizing is somewhat confusing: this setting does not mean the automatic adding of a storage location to the enterprise structure, but the automatic creation of a missing material view (storage location view) in the background—that is, the enhancement of a material master record.

Example

Let's explain the meaning of this setting with an example. Assume you have ordered a material for plant 1010. You now post the goods receipt with reference to the purchase order (movement type 101) to a storage location in plant 1010. The material does not yet have any stock in this storage location, and the corresponding storage location view of the material master record has not yet been created. The settings shown in Figure 3.6 (plant 1010 and movement type 101 selected) allow the system to automatically generate the view in the background, together with the goods receipt posting.

Purchase Order for Stock Material

There are several ways to create a purchase order. If you want to create a purchase order from scratch, you can use the Create Purchase Order, Create Purchase Order Advanced, and Manage Purchase Order apps.

You can also convert purchase requisitions manually or automatically into purchase orders. We will discuss these possibilities in detail in Chapter 6. Here we concentrate on the data that make up a stock material order.

You can create a stock purchase order item as follows:

- You first enter the header data: order type (only if you use the Create Purchase Order – Advanced app or Transaction ME21N), vendor, purchasing organization, purchasing group, company code. The order currency is derived from the vendor.

- You then enter the items. In a purchase order, you can basically include both stock material items and consumable material items.
- A purchase order item always contains precisely the required goods, the plant, the quantity, the purchasing price, and the desired delivery date.
- A stock material order item is characterized by:
  - A material with a material number, as shown in Figure 3.7
  - No account assignment category (account assignment category = blank), also shown in Figure 3.7
  - An expected valuated goods receipt, as shown in Figure 3.8; you cannot deselect the goods receipt indicator

Figure 3.7 Stock Material Purchase Order Item: Material with Material Number and Account Assignment Category Blank

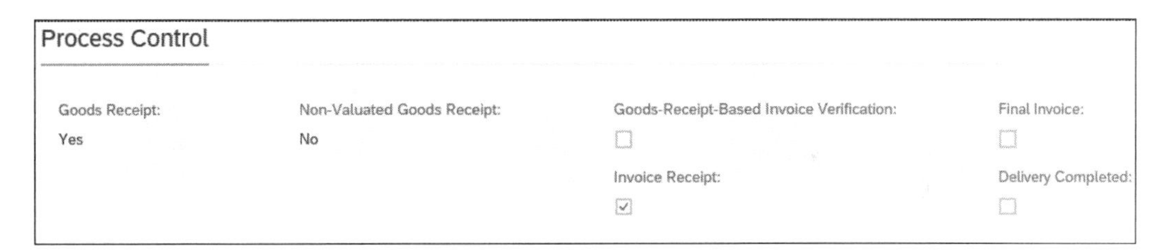

Figure 3.8 Stock Material Purchase Order Item: Valuated Goods Receipt Mandatory

Figure 3.8 shows the process control of a stock material purchase order item: the **Goods Receipt** indicator is marked with **Yes**, the **Non-Valuated Goods Receipt** indicator is marked with **No**, and neither indicator can be changed. It means that a valuated goods receipt is expected for a stock material purchase order item.

A storage location can but does not have to be e specified in the purchase order item. The storage location must be entered at the latest at goods receipt.

Goods Receipt with Reference to a Stock Material Purchase Order

In this section, we will explain how to post a goods receipt with reference to a stock material purchase order. We will then discuss the results of the goods receipt posting.

Posting a Goods Receipt

SAP S/4HANA offers several options for posting a goods receipt with reference to a purchase order, especially the Post Goods Receipt for Purchase Order app or transaction code, Transaction MIGO (Post Goods Movements).

In all cases, you must first enter the purchase order reference. The system then proposes the open purchase order items with the corresponding open quantities and other data from the purchase order item, such as storage location if available, or the stock type. This data can be overwritten.

Figure 3.9 and Figure 3.10 show Transaction MIGO (Post Goods Movements). Proceed as follows to start the entry of a goods receipt with reference to a purchase order:

1. Choose **Goods Receipt** in the **Transaction** field.
2. Choose **Purchase Order** in the **Reference** field.
3. Enter the purchase order number.
4. Verify that the movement type is **101**.
5. Choose **Execute**.

Figure 3.9 Entering the Purchase Order Reference in the Post Goods Movements App (Transaction MIGO)

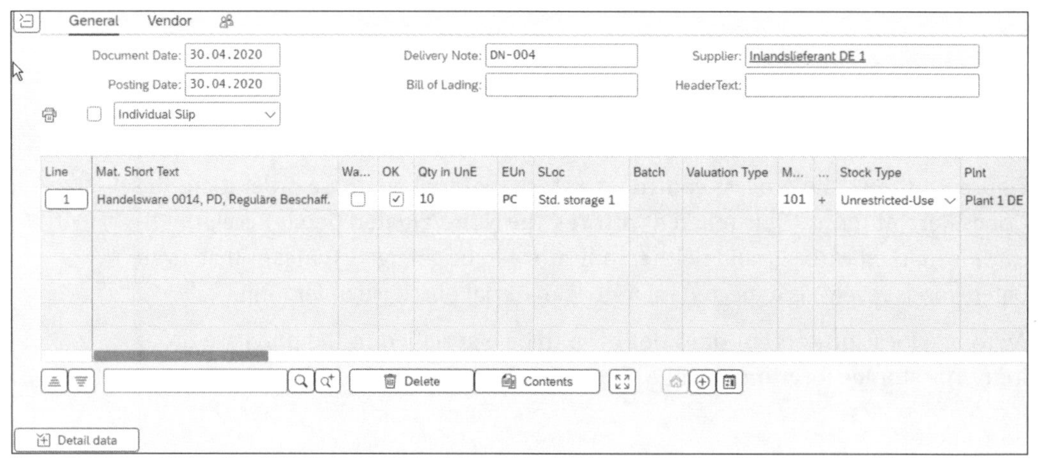

Figure 3.10 Proposed Items in the Post Goods Movements App (Transaction MIGO)

The system then proposes the purchase order data. Proceed as follows:

1. Enter a delivery note and check document and posting dates, as shown in Figure 3.10.

2. Check (and, if required, change) the proposed quantity, storage location, and stock type.
3. Flag the **OK** indicator.
4. Choose **Post**.

Figure 3.11 shows the Post Goods Receipt for Purchase Order app, which is intuitive and easy to use as follows:

1. Enter the purchase order number.
2. Enter a delivery note and check document and posting dates.
3. Select the relevant items, check (and, if required, change) the proposed quantity, storage location, and stock type.
4. Choose **Post**.

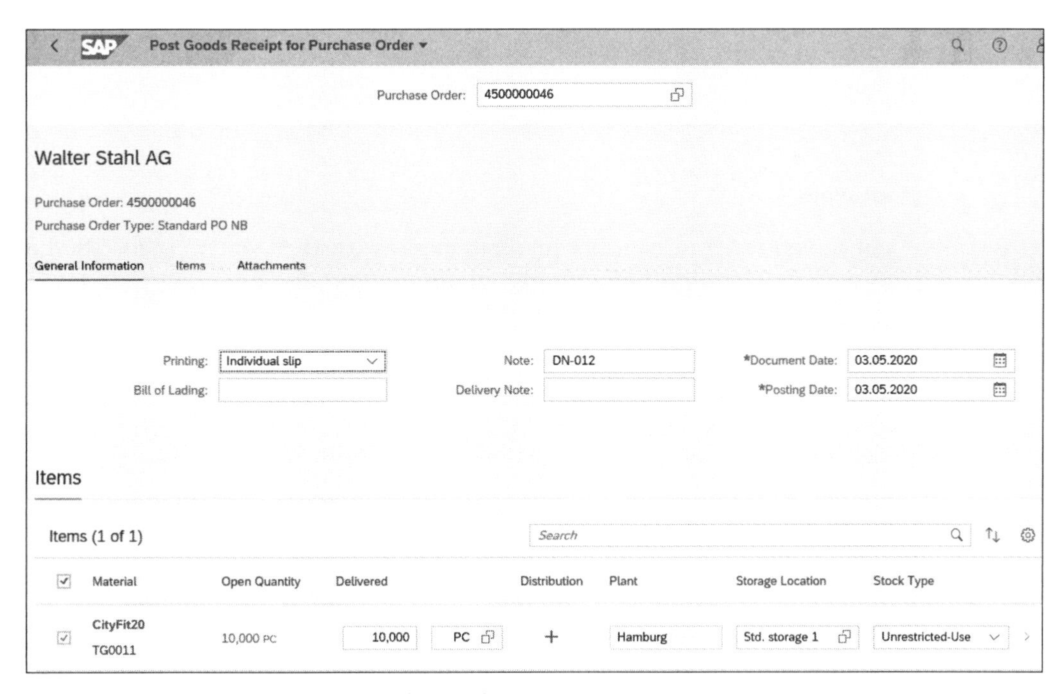

Figure 3.11 The Post Goods Receipt for Purchase Order App

The Post Goods Receipt for Purchase Order app has the following restrictions compared to Transaction MIGO:

- You can only enter the goods receipt for one purchase order and not for several.
- You cannot post a goods receipt for a purchase order item for a serialized material.
- You cannot post a goods receipt for a purchase order item with the delivery completed indicator.

When you post a goods receipt for a stock material purchase order, the quantities and values are usually updated.

Results of a Goods Receipt

The effects of a goods receipt for a stock material order are as follows:

- The material stock is increased in quantity and value.
- The purchase order history is updated. The purchase order history allows you to monitor the procurement process and shows all goods receipts, return deliveries, invoices, credit memos, and so on for each purchase order item.
- A material document is created. The material document provides information about the goods movement and is a source of information for all subsequent processes. The material document consists of a header and at least one item. The header contains general data, such as the date and the delivery note number. The items describe the individual movements (material number, quantity, movement type, plant and storage location, and so on).
- An accounting document is created. Using automatic account determination, the system updates the general ledger accounts that are affected by a goods movement.

Stock Types

When you enter a goods receipt for a purchase order item, you must specify the storage location and stock type for the quantity posting, as shown in Figure 3.12.

Figure 3.12 Storage Location and Stock Types

The stock type controls the usability of a quantity of material: While you can post all goods withdrawals from *unrestricted-use stock*, you can only scrap quantities in *quality inspection* or *blocked stock*, transfer them to another stock type, or take samples. Unrestricted-use stock and stock in quality inspection are relevant for material planning. You can define in Customizing whether or not blocked stock is relevant for materials planning.

If the purchase order item quantity is to be divided among different storage locations or stock types, you can split the proposed item, as shown in Figure 3.13.

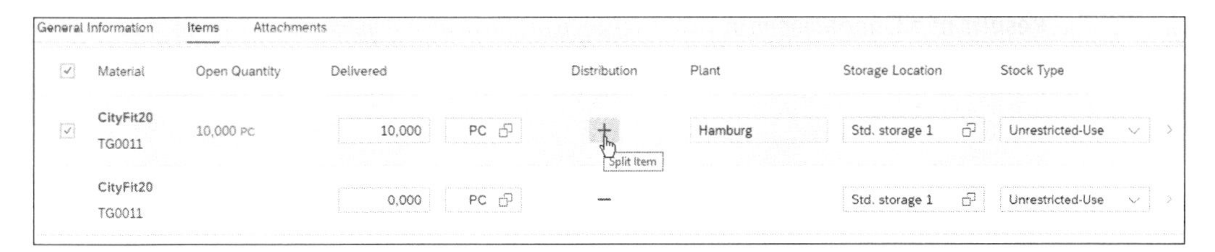

Figure 3.13 Split Item

It is possible to specify the stock type in the purchase order item. This is then proposed at goods receipt.

You can even stipulate the posting to inspection stock in the material master record: If you want to place a material in inspection stock first, for example, because it is a new material or because it is critical, set the **Post to Insp. Stock** indicator in the material master record at the plant level (Purchasing view), as shown in Figure 3.14. This indicator causes the stock type *inspection stock* to be set automatically in purchase order items and then in the goods receipt item. You can therefore plan the posting to inspection stock.

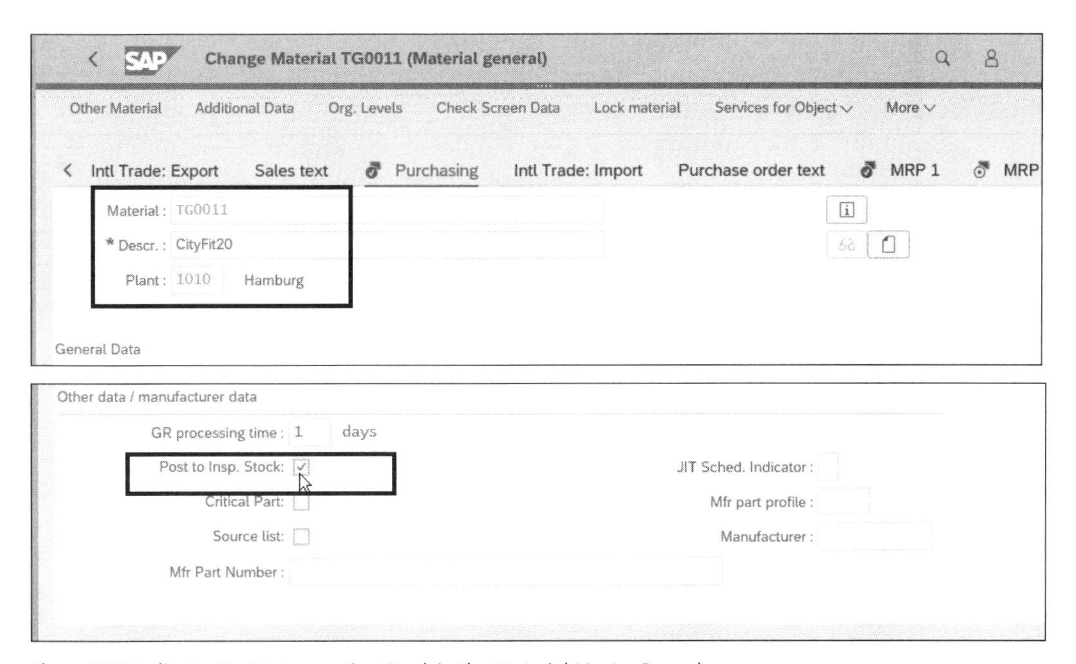

Figure 3.14 Indicator Post to Inspection Stock in the Material Master Record

Goods Receipt Blocked Stock

The purpose of the goods receipt (GR) blocked stock is to accept delivered goods under conditional acceptance. Unlike the other types of stock (unrestricted-use stock, stock in quality inspection, and blocked stock), GR blocked stock is not part of your regular stock in terms of quantity and value. It is not assigned to a storage location and is not valuated. Only the purchase order history is updated. The material document created at the time of the goods receipt serves as the document for the goods receipt.

You can see the GR blocked stock in several places:

- In the purchase order history
- In the stock overview (Transaction MMBE or the Stock Overview app)
- With the Overdue Materials—GR Blocked Stock app, which allows you to keep track of the quantities in GR blocked stock and to complete the process if necessary

If the conditions for accepting the delivery are met, you transfer the material to your regular stock. You transfer the materials from GR blocked stock either to unrestricted-use stock, to stock in quality inspection, or to blocked stock. This transfer posting is valuated, and an accounting document is created.

Figure 3.15 and Figure 3.16 show how you can monitor GR blocked stock using the Overdue Materials—GR Blocked Stock app and then transfer these materials to your regular stock using linked apps.

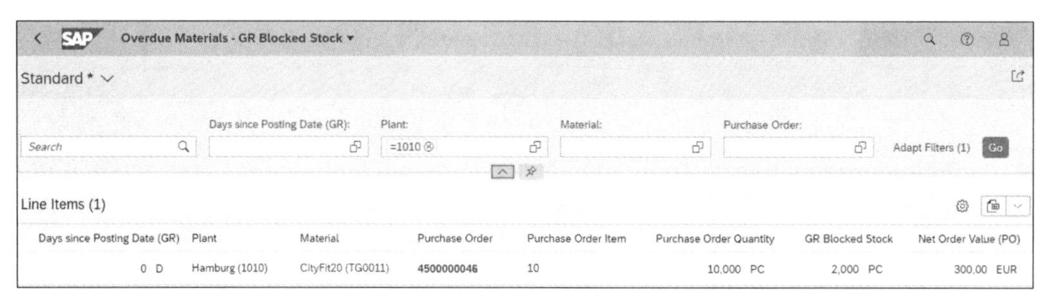

Figure 3.15 Overview of Quantities in GR Blocked Stock

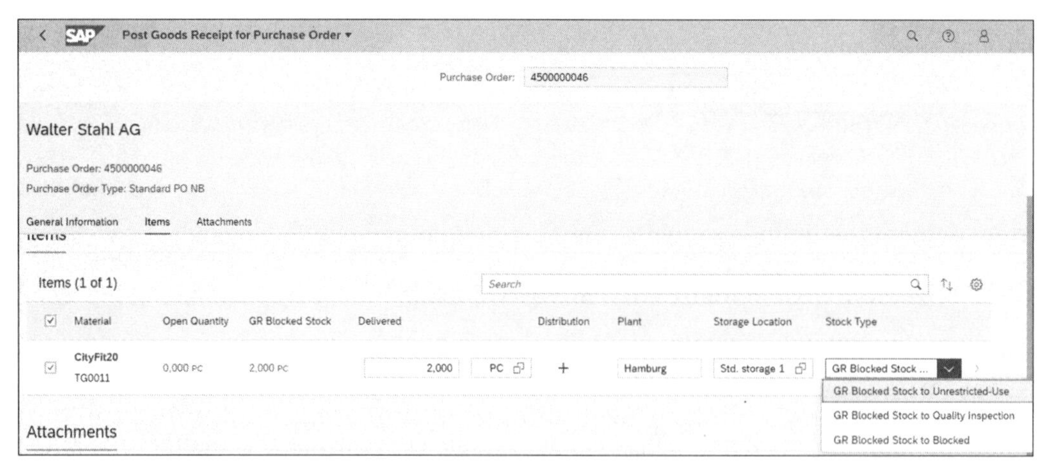

Figure 3.16 Transfer GR Blocked Stock to Own Regular Stock

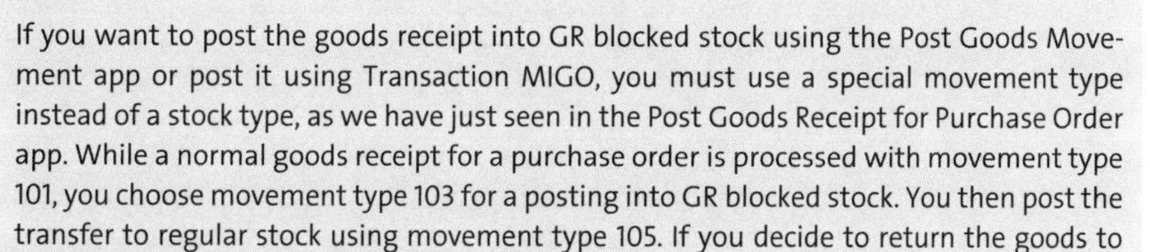

Note

If you want to post the goods receipt into GR blocked stock using the Post Goods Movement app or post it using Transaction MIGO, you must use a special movement type instead of a stock type, as we have just seen in the Post Goods Receipt for Purchase Order app. While a normal goods receipt for a purchase order is processed with movement type 101, you choose movement type 103 for a posting into GR blocked stock. You then post the transfer to regular stock using movement type 105. If you decide to return the goods to the vendor, choose movement type 124.

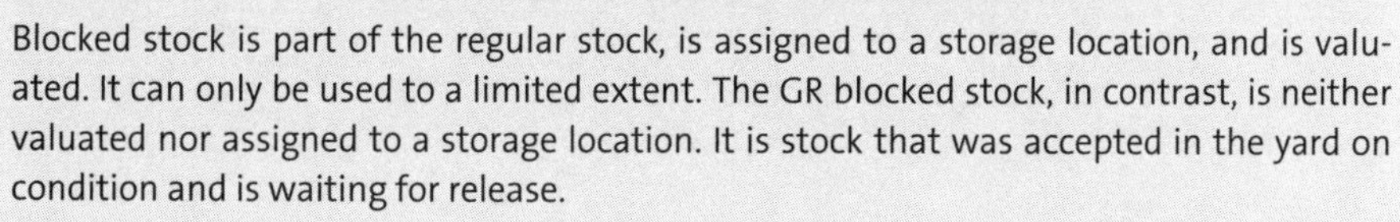

Warning

Do not confuse the stock type blocked stock with the GR blocked stock.

Blocked stock is part of the regular stock, is assigned to a storage location, and is valuated. It can only be used to a limited extent. The GR blocked stock, in contrast, is neither valuated nor assigned to a storage location. It is stock that was accepted in the yard on condition and is waiting for release.

Under-Deliveries and Over-Deliveries

When you post a goods receipt for a purchase order, the system proposes the open purchase order quantity. The open purchase order quantity is the difference between the quantity ordered and the quantity already delivered. You can, of course, overwrite this proposed quantity according to the information on the delivery note and the actual quantity delivered.

Under-deliveries or partial deliveries are always allowed. If you have defined an *under-delivery tolerance limit* in the purchase order item and if the sum of the quantities delivered so far and the quantities currently entered falls below this limit, then the system generates a warning message—and that is all.

Over-deliveries for a purchase order item are only permitted if this is indicated in the purchase order item. You have the option of setting the indicator for *unlimited over-delivery* in the purchase order item or specifying an *over-delivery tolerance* as a percentage. If the sum of the quantities already delivered and the quantities currently entered lies within the over-delivery tolerance, then SAP S/4HANA does not issue a message. If the sum exceeds the tolerance, the system issues an error message and you cannot post the goods receipt.

The over- and under-delivery tolerance limits can be defined in the following places:

- In the material master, in the Purchasing view via the purchasing value key
- In the purchasing info record for a combination supplier material
- In the purchase order item; values that are contained in the material master record or in the purchasing info record are proposed in the purchase order item and can be overwritten there

Figure 3.17 shows the default values in the material master record.

Material : TG0011
* Descr. : CityFit20
Plant : 1010 Hamburg

General Data

* Base Unit of Measure :	PC Piece	Order Unit :	
Purchasing Group :	002	* Material Group :	L001
Plant-Sp.Matl Status :		Valid From :	
Tax ind. f. material :	1	Qual.f.FreeGoodsDis. :	
Material freight grp :		Autom. PO:	☐
Batch management:	☐		
Batch management(Plant):	☐		

Purchasing values

Purchasing value key :	5	Shipping Instr. :	
1st Reminder/Exped. :	10 days	Underdel. Tolerance :	10,0 percent
2nd Reminder/Exped. :	20 days	Overdeliv. Tolerance :	10,0 percent
3rd Reminder/Exped. :	30 days	Min. Del. Qty in % :	0,0 percent
StdValueDelivDateVar :	0 days		
Unltd Overdelivery:	☐		
Acknowledgment Reqd:	☐		

Figure 3.17 Purchasing View of the Material Master Record

Figure 3.18 shows that the over- and under-delivery tolerance limits maintained in the material master record are passed on to a purchase order item, where they can be changed, if required.

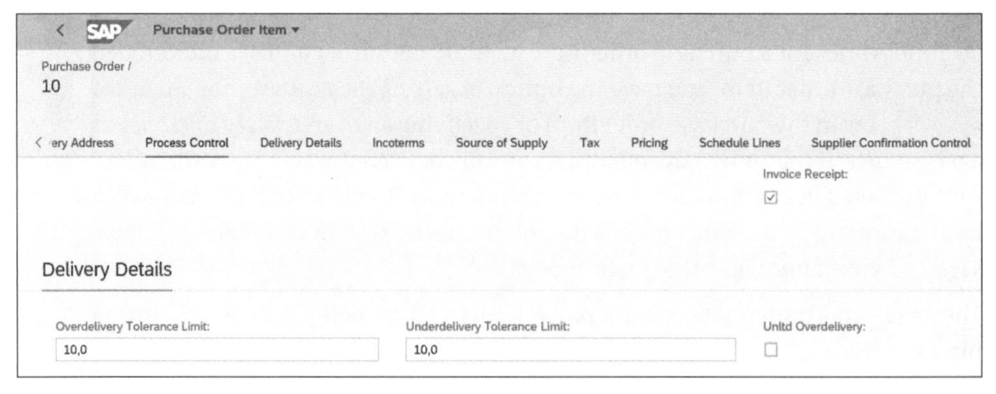

Figure 3.18 Purchase Order Item

Delivery Completed Indicator

When you post a goods receipt, you can manually set the **Delivery Completed** indicator if you do not expect any further delivery, even though the total order quantity has not been received.

Figure 3.19 and Figure 3.20 show how you can set the final delivery indicator manually when entering goods receipts with the Post Goods Receipt for Purchase Order app. In Figure 3.19, you can see a goods receipt item. Select the item detail by clicking the small arrow on the right end of the item.

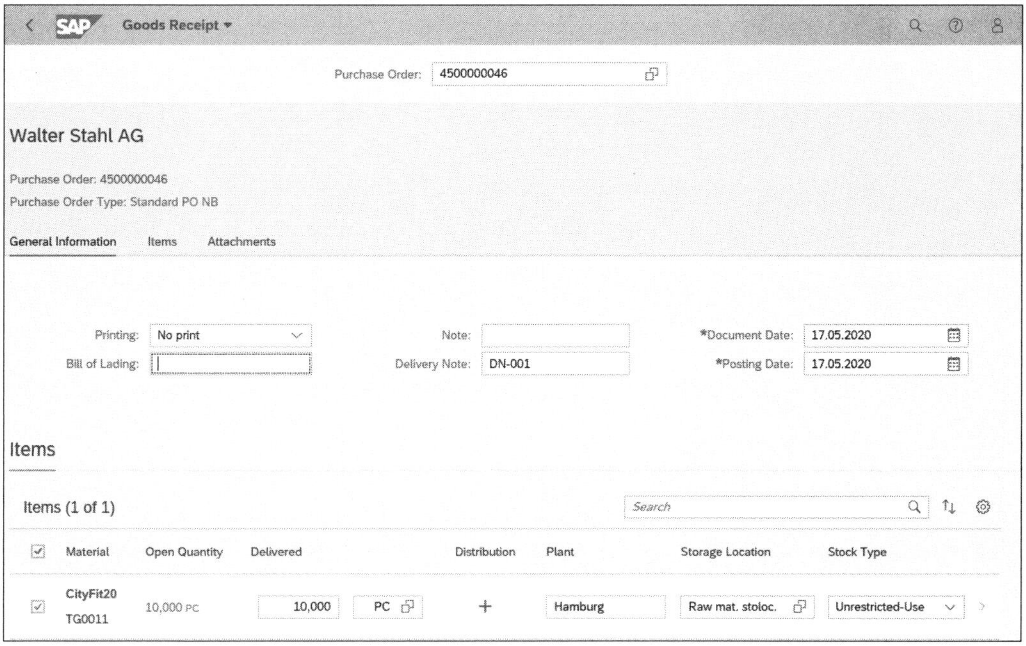

Figure 3.19 The Post Goods Receipt for Purchase Order App

In the item details, on the **Quantity** tab, you can set the **Delivery Completed** indicator shown in Figure 3.20.

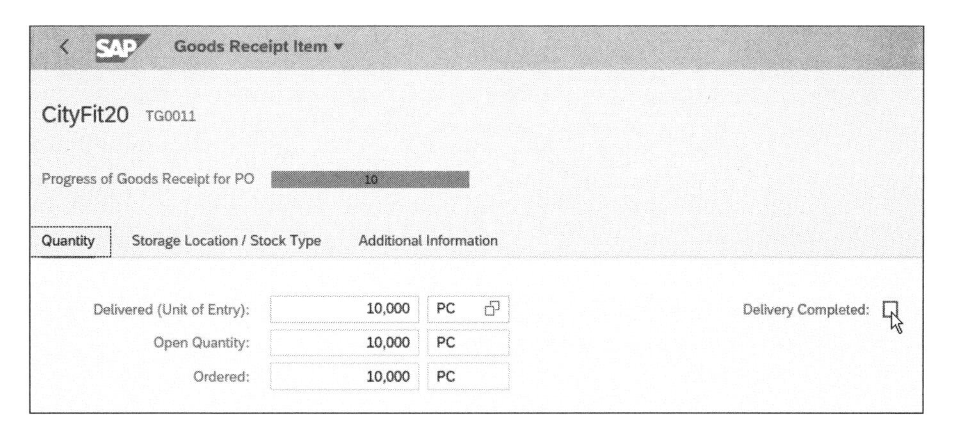

Figure 3.20 Setting the Delivery Completed Indicator Manually with the Post Goods Receipt for Purchase Order App

Figure 3.21 shows how you can manually set the final delivery indicator when entering goods receipts with the Post Goods Movement app (Transaction MIGO).

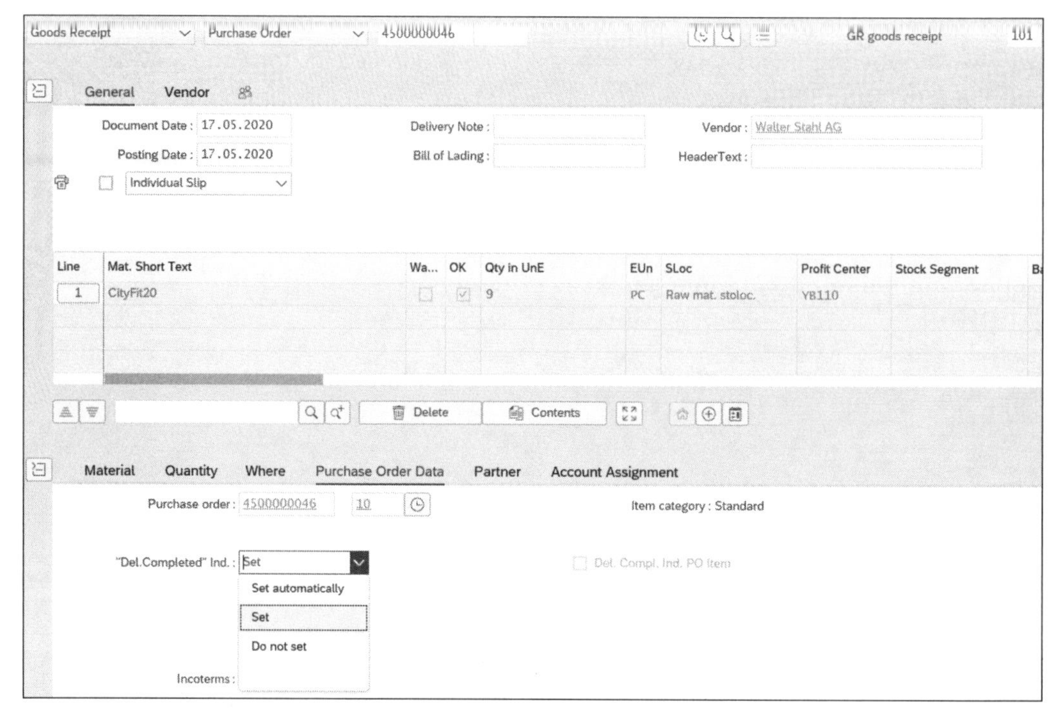

Figure 3.21 Setting the Delivery Completed Indicator Manually with the Post Goods Movement App (Transaction MIGO)

The **Delivery Completed** indicator is updated in the purchase order item and indicates that a purchase order item is completed. A goods receipt is no longer expected for the purchase order item, but it is still possible to post further goods receipts using the Post Goods Movements app (Transaction MIGO). Figure 3.22 shows a **Yes** value for the **Delivery Completed** indicator in the order item.

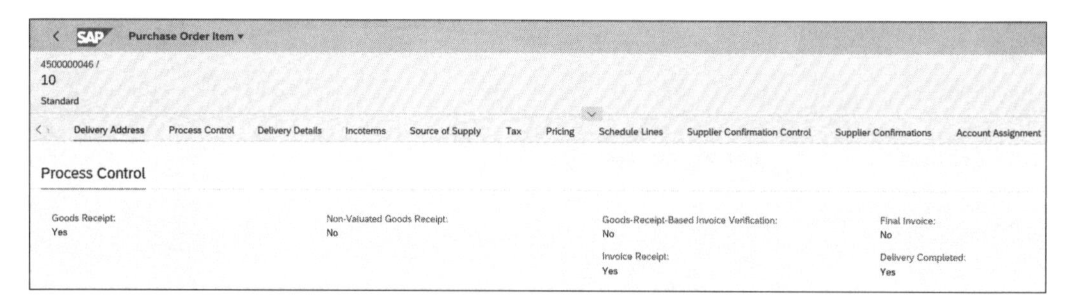

Figure 3.22 Delivery Completed Indicator in the Purchase Order Item

The delivery completed indicator has the following effects, even if the quantity delivered is less than the quantity ordered:

- No further goods receipt is expected for the purchase order item.
- The open order quantity is zero.
- The order item is no longer relevant for MRP.

- The commitment for the order item is canceled.
- No reminder is sent for the order item due to outstanding delivery.

We have seen that the delivery completed indicator can be set manually at goods receipt posting. There are other ways to set this indicator:

- Manually at goods receipt posting with the Post Goods Receipt for Purchase Order app
- Manually when entering any type of goods movement with reference to the purchase order item (goods receipt, return delivery, subsequent delivery, and cancellation) using the Post Goods Movement app or Transaction MIGO
- Manually when changing the purchase order item
- Automatically by the system when entering a goods receipt for the purchase order item

The system automatically sets the **Delivery Completed** indicator in the purchase order item when the goods receipt is posted under the following conditions:

- You have defined under-delivery tolerances in the purchase order item.
- The total goods receipt quantity in the purchase order item lies within the under-delivery tolerance limit.
- In Customizing, you have defined for each plant whether this automatic update is activated or not.

Invoice Receipt with Reference to a Stock Material Purchase Order

SAP S/4HANA offers several options for posting supplier invoices.

You use the Create Supplier Invoice app to enter manually and post a supplier invoice related to a purchase order or scheduling agreement. Simply enter the company code (if it is not automatically proposed), the gross invoice amount, the invoice date, the reference, and the purchase order number. The system determines all further data.

In the item list, SAP S/4HANA proposes all items that correspond to the referenced purchase order. It proposes the quantities and amounts of all items for which goods receipts were posted but not fully invoiced. If necessary, change the proposed data according to the received supplier invoice. The system compares the supplier invoice data with the related purchase order data. If the deviation is outside defined tolerances, the invoice is posted but the invoice is automatically blocked for payment.

Alternatively, you can use Transaction MIRO (Create Supplier Invoice—Advanced).

The effects of posting an invoice with reference to a purchase order are as follows:

- The purchase order history is updated. The purchase order history allows you to monitor the procurement process and shows all goods receipts, return deliveries, invoices, credit memos, and so on for each purchase order item.

- An invoice document is created.
- An accounting document is created. Using automatic account determination, the system updates the general ledger accounts that are affected by the invoice postings.

Value Flow and Valuation Basics

Now let's take a closer look at the accounting documents that are created with the goods receipt posting and the invoice posting. We will find out which values are updated to which accounts. Another lesson later in the book will show us how the system automatically determines the correct accounts.

The goods receipt for a purchase order is valuated at the purchase order price. This means that the value of a goods receipt for a purchase order is *goods receipt quantity × purchase order price*. However, the stock value being updated with exactly this value depends primarily on the price control defined for the material in the material master record. SAP S/4HANA offers two transaction-based valuation procedures:

- The moving average price procedure
- The standard price procedure

Note

The valuation procedure is usually defined per plant in the material master record, since the valuation level is usually the plant. This means that if the valuation level is the plant, you can valuate your individual stock materials differently per plant. For example, you can valuate a material with a standard price in a plant in which it is produced, while you valuate the same material in other plants with the moving average price.

In the moving average price procedure, a new material price is calculated after each receipt and after invoices that lead to a change in inventory value. This price is calculated by dividing the total stock value by the total quantity of the material in stock. In the moving average price method, the receipts are posted with the external receipt values and the issues are valuated with the current moving average price from the material master. With this procedure, the stock value is increased by exactly the value of the *Goods receipt quantity × Purchase order price* at the time of the goods receipt for the purchase order.

If a material is valuated at a standard price, the standard price remains constant for at least one period. The standard price is usually the result of a standard cost estimate for the material. If you valuate a material at a standard price, then all inventory postings are made with this constant standard price. The current inventory value is the product of the standard price and the current inventory quantity. In this case, the stock value at goods receipt for the purchase order is exactly the value of the *Receipt quantity × Standard price*. The difference to the purchase order price is posted to a price difference account.

To understand exactly which values are posted to which accounts at goods receipt and then at invoice receipt for the purchase order, we analyze two procurement process examples with two materials—one with standard price and the other with moving average price.

Material with Standard Price

The initial situation is that you have a material valuated at the standard price of EUR 10 whose current stock is 100 pieces with a value of EUR 1000.

We consider the following scenario, illustrated in Figure 3.23.

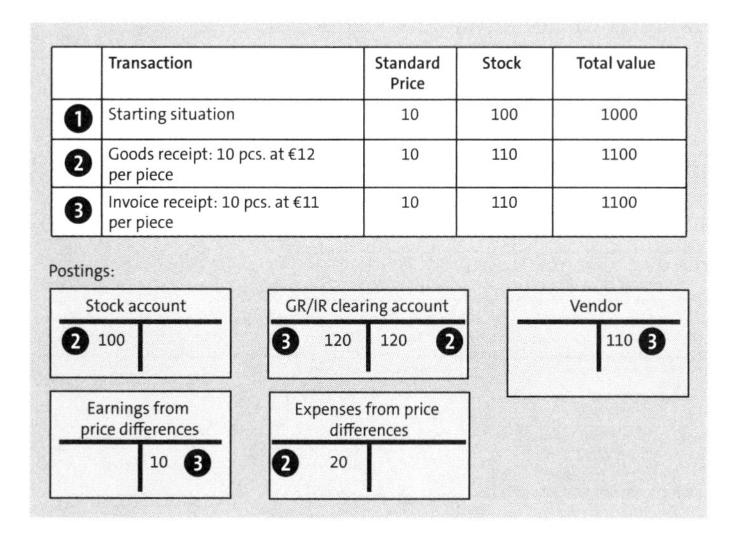

	Transaction	Standard Price	Stock	Total value
1	Starting situation	10	100	1000
2	Goods receipt: 10 pcs. at €12 per piece	10	110	1100
3	Invoice receipt: 10 pcs. at €11 per piece	10	110	1100

Figure 3.23 Postings with Standard Price

Let's first analyze the effects of the goods receipt:

- The stock increases by 10 pieces and is now 110 pieces.
- The stock value is now 1100 EUR (valuation at standard price).

Which account postings have been made?

- The stock account is debited with the value 100: *goods receipt quantity × standard price.*
- The difference to the order value (*2 × 10*) is charged to the price difference account.
- The offsetting entry is posted to the goods receipt/invoice receipt (GR/IR) clearing account at purchase order value 120 EUR.

Posting Logic of the GR/IR Clearing Account

If the goods receipt is posted before the invoice, then the GR/IR clearing account is credited with the order value and cleared when the invoice is posted. If the invoice is posted before the goods receipt, then the GR/IR clearing account is debited with the invoice value and cleared with this value when the goods receipt is posted.

Let's now consider the effects of posting the invoice.

- Quantitatively, nothing happens anymore.
- The difference to the purchase order price (10 EUR) is credited to the price differences account.

Material with Moving Average Price

Let's turn to our second example. The initial situation is that you have a material valuated at the moving average price of EUR 10, whose current stock is 100 pieces for a total value of EUR 1000.

We consider the following scenario, illustrated in Figure 3.24.

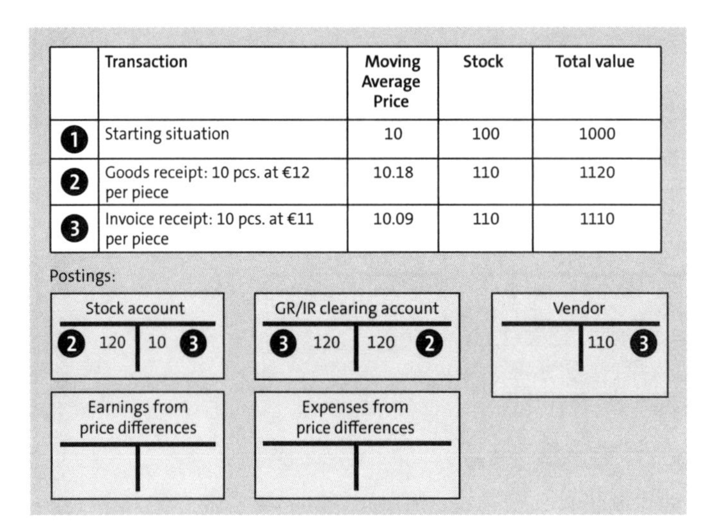

	Transaction	Moving Average Price	Stock	Total value
1	Starting situation	10	100	1000
2	Goods receipt: 10 pcs. at €12 per piece	10.18	110	1120
3	Invoice receipt: 10 pcs. at €11 per piece	10.09	110	1110

Figure 3.24 Postings with Moving Average Price

Let's first analyze the effects of the goods receipt:

- The stock increases by 10 pieces and is now 110 pieces.
- The stock value is now 1120 EUR (1000 + 120).
- The moving average price is calculated by dividing the total stock value by the total stock quantity.

Which account postings have been made?

- The stock account is debited with the value 120 EUR: *Goods receipt quantity × Purchase order price.*
- The offsetting entry is posted to the GR/IR clearing account at purchase order value 120 EUR.

Now, let's consider the effects of posting the invoice:

- Quantitatively, nothing happens anymore.
- The difference to the purchase order price (10 EUR) is credited to the stock account.

- The stock value is now 1110 EUR.
- The moving average price is calculated by dividing the total stock value by the total stock quantity.

If the invoice shows no variances from the purchase order price, the GR/IR clearing account is simply cleared in addition to the posting to the vendor account and the tax posting. However, if the invoice contains variances from the purchase order price, either an inventory value adjustment is posted (in the case of the moving average price) or a price difference (in the case of the standard price).

In cases such as the following, a price difference posting can also be made for a material with a moving average price:

- A goods receipt is posted with reference to a stock material purchase order
- Goods issues are posted in the time between goods receipt and invoice receipt
- When the invoice is posted, the stock quantity of the material is less than the invoice quantity
- The invoice price differs from the order price

In these cases, the difference amount is posted to two accounts when the invoice is posted: the stock value is adjusted proportionally for the remaining stock quantity, and the rest of the amount is posted to a price difference account.

Let's look at an example. The initial situation is that you have a material valuated at the moving average price of EUR 10, whose current stock is 100 pieces with a value of EUR 1000.

We consider the following scenario, illustrated in Figure 3.25:

❶ You order 100 pieces from a supplier at a purchase order price of 11 EUR.

❷ You post the goods receipt for this purchase order.

❸ You withdraw 150 pieces for a cost center.

❹ On the vendor invoice, the price differs from the purchase order price: instead of 11 EUR, the vendor invoices 12 EUR per unit. You post the vendor invoice.

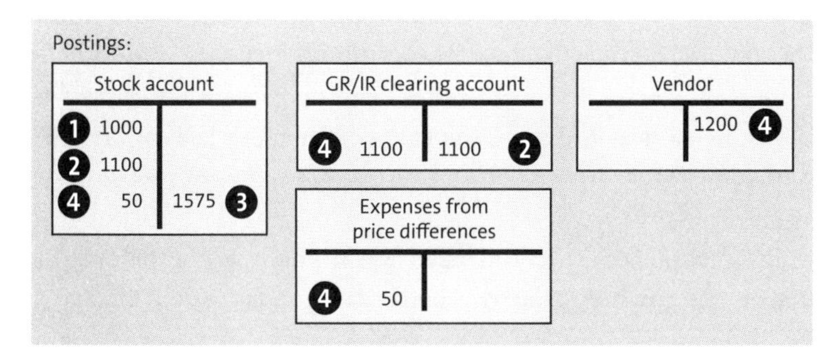

Figure 3.25 Postings with Moving Average Price—Insufficient Stock Coverage at the Time of Invoice Entry

Note

With SAP S/4HANA, the material ledger is active and cannot be deactivated. The material ledger is a subledger for materials. It collects all information relevant to stock and valuation. Based on this data, the material ledger determines prices for the valuation of these materials. It forms the basis for actual costing and provides several methods of price determination in addition to multiple currencies and valuations:

- **Transaction-based material price determination (option 2 in the material master)**
  The price controls for standard price and moving average price are available as already described.
- **Single-level/multi-level price determination through actual costing (option 3 in the material master)**
  The material is valuated with a preliminary valuation price (standard price). All variances are collected during a posting period. At the end of the period, these variances are assigned to the ending inventory quantity and the material consumption of the period. In multi-level price determination, additional price differences (e.g., those of raw materials) are rolled up to the semi-finished products and, in the next step, to the finished products. When the closing in the material ledger is complete, these results can be transferred to inventory valuation.

Important Terminology

The following terminology was used in this chapter:

- **Purchasing value key**
  The purchasing value key contains default values for the purchasing process and is entered in the material master record.
- **Stock type**
  There are three types of inventory in inventory management: unrestricted stock, stock for quality inspections, and blocked stock. Each of these stock types indicates how a quantity of material can be used.
- **Moving average price**
  A moving average price is an inventory valuation method in which the valuation price of the material is recalculated after each goods receipt.
- **Standard price**
  If a material is assigned a standard price, the value of the material is always calculated at this price.
- **GR/IR clearing account**
  A GR/IR clearing account is an account used when goods arrive before the invoice is issued or when an invoice arrives before the goods are delivered.

Practice Questions

These questions will help you evaluate your understanding of the topics covered in this chapter. They are similar in nature to those on the certification examination. Although none of these questions will be found in the exam itself, they will allow you to review your knowledge of the subject.

Select the correct answers, and then check the completeness of your answers in the next section. Remember that on the exam, you must select all correct answers—and only correct answers—to receive credit for the question.

1. What must a stock material purchase order item include?

☐ **A.** A plant and a storage location

☐ **B.** A material number and any account assignment category

☐ **C.** A material number and an account assignment category blank

☐ **D.** A storage location and an account assignment category blank

2. What requirements must be met before you can perform a stock material procurement process?

☐ **A.** A material with the Purchasing and Plant Data/Storage 1 views must have been created.

☐ **B.** A material with the Purchasing and Accounting 1 views must have been created.

☐ **C.** A purchasing info record must have been created.

☐ **D.** A material with the Purchasing and MRP 1 views must have been created.

3. Which of the following material master views can be created automatically by the system when you post the first goods receipt?

☐ **A.** Purchasing

☐ **B.** MRP 1

☐ **C.** Accounting 1

☐ **D.** Plant Data/Storage 1

4. True or false: The GR blocked stock is part of the valuated stock of a plant. Its usage is merely restricted.

☐ **A.** True

☐ **B.** False

5. Which documents can be created with a goods receipt posting? (There are two correct answers.)

☐ **A.** A material document

☐ **B.** An invoice document

☐ **C.** An inbound delivery

☐ **D.** An accounting document

6. True or false: When you enter a goods receipt for a purchase order, you must manually enter the purchase order number and all the necessary data for each item: material, quantity, storage location, stock type, and so on.

☐ **A.** True

☐ **B.** False

7. You have ordered a material for storage. This material is valuated with a standard price. Which of the following happens if you now post the goods receipt for this purchase order?

☐ **A.** If the purchase order price differs from the standard price, this difference is posted directly to the stock account.

☐ **B.** The standard price in the material master record is adjusted to the purchase order price.

☐ **C.** The total stock value is increased by the goods receipt quantity multiplied by the standard price.

☐ **D.** The posting to the GR/IR account is made at the value of the goods receipt quantity multiplied by the standard price.

8. True or false: In SAP S/4HANA, you can deactivate the material ledger.

☐ **A.** True

☐ **B.** False

9. You have ordered a material for storage. This material is valuated with a moving average price. Which of the following happens when you now post the goods receipt for this purchase order? (Here we assume that the goods receipt is posted before the invoice.) (There are three correct answers.)

☐ **A.** The total stock value is increased by the goods receipt quantity multiplied by the purchase order price.

☐ **B.** The total stock of the material is increased by the goods receipt quantity.

☐ **C.** The offsetting entry for the stock posting is made to a price difference account.

☐ **D.** The moving average price of the material is recalculated based on the new total stock value and the new total stock quantity.

10. True or false: A material is valued uniformly in a company group.

☐ **A.** True

☐ **B.** False

11. When is a purchase order item considered as fully delivered? (There are two correct answers.)

☐ **A.** If the goods receipt quantity is greater than or equal to the ordered quantity

☐ **B.** If the **Delivery Completed** indicator is set and the total goods receipt quantity is within the under-delivery tolerance

☐ **C.** If the **Delivery Completed** indicator is set, regardless of the total goods receipt quantity

☐ **D.** If the total goods receipt quantity was invoiced

12. True or false: Only when posting the goods receipt can you decide whether a material should be posted to inspection stock, blocked stock or unrestricted-use stock.

☐ **A.** True

☐ **B.** False

13. Which accounts can possibly be booked when posting an invoice for a stock material purchase order if the material is valuated at the moving average price? (There are three correct answers.)

☐ **A.** Stock account

☐ **B.** GR/IR clearing account

☐ **C.** Price difference account

☐ **D.** Bank account

14. True or false: A purchase order item with the **Delivery Completed** indicator is no longer relevant for MRP, even if the goods receipt quantity is less than the purchase order quantity.

☐ **A.** True

☐ **B.** False

15. Which of the following statements about under- and over-deliveries are correct? (There are three correct answers.)

☐ **A.** If the total goods receipt quantity falls below the under-delivery tolerance, the system generates an error message.

☐ **B.** If the total goods receipt quantity exceeds the over-delivery tolerance, the system generates an error message.

☐ **C.** Under-deliveries are generally permitted. This makes partial deliveries possible.

☐ **D.** It is possible to allow unlimited over-deliveries.

16. Which organizational units can be left empty in the purchase order and entered at goods receipt?

☐ **A.** The purchasing organization

☐ **B.** The plant

☐ **C.** The storage location

☐ **D.** The purchasing group

17. True or false: You can only post invoices with an invoice price that differs from the purchase order price for materials with a moving average price.

☐ **A.** True

☐ **B.** False

18. True or false: The stock types unrestricted-use stock, quality inspection, and blocked stock belong to the physical and valuated stock of a plant.

☐ **A.** True

☐ **B.** False

Practice Question Answers and Explanations

1. Correct answer: **C**

   A stock material order item must contain a material number and the account assignment category must be blank. You can specify the storage location in the purchase order item, but it is not a must.

2. Correct answer: **B**

   For the procurement of stock material, a material with the Purchasing and Accounting 1 views must have been created. It enables you to order the material and then post a valuated goods receipt. Depending on the system configuration, it is not necessary to have created the storage location view of the material master record beforehand. The system can automatically add the missing storage location view to the material master record in the background when the first goods receipt is posted to this storage location. The MRP view is only necessary if you want to plan the material. This is useful but not mandatory for stock materials.

3. Correct answer: **D**
   The storage location view (Plant Data/Storage 1) of a material can be generated automatically in the background during the first goods receipt posting, if the necessary settings have been made in Customizing. The Purchasing view and the posting view are prerequisites for procurement. MRP 1 is only required if you want to plan the material.
4. Correct answer: **B**
   False. The GR blocked stock is neither valuated nor assigned to a storage location. Do not confuse this with the stock type blocked stock, which means that the usability of a material quantity is restricted but is both valuated and assigned to a storage location.
5. Correct answers: **A, D**
   When you post a goods receipt, the system creates a material document. This document contains information about the material delivered, the quantity delivered, and the plant and storage location where the material is put away. If the goods receipt is valuated—which is the rule for stock materials—the system also creates an accounting document. This document contains details of the accounting effects of the material movement, such as general ledger accounts and values.
6. Correct answer: **B**
   False. When you enter a goods receipt for a purchase order, the system proposes all possible data: the material number, the open quantity (ordered and not yet delivered), and the storage location, if available. As far as the stock type is concerned, the system defaults to unrestricted-use stock, unless stock in quality inspection has already been predefined in the purchase order item. All proposed values can be overwritten at goods receipt posting, according to the delivery note and the actual delivery.
7. Correct answer: **C**
   If a material is valuated using the standard price procedure, all postings to the stock account are made at a standard price. When goods are received for a purchase order, the posting to the stock account is therefore made at the value of the goods receipt quantity multiplied by the standard price. The stock value increases accordingly.

   The standard price is not adjusted to the purchase order price at goods receipt: the standard price is set for a certain period and is often the result of a product cost calculation.

The value resulting from a difference between the purchase order price and the standard price is posted to a price difference account.

If the goods receipt is posted before the invoice, the posting is made to the GR/IR account at the value of the goods receipt quantity multiplied by the purchase order price. If the invoice is posted first, the posting to the GR/IR account is made at the value of the invoice value. This value is cleared at goods receipt. This posting rule applies regardless of the valuation procedure.

8. Correct answer: **B**

   False. In SAP S/4HANA, the material ledger is active and cannot be deactivated.

9. Correct answers: **A, B, D**

   At goods receipt, the stock quantity is increased by the goods receipt quantity. At the same time, the stock value is increased. For a material with a moving average price, the stock value is incremented by the amount of the goods receipt quantity multiplied by the purchase order price. Together with the goods receipt, the moving average price is recalculated using the formula *total value × total quantity*.

   The offsetting entry for the stock posting is made to a GR/IR clearing account.

10. Correct answer: **B**

    False. A material can be valuated differently for each valuation area. Since the valuation area usually corresponds to the plant, you can define different valuation procedures, prices, and valuation classes for each material per plant.

11. Correct answers: **A, C**

    From the perspective of goods receipt, a purchase order item is complete if the quantity delivered is greater than or equal to the order quantity or if the **Delivery Completed** indicator has been set. For the system to set the **Delivery Completed** indicator automatically, the total quantity of goods received must be within the under-delivery tolerance, but the indicator can be set manually at any time, regardless of the quantity delivered. Once it has been set, the purchase order item is fully delivered, regardless of the actual quantity of goods received.

12. Correct answer: **B**

    False. When you post a goods receipt, you can decide whether a material is to be posted to inspection stock, blocked stock, or unrestricted-use stock. However, you can specify this already in the purchase order item. You can also plan the posting to inspection stock by setting an indicator in the material master record.

13. Correct answers: **A, B, C**

    When you post an invoice for a stock material purchase order, the GR/IR clearing account is always updated. If the invoice price differs from the purchase order price, the stock value is corrected (posting from a stock account). If there is insufficient stock coverage, the difference value is posted to a price difference account in proportion to the missing quantity. A bank account is posted first upon payment.

14. Correct answer: **A**
    True. A purchase order item with the delivery completed indicator has an open quantity equal to zero. A goods receipt is no longer expected, and this purchase order item is no longer relevant for requirements planning.
15. Correct answers: **B, C, D**
    In any case, under-deliveries (partial deliveries) are possible. If the quantity delivered falls below the under-delivery limit, a warning message is generated, not an error message. However, an error message is generated if the over-delivery tolerance limit is exceeded. But you can accept an unlimited over-delivery in the material master and then in the purchase order item.
16. Correct answer: **B**
    Purchasing organization and purchasing group must be entered at the purchase order header level. You must enter a plant for each item. Only the storage location can be left blank and is entered at the latest at goods receipt.
17. Correct answer: **B**
    False. You can always post an invoice at an invoice price that differs from the purchase order price. The difference is posted either to the stock account (in the case of a moving average price) or to a price difference account (in the case of a standard price or in the case of a moving average price and insufficient stock coverage). The invoice may be blocked for payment if defined tolerance limits are exceeded.
18. Correct answer: **A**
    True. The unrestricted-use stock types, quality inspection, and blocked stock are valuated and assigned to a storage location. Only the usability and availability are restricted.

Takeaway

You should now understand the procurement process for stock material and be able to explain both how to maintain a stock material and how to create a purchase order for it. You should understand the effects of a goods receipt posting with reference to a purchase order for a stock material. You should also be able to explain the postings resulting from both a goods receipt and the entry of a supplier invoice. Finally, you should be familiar with the term *stock type* and be able to explain how to enter a goods receipt with conditional acceptance.

Summary

It is important that you understand the characteristics and special features of a simple stock procurement process: What is a stock material, what does a stock material order item look like, what happens when we post a goods receipt to the warehouse with reference to an order item, what does the value flow look like, and what can happen when an invoice is posted?

In the following chapters, we will compare the procurement of stock material with the procurement of consumables. Later in this book, we will also explore how to optimize the procurement processes, particularly the procurement of stock material.

This basic knowledge, which we have refreshed in this chapter, is the prerequisite for designing a lean, optimized, and efficient warehouse procurement process in your company or for your customers. Mastering this basic knowledge is therefore of utmost importance.

Chapter 4
Procurement for Direct Consumption

4

Techniques You'll Master

- Explain the differences between storage material and consumable material.
- Create a purchase order for direct consumption.
- Understand the control options of an account assignment category and configure your own account assignment category if required.
- Post a goods receipt with reference to a purchase order for direct consumption and explain the consequences of this goods receipt.
- Post a vendor invoice with reference to a purchase order for direct consumption and explain the consequences of this posting.
- Describe a self-service procurement process.
- Explain the features and advantages of a blanket purchase order.

In this chapter, we clarify the term *consumables* and explain the special features of procurement for direct consumption: What constitutes a purchase order with account assignment, and what are the effects of the goods receipt posting and the subsequent invoice posting? We introduce the blanket order, which is a special form of a purchase order with an account assignment, and we highlight the self-service procurement.

Real-World Scenario

In every company, the procurement of materials for direct consumption is a relevant issue: brochures for a trade fair, office supplies for a cost center, special auxiliary materials for a production order, special spare parts for a repair order. Procurement for direct consumption is of universal importance, regardless of the industry in which a company operates, whether it is a production company, a retailer, or a service provider. Highly automated self-service procurement is now also very widespread, where the requester personally creates a specific order request and sends it off for approval, and then initiates the goods receipt posting via confirmation of receipt. The ability to understand, implement, and configure the consumption procurement process is part of the basic skills of the sourcing and procurement consultant.

Objectives of This Portion of the Test

The purpose of this part of the certification exam is to test your knowledge of procurement for direct consumption. For the certification exam, you must be able to do the following:

- Explain the term *consumable material* and list what differentiates this from a stock material.
- Describe what constitutes an order for direct consumption and specify the important mandatory and optional input data.
- Understand the control options of the account assignment category.
- Describe the various options that a purchase order with account assignment allows for goods receipt.
- Describe the various posting scenarios of an incoming invoice for a purchase order item assigned to an account.

Key Concept Refresher

In this section, we will go through the main steps of a procurement process for direct consumption—purchase order, goods receipt, and invoice entry—and will examine their features and effects. We will also observe the impacts in terms of value.

Let's start by clarifying the concept of procurement for direct consumption.

Procurement for Direct Consumption

We speak of procurement for direct consumption when the value of the goods purchased is posted to a cost element account or an asset account. The purchase order items are allocated directly to an account assignment object (assets, cost centers, and so on). The ordered materials are not posted to the warehouse but are assigned directly to the cost objects that consume the materials (office supplies, computer systems, and so on).

What types of materials or services can you purchase for direct consumption?

- Stock materials, which are usually purchased for the warehouse, can be procured directly for a cost center, project, or order if required.
- Pure consumable material without a material master record or services without a master record can be purchased. Here, instead of a material number, you enter a description, a suitable material group, and a unit of measure when you create the purchase order item.
- You can purchase pure consumable material with a material master record. Two standard material types are suitable for consumable materials:
  - Type NLAG for materials that are stored neither by quantity nor by value
  - Type UNBW for materials that are only stored by quantity only
- You can purchase services with a material master record. SAP S/4HANA offers a new standard material type, SERV, suitable for creating services.

Purchase Order for Direct Consumption

There are several ways to create a purchase order. If you want to create a purchase order from scratch, you can use the Create Purchase Order, Create Purchase Order – Advanced, and Manage Purchase Order apps. (You can also convert purchase requisitions manually or automatically into purchase orders. We will discuss these possibilities in detail in Chapter 6. Here we concentrate on the data that make up a purchase order for direct consumption.)

Creating a purchase order item for direct consumption is quite straightforward:

- You first enter the header data: order type (only if you use the Create Purchase Order – Advanced app or Transaction ME21N), vendor, purchasing organization, purchasing group, and company code. The order currency is derived from the vendor.
- You then enter the items. In a purchase order, you can basically include both stock material items and items for direct consumption.

- A purchase order item always contains precisely the required goods, the plant, the quantity, the purchasing price, and the desired delivery date.
- Here are important characteristics of a purchase order item for direct consumption:
  - You must specify an account assignment category (account assignment category must not be blank).
  - You can order a material or a service with or without a material number.
  - A goods receipt is not necessary, and if you decide on a goods receipt, this goods receipt can be non-valuated.

Note
The account assignment category determines whether a goods receipt can or must be posted and, if so, whether the goods receipt must or can be valuated or non-valuated.

Figure 4.1 shows a purchase order item for direct consumption. In this example, the ordered material does not have a material master record; you must enter the description (short text), quantity, unit of measure, price, price unit, and material group manually.

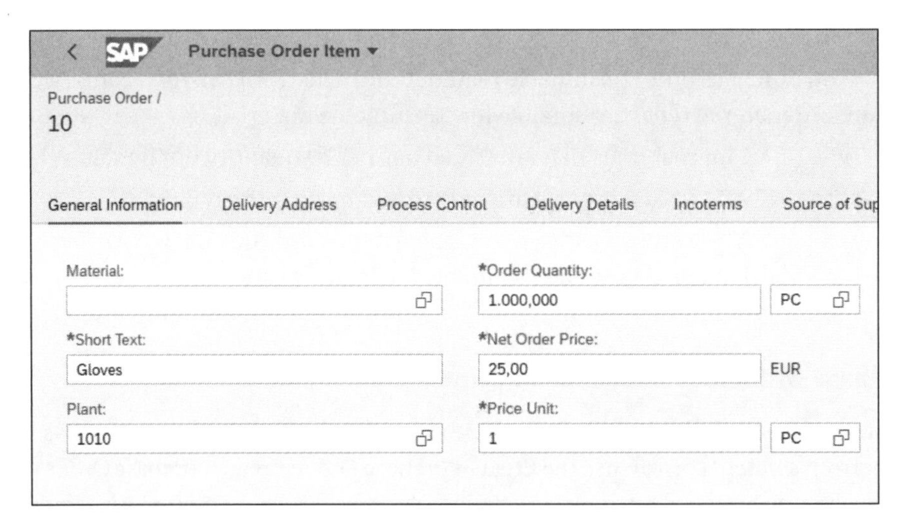

Figure 4.1 Item for Direct Consumption in the Manage Purchase Order App

A material master record has the advantage of allowing you to enter less data. You must always enter an account assignment category (as seen in Figure 4.2). The account assignment category turns a purchase order item into a purchase order item for direct consumption.

Figure 4.2 shows the process control of a purchase order item for the cost center: After you have selected the account assignment category, you can specify the process control. Depending on the account assignment category, you can decide whether you want to post a valuated or non-valuated goods receipt or no goods receipt at all.

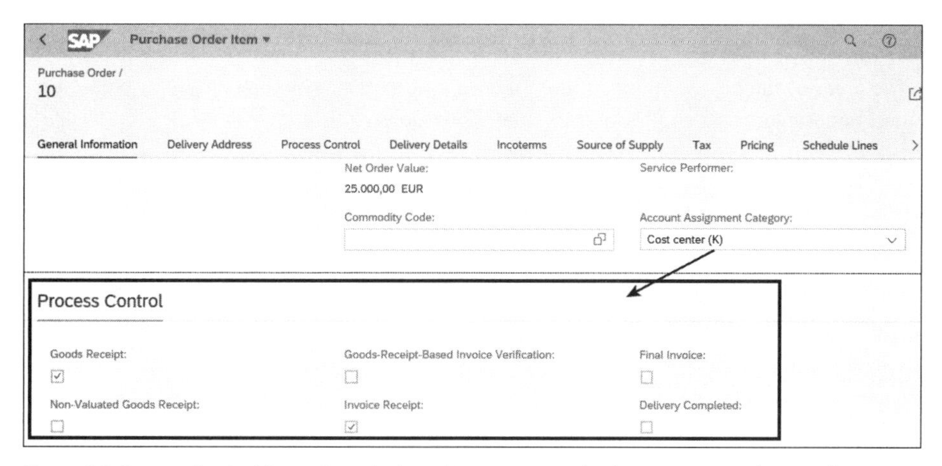

Figure 4.2 Process Control for an Item Assigned to an Account in the Manage Purchase Order App

As already mentioned, if you want to request materials or services directly for an account assignment object, you must specify an account assignment category for the item. As a result, you must enter additional account assignment data in the item details on the **Account Assignment** tab.

To access the details of the account assignment data, select the small arrow at the right of the account assignment row, as shown in Figure 4.3. For account assignment category K (cost center), you must enter an account and a cost center, as shown in Figure 4.4.

Figure 4.3 Account Assignment Tab of a Purchase Order Item in the Manage Purchase Order App

Purchase Order Item Account Assignment
Purchase Order / 00010 /
1
Delete
General Information
Project
Sales and Distribution Information
Network
Asset Management
Service Order
*G/L Account:
51100000
Functional Area:
YB75
Controlling Area:
A000
Cost Center:
4891
Profit Center:
YB600
Unloading Point:

Figure 4.4 Account Assignment Details of a Purchase Order Item in the Manage Purchase Order App

Note

The account can be determined and proposed automatically by the system based on the account assignment category and the material group.

Account Assignment Category

The account assignment category determines which account assignment object type is to be charged and which account assignment data you must enter. In addition, the account assignment category performs several other important control functions. You can create as many account assignment categories as you require in Customizing. Figure 4.5 shows the path to the account assignment category in Customizing.

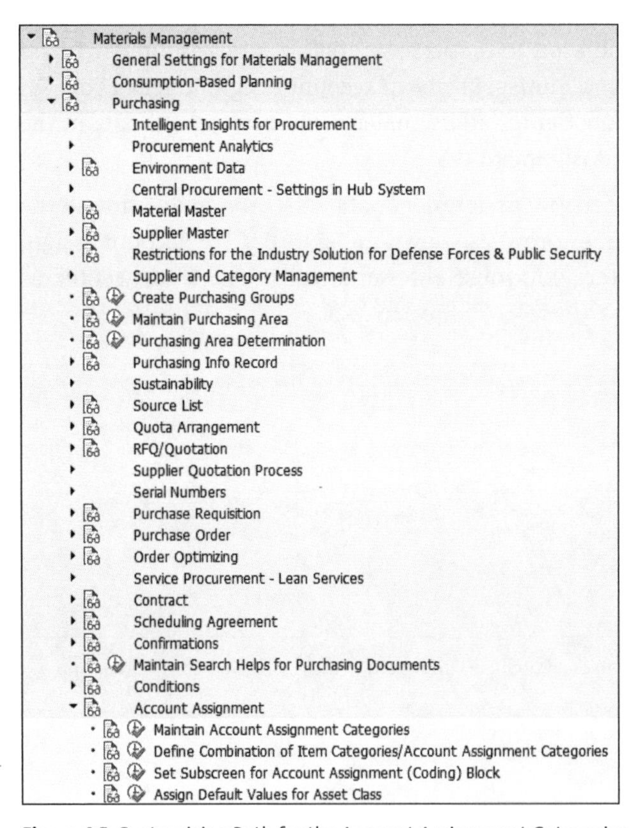

Figure 4.5 Customizing Path for the Account Assignment Categories

If you select **Maintain Account Assignment Categories**, you get a list of the available account assignment categories. You can create a new account assignment category, delete an existing one, or change it. For example, select **Cost Center** as shown in Figure 4.6.

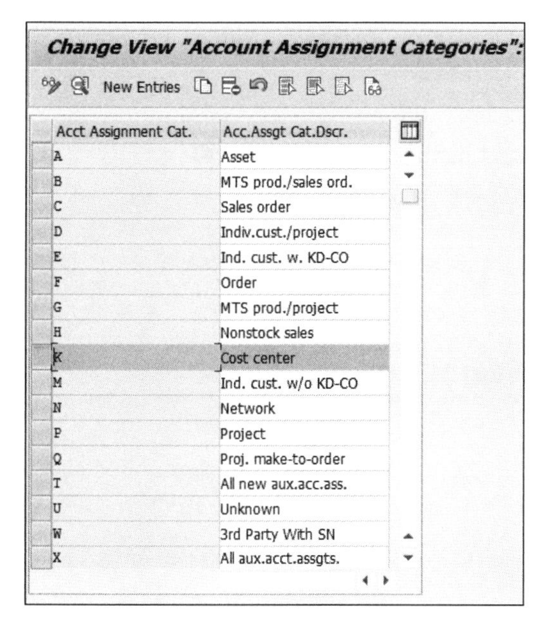

Figure 4.6 Customizing Account Categories

Figure 4.7 shows the parameters of account assignment category K, which we will look at more closely in the following section.

Acct Assignment Cat. K Cost center

Detailed information

☐ Acct.assg.changeable	Consumption posting	V	Distribution	1
☑ AA Chgable at IR	Account Modification	VBR	Partial invoice	2
☐ Derive acct. assgt.	ID: AcctAssgt Scrn	1	Single account assig	
☐ Del.CstsSep.	Special Stock			

☑ Goods receipt	☐ GR non-valuated	☑ Invoice receipt
☐ GR Ind. Firm	☐ GR NonVal. Firm	☐ IR Ind. Firm

Fields

Field Label	Mand.Ent.	Opt.Entry	Display	Hidden
Asset	○	○	○	◉
Asset Subnumber	○	○	○	◉
Business Area	○	○	○	◉
Business process	○	○	○	◉
CO/PP Order	○	○	○	◉
Commitment Item	○	○	○	◉
Cost Center	◉	○	○	○
Cost Object	○	○	○	◉
Cost Type	○	○	○	◉
Deletion indicator	○	◉	○	○

Figure 4.7 Customizing for Account Assignment Category K (Cost Center)

Let's examine the most important control parameters of the account assignment category, using Figure 4.8 as a reference:

- Field selection control ❻
- **Distribution** and **Partial Invoice** indicators ❶
- **Account Modification** field ❷
- Account assignment changeable (**AA Chgable at IR**) indicators ❸
- Indicators related to the goods receipt postings (❹ and ❺)

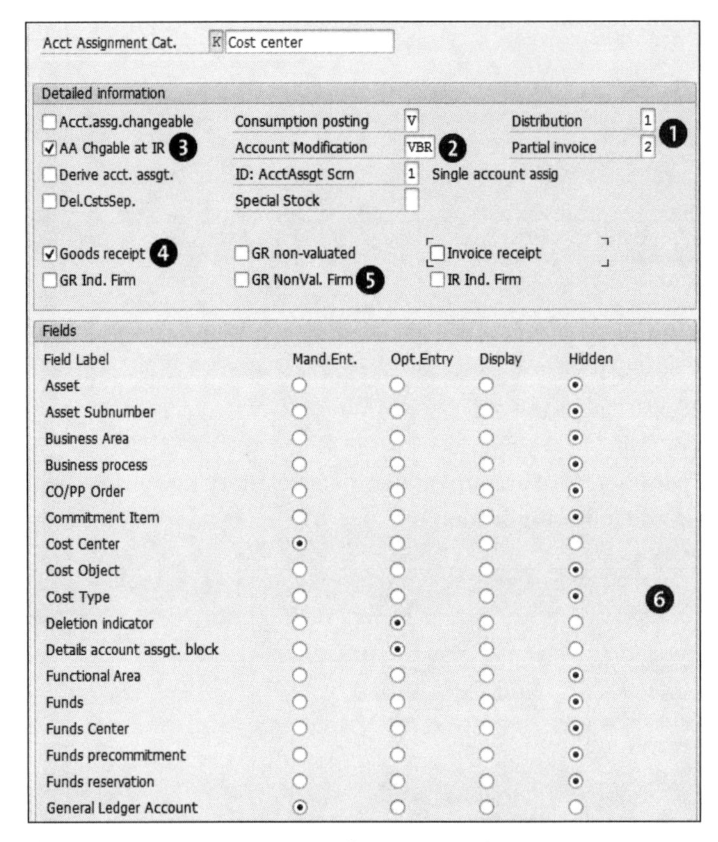

Figure 4.8 Customizing Parameters for Account Assignment Category K

Field Selection Control

The area marked ❻ in Figure 4.8 for account assignment category K (cost center) shows the fields you can control for the account assignment category. **General Ledger Account** and **Cost Center** are required entry fields for account assignment category K. If you also require the entry of an order for account assignment to the cost center, you can create your own additional account assignment category for which the **CO/PP Order** field is also mandatory. Alternatively, you can change the account assignment category K.

Distribution and Partial Invoice Indicator

If you wish to order the same material or the same service for several cost centers, for example, you do not necessarily need to create several purchase order items. You can create a single purchase order item and opt for multiple account assignment. This means that you distribute the purchase order quantity among different cost centers. Distribution to different account assignment objects is only valid within one account assignment category. If you need part of the quantity for a cost center and another part for a work breakdown structure (WBS) element, you must create two items, each with the correct account assignment category.

In connection with multiple account assignments, two indicators can be maintained for the account assignment category in Customizing: **Distribution** and **Partial Invoice**, shown on the right side of Figure 4.9.

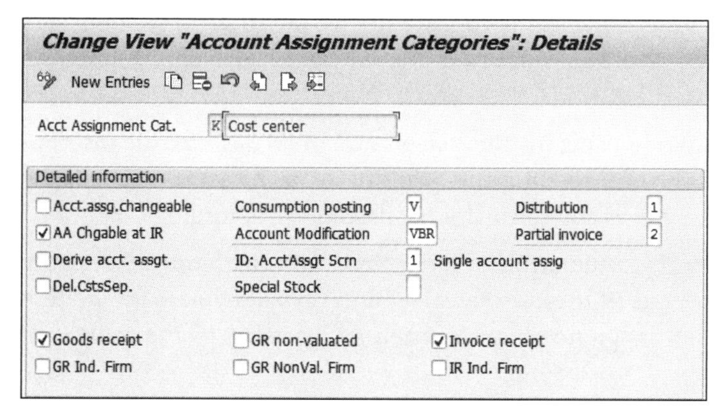

Figure 4.9 Distribution and Partial Invoice Indicators

Figure 4.10 shows the possible values for the **Distribution** indicator, and Figure 4.11 shows the possible values for the **Partial Invoice** indicator.

Distribut. indicator	Short Descript.
	Single Account Assignment
1	Distrib. on Quantity Basis
2	Distribution by Percentage
3	Distribution by Amount

Figure 4.10 Possible Values for the Distribution Indicator

Partial invoice	Short Descript.
	No multiple account assignment
1	Apportion IR quantities to GR quantities one after another
2	Apportion IR quantities to GR quantities proportionately

Figure 4.11 Possible Values for the Partial Invoice Indicator

Figure 4.12 shows the **Distribution** and **Partial Invoice** indicators on the purchase order item, copied as default values from the account assignment category Customizing.

Material Data | Quantities/Weights | Delivery Schedule | Delivery | Invoice | Conditions | Account Assignment

AccAssCat: Cost center | Distribution: Distrib. on Quantity Basis | CoCode: Company Code..

Partial Inv.: Distribute Proportionally

S...	A...	Quantity	Perce	Net Value	Cost Ctr	G/L Acct	Unloading Point	Recipient	D
	1	500,000	50,0	1.000,00	4110	51100000			
	2	300,000	30,0	600,00	4891	51100000			
	3	200,000	20,0	400,00	4300	51100000			

Figure 4.12 Distribution and Partial Invoice Indicators in the Purchase Order Item

The indicator **Distribution** determines how quantity and value are to be distributed among the individual account assignment items when a purchase order item with multiple account assignment is created: by percentage, by quantity, or by amount. This **Distribution** indicator is copied into the purchase order item as a default value when it is created and can be changed by the user. It simply provides an entry aid for the user.

As shown in Figure 4.12, choosing **Distribution on a Quantity Basis** for the **Distribution** indicator allows the user to enter the quantity for each cost center, and the system distributes the values and calculates the percentage accordingly.

The **Partial Invoice** indicator determines for each account assignment category how partial goods receipts (if they are valuated) and partial invoices for an item with multiple account assignments are distributed by value to the individual account assignments.

There are two ways of distributing valuated partial deliveries and partial invoices for an item with multiple account assignment to the individual account assignment elements by value:

- On a progressive refill basis
- Proportional to distribution in the order item

In the purchase order item, you can explicitly choose one of these two options or you can adopt the setting of the account assignment category in Customizing (the first option shown in Figure 4.13).

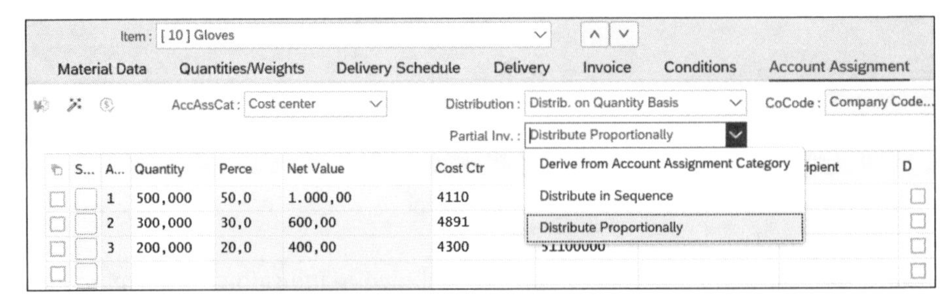

Figure 4.13 Partial Invoice Indicator

Example

Consider the example shown in Figure 4.13. One thousand pieces of a material are ordered for different cost centers. Five hundred pieces are assigned to cost center 4110, 300 pieces are assigned to cost center 4891, and 200 pieces are assigned to cost center 4300.

A partial goods receipt of 700 pieces at 1,400 EUR is posted for this purchase order. Depending on whether you selected **Distribute in Sequence** or **Distribute Proportionally** for the **Partial Invoice** indicator, the invoice amount is distributed as follows:

- **Distribute proportionally**
  - Cost center 4110: $700, corresponding to 50% (350 pieces)
  - Cost center 4891: $420, corresponding to 30% (210 pieces)
  - Cost center 4300: $280, corresponding to 20% (140 pieces)
- **Distribute in sequence**
  - Cost center 4110: $1,000 (the value assigned to the first cost center in the purchase order item is filled up)
  - Cost center 4891: $600 (the value assigned to the second cost center in the purchase order item is filled up)
  - Cost center 4300: $100 (remaining amount)

Note

Although the indicator is called *partial invoice*, the distribution rule applies to a partial goods receipt if it is valuated. If the goods receipt is not valuated or no goods receipt is posted at all, then the distribution rule applies to the partial invoice.

Account Modification

The system uses the *account modification* of the account assignment category to propose a general ledger account when creating a purchase order item with account assignment. The account modification is one of the keys of the account determination table, which we will investigate in Chapter 10.

Account Assignment Changeable at Invoice Receipt

With the **AA Chgable at IR** indicator, you allow the account assignment that you entered in the purchase order item to be changed during invoice entry. A prerequisite for this is that you are working with non-valuated goods receipt or no goods receipt at all. In the case of a valuated goods receipt, the costs are updated at goods receipt. In this case, therefore, you are not allowed to change the account assignment when you post the invoice.

Indicators Related to the Goods Receipt Postings

Now it is time to talk about these important indicators, which control whether a goods receipt is to be posted for a purchase order item assigned to an account and, if so, whether it is valuated or non-valuated.

For a purchase order item assigned to an account, it is possible to post a valuated or non-valuated goods receipt or no goods receipt at all. The prerequisites for this are set in Customizing for the account assignment category.

Reminder

The settings in Customizing for the account assignment category are copied to each purchase order item created with this account assignment category. They are either only default values or binding, depending on whether the binding indicator is set.

Let's look at some different settings combinations and their effects so that you can easily understand the logic of this approach.

- **Goods receipt binding, valuated proposed, as shown in Figure 4.14**
  Based on this setting in Customizing for the account assignment category, a goods receipt *must* be posted (**Goods Receipt** and **GR Ind. Firm** indicators set). The default is that the goods receipt is valuated (**GR Non-Valuated** indicator not set), but this setting can be adjusted in the purchase order item. (The **GR NonVal. Firm** indicator is not set.)

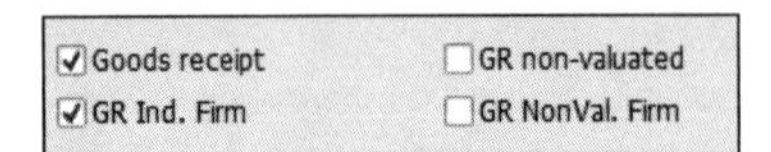

Figure 4.14 Goods Receipt Firm

- **Goods receipt valuated proposed but changeable in the purchase order, as shown in Figure 4.15**
  Based on this setting in Customizing for the account assignment category, a goods receipt valuated posting is defaulted in the purchase order item (the **Goods Receipt** indicator is set and the **GR Non-Valuated** indicator is not set), but this proposal can be changed in the purchase order item (no **Firm** indicator is set).

  The setting shown in Figure 4.16 offers the same options, but instead of valuated, the non-valuated goods receipt is proposed.

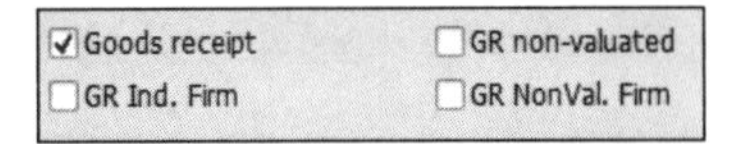

Figure 4.15 Goods Receipt Valuated Proposed

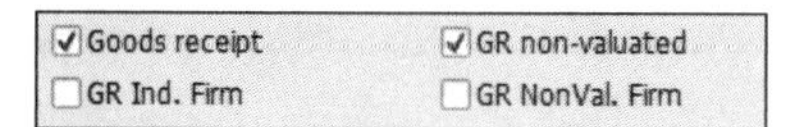

Figure 4.16 Goods Receipt Non-Valuated Default

- **If goods receipt posting, then non-valuated, as shown in Figure 4.17**
  With this combination of settings, the goods receipt posting is proposed (the **Goods Receipt** indicator is set), but you can choose not to do so (the **GR Ind. Firm** indicator is not set). However, if the goods receipt is posted, it can only be posted non-valuated (both the **GR Non-Valuated** and **GR NonVal. Firm** indicators are set).

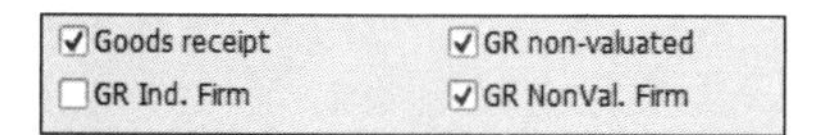

Figure 4.17 If Goods Receipt, Then Non-Valuated

Combination of Item Categories and Account Assignment Categories

You can assign account assignment categories to the item categories. This enables you to permit only certain account assignment categories for procurement processes.

As a reminder, the item categories specify the procurement process you want to trigger. By assigning item categories and account assignment categories, you determine which account assignments are allowed in which procurement processes.

Figure 4.18 shows the Customizing path for combining item categories with account assignment categories.

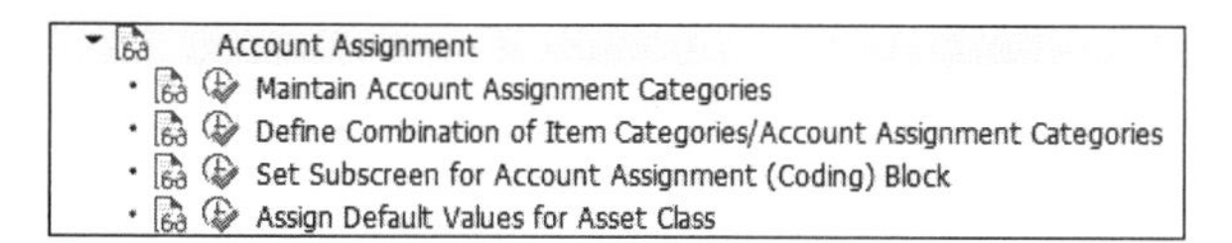

Figure 4.18 Customizing: Path to the Assignment of Account Assignment Categories to Item Categories

Figure 4.19 shows which account assignment categories are allowed for the normal procurement process (with the item category blank).

Figure 4.20 shows which account assignment categories are allowed for the subcontracting process (item category L). If, for example, you only want to use the subcontracting process for the warehouse, simply do not allow any account assignment category for item category L.

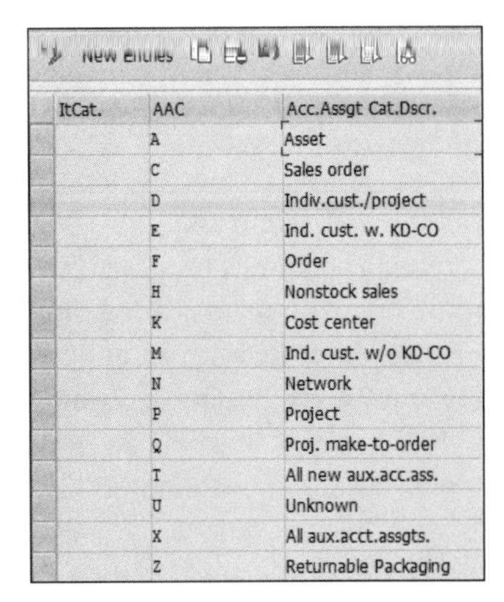

ItCat.	AAC	Acc.Assgt Cat.Dscr.
	A	Asset
	C	Sales order
	D	Indiv.cust./project
	E	Ind. cust. w. KD-CO
	F	Order
	H	Nonstock sales
	K	Cost center
	M	Ind. cust. w/o KD-CO
	N	Network
	P	Project
	Q	Proj. make-to-order
	T	All new aux.acc.ass.
	U	Unknown
	X	All aux.acct.assgts.
	Z	Returnable Packaging

Figure 4.19 Customizing: Combining Item Categories with Account Assignment Categories

L	E	Ind. cust. w. KD-CO
L	F	Order
L	K	Cost center
L	Q	Proj. make-to-order
L	T	All new aux.acc.ass.
L	X	All aux.acct.assgts.

Figure 4.20 Customizing: Combining Item Categories with Account Assignment Categories

Goods Receipt with Reference to a Purchase Order for Direct Consumption

In this section, we explain how to post a goods receipt with reference to a purchase order item assigned to an account. We will then discuss the results of the goods receipt posting.

We have already discussed that, in contrast to the procurement of stock material, there are basically three options for account assigned procurement:

- Goods receipt valuated
- Goods receipt non-valuated
- No goods receipt

In the next subsection, we will explain how to post a goods receipt with reference to a purchase order item assigned to an account. We will then discuss the results of the goods receipt posting, depending on whether the goods receipt is valuated or not valuated.

Posting a Goods Receipt

SAP S/4HANA offers several options for posting a goods receipt with reference to a purchase order, especially the Post Goods Receipt for Purchase Order app or Transaction MIGO (Post Goods Movements).

In all cases, you must first enter the purchase order reference. The system then proposes the open purchase order items with the corresponding open quantity. This data can be overwritten.

Figure 4.21 and Figure 4.22 show Transaction MIGO (Post Goods Movement). You can proceed as follows to start the entry of a goods receipt with reference to a purchase order:

1. Choose **Goods Receipt** in the **Transaction** field.
2. Choose **Purchase Order** in the **Reference** field.
3. Enter the purchase order number.
4. Verify that the movement type is **101**.
5. Choose **Execute**.

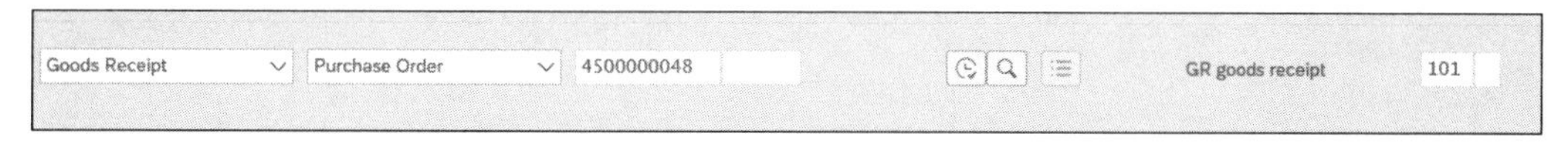

Figure 4.21 Entering the Purchase Order Reference in the Post Goods Movement App (Transaction MIGO)

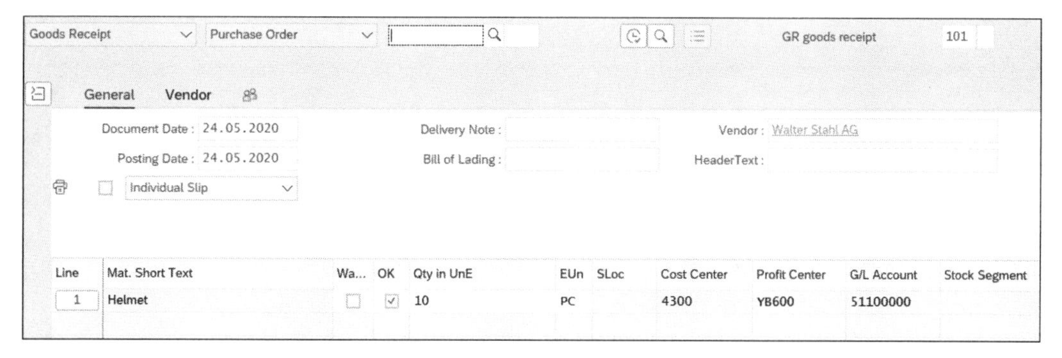

Figure 4.22 General Data and Items in the Post Goods Movement App (Transaction MIGO)

The system then proposes the purchase order data. Proceed as follows:

1. Enter a delivery note and check document and posting dates.
2. Check (and, if required, change) the proposed quantity, storage location, and stock type.
3. Flag the **OK** indicator.
4. Choose **Post**.

The Post Goods Receipt for Purchase Order app is intuitive and easy to use, as follows:

1. Enter the purchase order number, as shown in Figure 4.23.
2. Enter a delivery note and check document and posting dates, as shown in Figure 4.24.

3. Select the relevant items, check, and (if required) change the proposed quantity.
4. Choose **Post**.

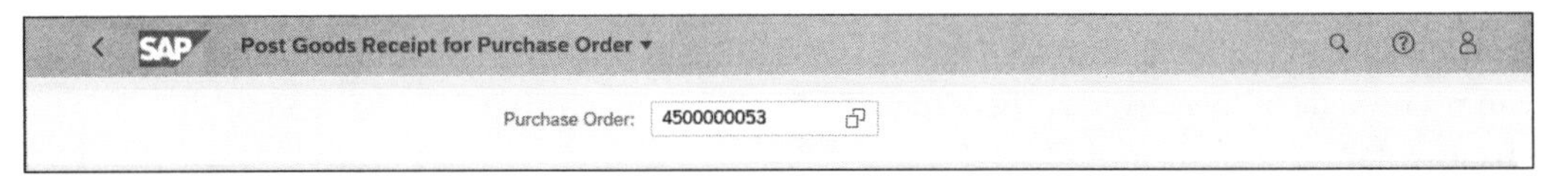

Figure 4.23 Entering the Purchase Order Number in the Post Goods Receipt for Purchase Order App

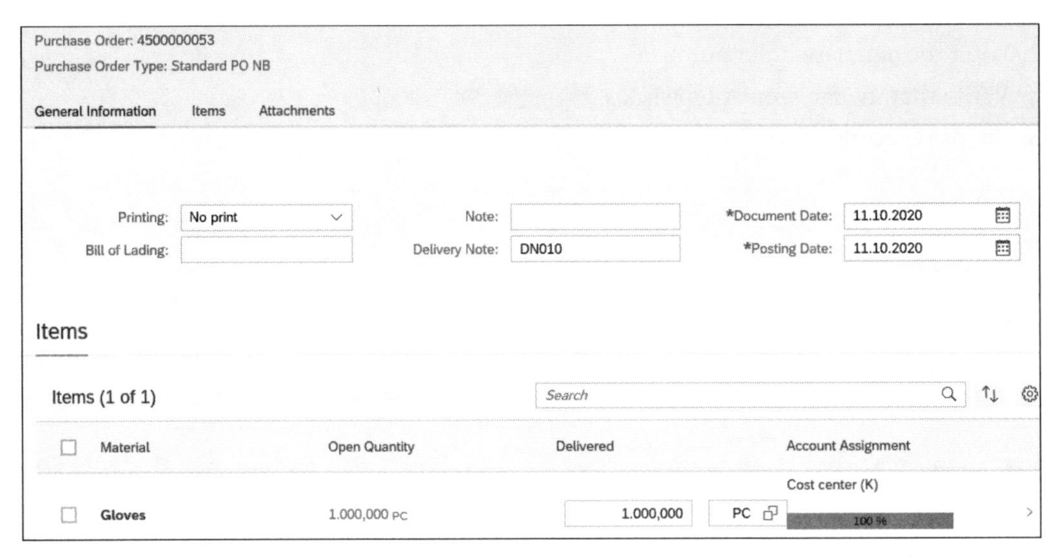

Figure 4.24 General Information and Items in the Post Goods Receipt for Purchase Order App

The Post Goods Receipt for Purchase Order app has the following restrictions compared to Transaction MIGO:

- You can only enter the goods receipt for one purchase order and not for several.
- You cannot post a goods receipt for a purchase order item with the delivery completed indicator.

When you post a goods receipt for a purchase order assigned to an account, the value updates depend on whether you post a valuated or non-valuated goods receipt.

Results of a Goods Receipt

The effects of a goods receipt for a purchase order item assigned to an account are as follows:

- The purchase order history like the one shown in Figure 4.25 is updated. The purchase order history allows you to monitor the procurement process and shows all goods receipts, returns, invoices, credit memos, and so on for each purchase order item. It is also known as a *process flow*.

- A material document is created. The material document provides information about the goods receipt and is a source of information for all subsequent processes. The material document consists of a header and at least one item.
- An accounting document is only created if the goods receipt is posted as valuated. Among others, the general ledger account entered in the purchase order item is updated.

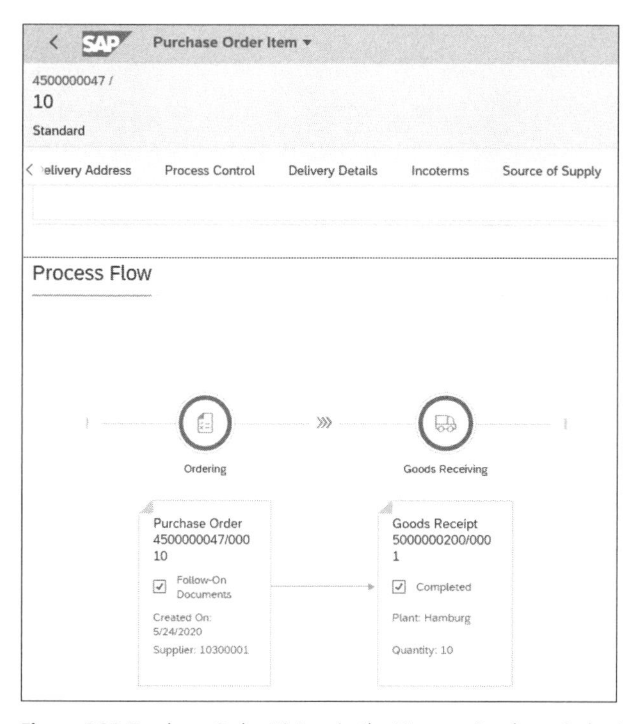

Figure 4.25 Purchase Order History in the Manage Purchase Order App

Under-Deliveries and Over-Deliveries

When you post a goods receipt for a purchase order, the system proposes the open purchase order quantity. The open purchase order quantity is the difference between the quantity ordered and the quantity already delivered. You can, of course, overwrite this proposed quantity according to the information on the delivery note and the actual quantity delivered.

The option of posting over- or under-deliveries is the same as for a stock purchase order item. The logic of the **Delivery Completed** indicator is also the same (see Chapter 3).

Invoice Receipt with Reference to a Purchase Order Item Assigned to an Account

The invoice entry procedure is also very similar for stock material purchase orders and purchase orders with account assignment.

You use the Create Supplier Invoice app (F0859) to manually enter and post a supplier invoice related to a purchase order. Alternatively, you can use Transaction MIRO (Create Supplier Invoice—Advanced).

The effects of posting an invoice with reference to a purchase order are as follows:

- The purchase order history is updated. The purchase order history allows you to monitor the procurement process and shows all goods receipts, return deliveries, invoices, credit memos, and so on for each purchase order item.
- An invoice document is created.
- An accounting document is created. Using automatic account determination, the system updates the general ledger accounts that are affected by the invoice postings.

Under certain conditions, it is possible to change the account assignment during invoice entry. To change the account assignment with the Create Supplier Invoice app, follow these steps:

1. In the item list, select the small arrow to the right of the relevant item to access the details, as shown in Figure 4.26.
2. In the item details, within the **Account Assignment** tab, you can adjust the account assignment elements, as shown in Figure 4.27.

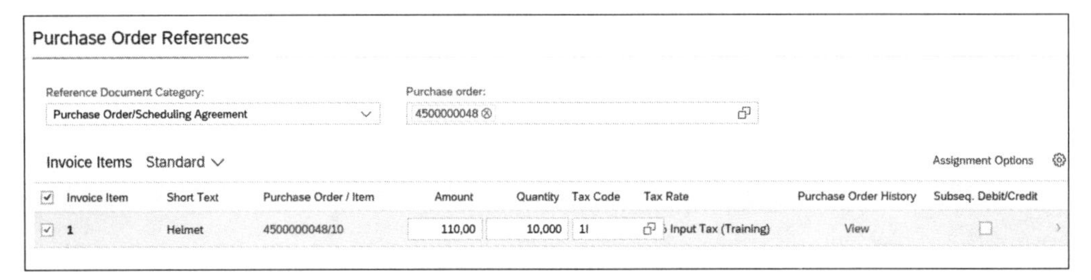

Figure 4.26 Changing an Account Assignment in the Create Supplier Invoice App (I)

Figure 4.27 Changing an Account Assignment in the Create Supplier Invoice App (II)

The prerequisites for changing the account assignment in invoice verification are as follows:

- In Customizing, for the account assignment category, you must define that the account assignment can be changed in invoice verification (the **AA chgable at IR** checkbox).
- A non-valuated or no goods receipt must be planned for the purchase order item.

Value Flow

Now let's take a closer look at the accounting documents that are created with the goods receipt posting and the invoice posting with reference to a purchase order item assigned to an account. For a purchase order item assigned to an account, the postings at goods receipt/invoice receipt (GR/IR) obviously depend on whether the goods receipt is valuated or non-valuated.

As for the procurement of stock material, we examine the booking logic using examples.

Valuated Goods Receipt

We consider the following example scenario:

1. You order 10 helmets for a cost center at a purchase order price of 12 EUR. The **Goods Receipt** indicator is set, whereas the **GR Non-Valuated** indicator is not set, as shown in Figure 4.28.

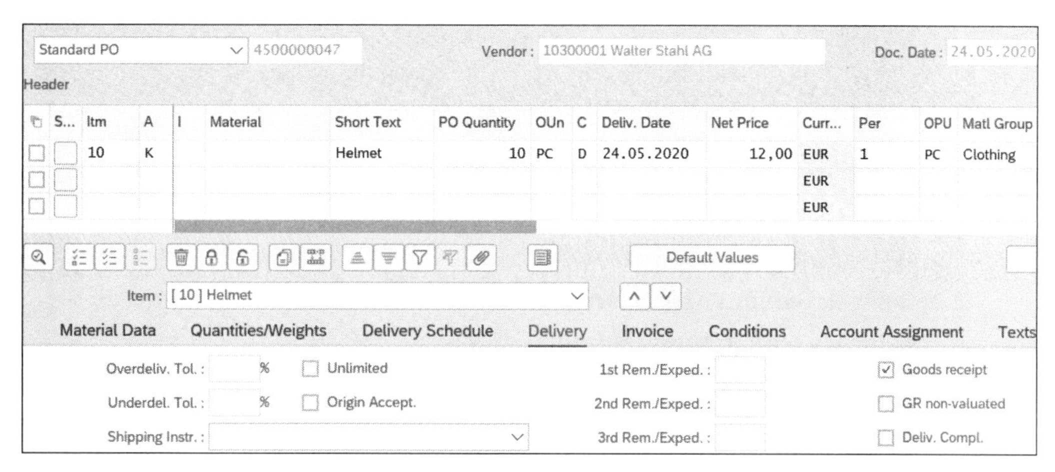

Figure 4.28 Purchase Order Item Assigned to an Account: Goods Receipt Valuated in the Create Purchase Order—Advanced App (Transaction ME21N)

2. You post the goods receipt for this purchase order.

3. On the vendor invoice, the price differs from the purchase order price: instead of 12 EUR, the vendor invoices 11 EUR per unit, as shown in Figure 4.29. You post the vendor invoice.

The following account postings are made at goods receipt:

- The cost account is debited with the value 120 EUR: *goods receipt quantity × purchase order price*.
- The offsetting entry is posted to the GR/IR clearing account at purchase order value 120 EUR.

Now let's consider the effects of posting the invoice, disregarding the tax posting for simplification purposes:

- The vendor account is credited by the invoice value.
- The GR/IR clearing account is cleared.
- The difference to the purchase order price (10 EUR) is credited to the cost account.

Note

If the invoice does not contain any variances from the purchase order price, only the GR/IR clearing account is cleared, in addition to posting to the vendor account and the tax posting.

However, if the invoice contains variances from the purchase order price, a cost value adjustment is posted.

Posting Logic of the GR/IR Clearing Account

If the goods receipt is posted before the invoice, the GR/IR clearing account is credited with the order value and cleared when the invoice is posted. If the invoice is posted before the goods receipt, the GR/IR clearing account is debited with the invoice value and cleared with this value when the goods receipt is posted.

Here we can see the account movements for a PO item assigned to each account, goods receipt valuated:

❶ Purchase order: 10 pieces at €12 per piece

❷ Goods receipt: 10 pieces at €12 per piece

❸ Invoice receipt: 10 pieces at €11 per piece

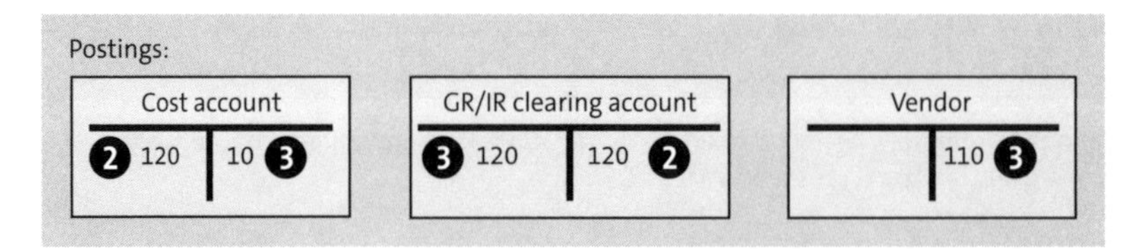

Figure 4.29 Goods Receipt Valuated—Postings

Non-Valuated Goods Receipt or No Goods Receipt

We consider the following example scenario:

1. You order 10 helmets for a cost center at a purchase order price of 12 EUR, as shown in Figure 4.30. The **Goods Receipt** indicator is set; the **GR Non-Valuated** indicator is also set, as shown in Figure 4.31.

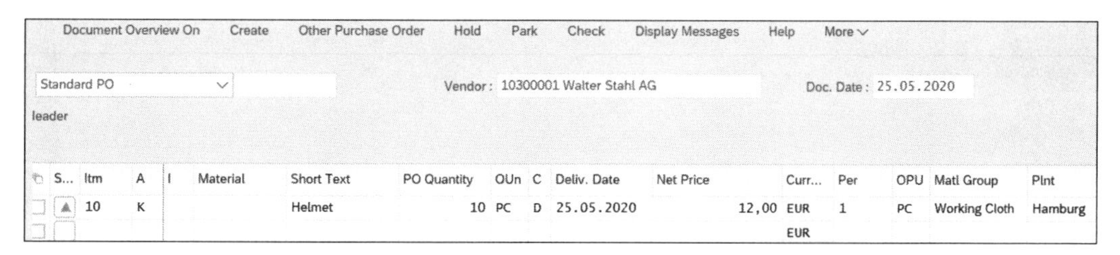

Figure 4.30 Purchase Order Item Assigned to an Account in the Create Purchase Order—Advanced App (Transaction ME21N)

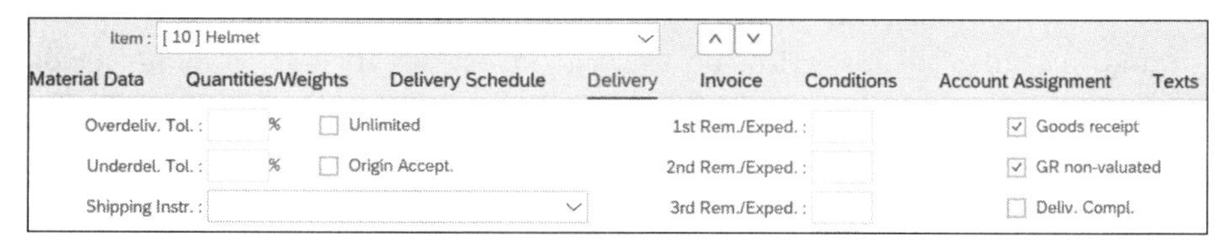

Figure 4.31 Purchase Order Item Assigned to an Account, Goods Receipt Non-Valuated in the Create Purchase Order—Advanced App (Transaction ME21N)

2. You post the goods receipt for this purchase order.
3. On the vendor invoice, the price differs from the purchase order price: instead of 12 EUR, the vendor invoices 11 EUR per unit, as shown in Figure 4.32. You post the vendor invoice.

Now let's consider the effects of posting the goods receipt and the invoice, again disregarding the tax posting for simplification purposes:

- No account postings are made at goods receipt.
- At invoice receipt, the vendor account is credited by the invoice value.
- The invoice value is debited to the cost account.

> **Note**
>
> Since no posting is made in financial accounting at the time of goods receipt, the posting to the GR/IR clearing account is not applicable.

Here we can see the account movements for a PO item assigned to an account, goods receipt non-valuated:

❶ Purchase order: 10 pieces at €12 per piece

❷ Goods receipt: 10 pieces

❸ Invoice receipt: 10 pieces at €11 per piece

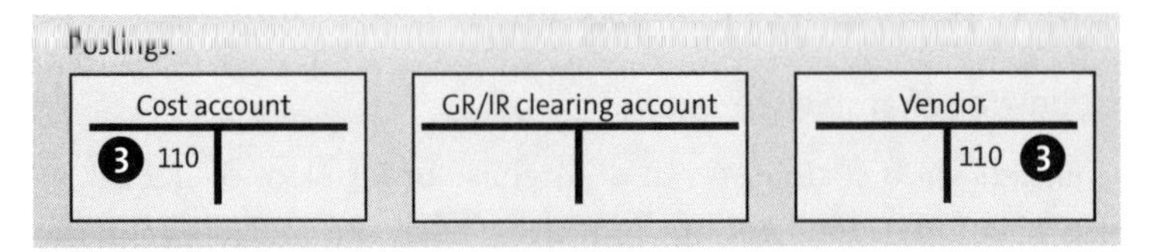

Figure 4.32 Goods Receipt Non-valuated—Postings

Note

The postings are the same if no goods receipt is posted.

Self-Service Procurement

Self-service procurement offers users the option to order items for their own use in a simplified and user-friendly form. The Create Purchase Requisition app allows users to easily order items while a purchase requisition is created in the background. Instead of spending time posting a goods receipt, the recipient only needs to confirm receipt of the ordered goods. The actual goods receipt posting takes place in the background.

Figure 4.33 shows the apps that users can utilize to order goods and services for their own needs.

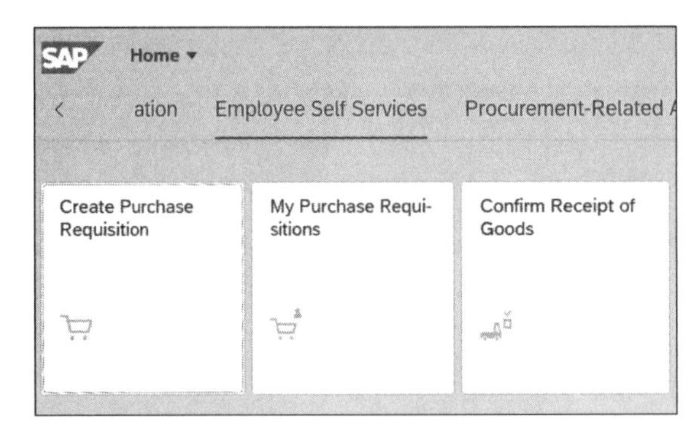

Figure 4.33 Main Self-Service Procurement Apps

A user can maintain default values so that they can order products with a minimum of effort and the system still have all the necessary data to create a complete purchase requisition. Users can maintain default values, for example, for plant, company code, delivery date, purchasing group, purchasing organization, account assignment category, cost center, and so forth, and assign catalogs to themselves.

If you want to maintain your default settings, start the Create Purchase Requisition app shown in Figure 4.34 and then choose the personalization icon in the upper-right corner of the screen.

You can then select **Default Settings for User** (see Figure 4.35) and maintain personal default values.

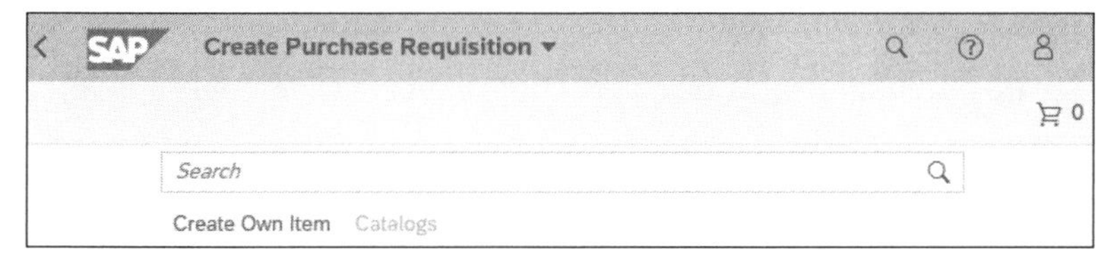

Figure 4.34 The Create Purchase Requisition App

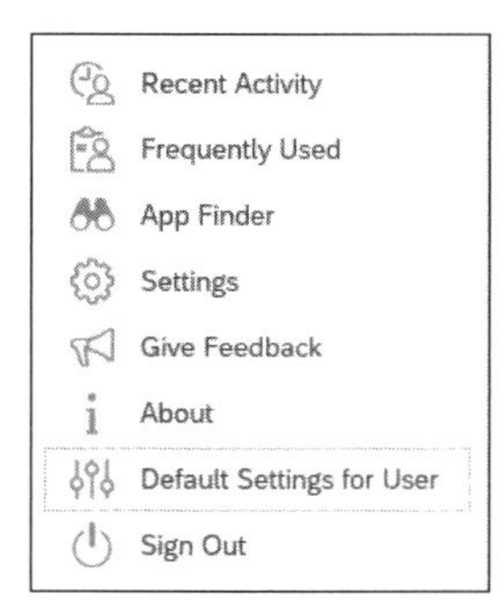

Figure 4.35 Default Settings for Users

If all the required organizational units, account assignment category, and cost elements (e.g., cost center and general ledger account) are maintained in the default values, then within the Create Purchase Requisition app, the user only needs to enter a small amount of data:

- Material or short description
- Material group if no material number is specified
- Valuation price, if no or a non-valuated material was entered
- Quantity
- Desired delivery date

When you click on the **Add to Cart** button, the system updates the shopping cart. In the next step, you click on **Order** and the system creates a purchase requisition in the background.

Note

When you create a purchase requisition item for valuated material, the valuation price is taken from the material master record. In the case of non-valuated material or material without a master record, the person entering the item must enter the valuation price manually. This valuation price can be used for a value-based release procedure.

Note that this is not necessarily the purchase order price! However, when the item is converted into a purchase order, this price is adopted for materials without a master record and without a source of supply. It must be checked and changed by the buyer.

In the case of a material with a master record, this price is *not* adopted in the purchase order from the purchase requisition.

To place an order for your own needs, take the following steps.

❶ Start the Create Purchase Requisition app, choose **Create Own Item**, and enter data, as shown in Figure 4.36.

❷ Select the **Add to Cart** button in the lower right corner of the screen, as shown in Figure 4.36.

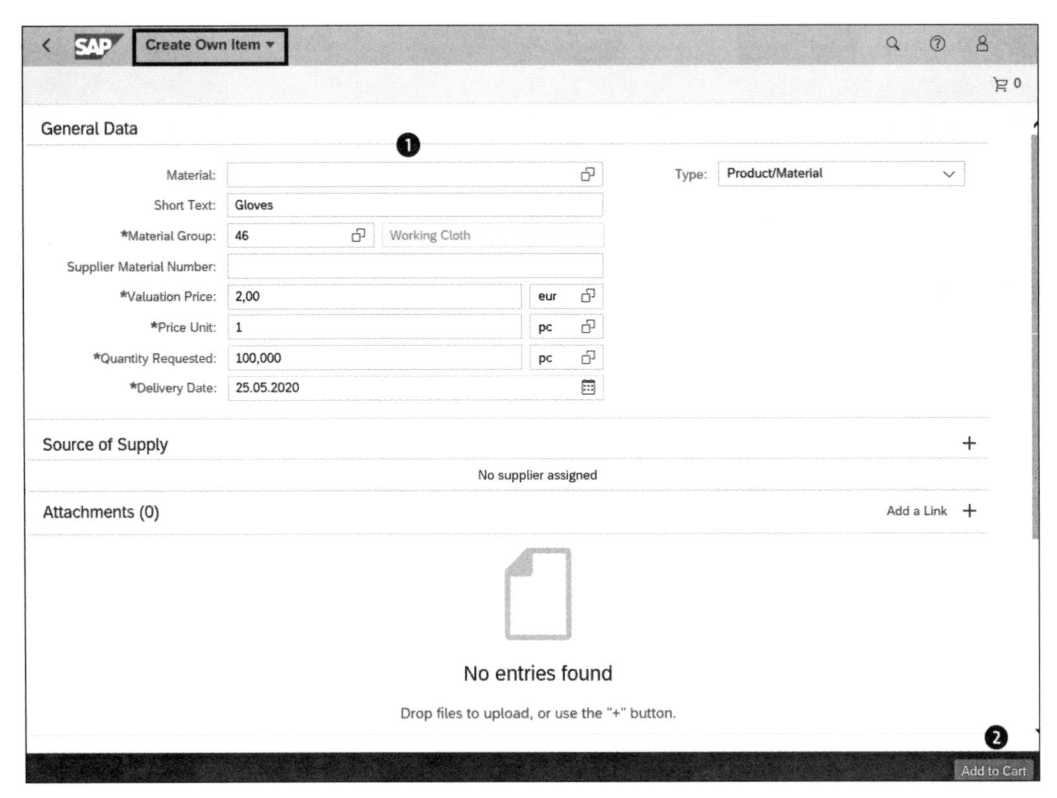

Figure 4.36 Creating an Item and Adding Items to Cart in the Create Purchase Requisition App

❸ Your shopping cart will then be filled. You can access it in the upper-right corner of the screen and press the **Order** button shown in Figure 4.37 when you are finished filling it.

Figure 4.37 Ordering from Your Shopping Cart in the Create Purchase Requisition App

This purchase requisition may have to be approved; it is then converted into a purchase order. The user can monitor the status of a request with the My Purchase Requisitions app. When the user receives the ordered goods, they confirm the goods receipt with the Confirm Receipt of Goods app, and the system posts the goods receipt in the background.

Blanket Purchase Order

Process costs for procurement sometimes bear no relation to the value of the materials or services procured. You can reduce these process costs by using a blanket purchase order instead of a standard purchase order. A blanket purchase order is a long-term framework order that is used to procure many materials or services whose low value does not justify the high processing costs for the procurement of individual items.

The main advantage of using blanket purchase orders is the reduction of processing costs. These costs are significantly lower for the following reasons:

- The blanket purchase order is valid for a longer period.
- Goods receipt is not posted for a blanket purchase order.
- You do not need to create purchase order items for single items procured.
- You usually agree on collective invoices (e.g., monthly invoices) with your supplier.

The characteristics of a blanket purchase order are as follows:

- **Item category B (limit) or E (extended limit)**
  These item categories control the flow of the process: You do not enter a purchase order quantity or a net price in the item, but only a rough description of the intention to buy (e.g., office supplies). You enter the expected value and the total limit, which represents the maximum value up to which you can post invoices with reference to the purchase order item. The expected value and the overall limit, of course, can be the same.
- **Account assignment category required**
  In Customizing, you can specify which account assignment categories you allow in combination with the item category limit or the item category extended limit. If you do not know the exact account assignment when you create the purchase order, you can use account assignment category U (unknown), which is allowed in the standard SAP S/4HANA system for purchase order items of item categories B and E. You do not have to specify the account assignment until you enter the invoice.

SAP S/4HANA offers two options for creating a blanket purchase order.

The Create Purchase Order—Advanced App

When you start the Create Purchase Order—Advanced app, you should use the document category framework order (FO). With this document type, the **Validity Start** and **Validity End** fields are displayed automatically, and the **Validity Start** field is defined as a required entry field, as shown in Figure 4.38.

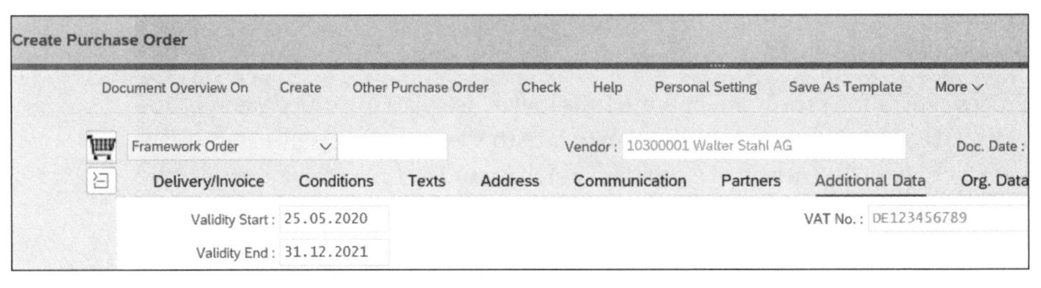

Figure 4.38 Creating a Blanket Purchase Order with Order Type FO

You enter the B (limit) item category and an account assignment category in the purchase order item overview (see Figure 4.39). Entering this item category ensures that the **Limits** tab appears on the **Item Details** screen. You then enter an overall limit and an expected value, as shown in Figure 4.40.

S...	Itm	A	I	Material	Short Text	PO Quantity	OUn	C	Deliv. Date	Net Price	Curr...	Per	OPU	Matl
	10	K	B		Office supplies	1	EA	D	25.05.2020	1.000,00	EUR	1	EA	Offic
											EUR			

Figure 4.39 Blanket Purchase Order with Item Category B

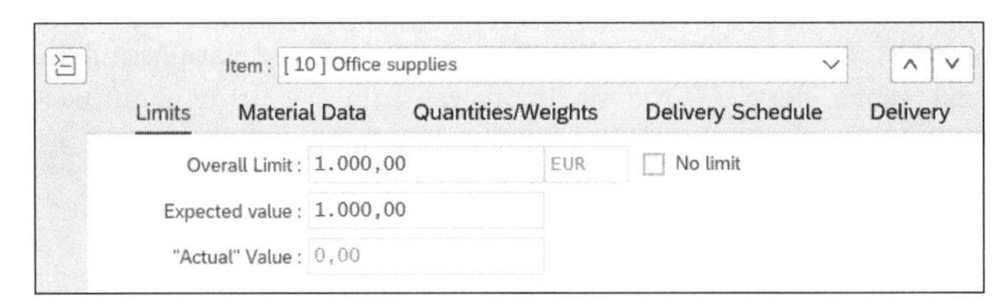

Figure 4.40 Creating a Blanket Purchase Order: Limits

The Manage Purchase Order App

The second option is to create a limit item with the Manage Purchase Order app: It has a description, an expected value, and an overall limit, as shown in Figure 4.41. In the item details in the **General Information** tab in Figure 4.42, you can define the account assignment category; on the **Schedule Lines** tab in Figure 4.43, you can define the validity period. When you save the purchase order, the system creates a purchase order item with item category E for it.

Limit Items

Purchase Order Limit Items (1)

Search Delete

Item	Short Text	Expected Value		Overall Limit		Acct Assignment Cat.	Product Type Group	Plant
10	Office supplies	10.000,00	EUR	10.000,00	EUR	Cost center (K)	Material (1)	1010

Figure 4.41 Creating a Limit Item with the Manage Purchase Order App

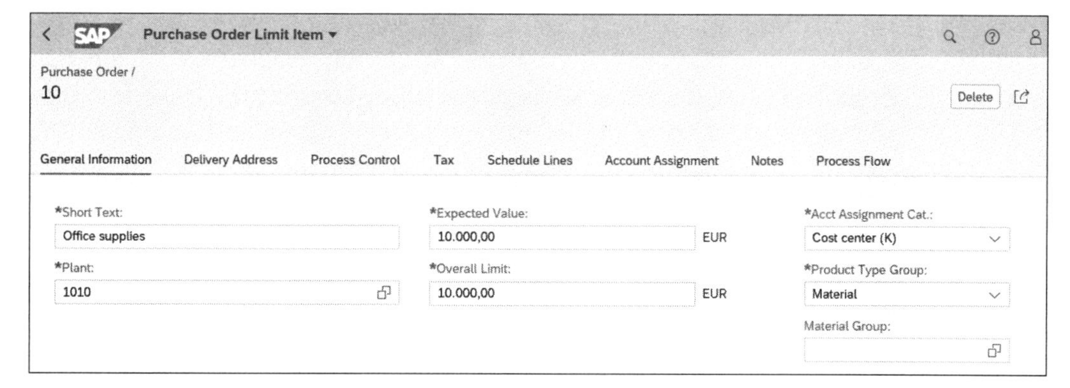

Figure 4.42 Using the General Information Tab to Create a Limit Item

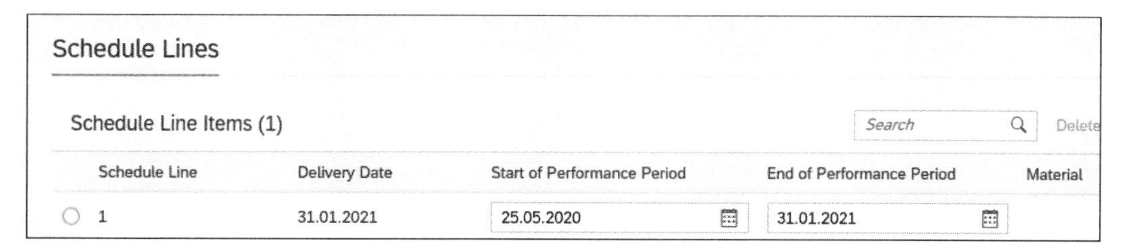

Figure 4.43 Using the Schedule Lines Tab to Create a Limit Item

The next process step in a blanket purchase order is the invoice posting at regular intervals. The system checks the overall limit and validity. If the overall limit or the validity range is exceeded when an invoice is posted, the system issues a warning or error message, depending on the settings in Customizing, and the invoice may be blocked for payment.

Important Terminology

The following terminology was used in this chapter:

- **Account assignment category**
  The account assignment category is a key that determines whether a purchase order item is intended for consumption or for stock. This is configured in Customizing; you can adapt existing account assignment categories or add new ones.

- **Distribution indicator**
  The **Distribution** indicator is used in purchase order items with multiple account assignments and determines how the item quantities and values should be distributed to the different account assignments. It is defined as a default value in Customizing of the account assignment category and can be adjusted in the purchase order item.
- **Partial Invoice indicator**
  The **Partial Invoice** indicator is used in purchase order items with multiple account assignments and determines how a partial goods receipt or a partial invoice is to be assigned to the various account assignments—filling in or proportionally distributed. It is defined as a default value in Customizing for the account assignment category and can be adjusted in the purchase order item.
- **Account modification**
  The account modification is used for account determination. An account modification in Customizing for the account assignment category allows a general ledger account to be proposed when creating a purchase order item with account assignment.
- **Item category**
  The item category is a one-character key, is entered at the purchase order item level, and determines the procurement process. It is given by SAP and cannot be configured. Only the external display and the description can be customized.
- **GR/IR clearing account**
  A GR/IR clearing account is an account used when goods arrive before the invoice is issued or when an invoice arrives before the goods are delivered.
- **Non-valuated goods receipt**
  For a purchase order item with account assignment, you can decide—depending on the account assignment category—whether the goods receipt is valuated or not (i.e. whether the costs are updated at goods receipt or only when the invoice is posted).
- **Blanket purchase order**
  A framework purchase order is used for the procurement of materials or services whose small value does not justify the high handling costs for the procurement of individual articles. A framework order item is created with item category B or E and assigned to an account.
- **Self-service procurement**
  Self-service procurement consists of SAP Fiori apps that allow users to order items for their own use in a streamlined and user-friendly form.

Practice Questions

These questions will help you evaluate your understanding of the topics covered in this chapter. They are similar in nature to those on the certification examination. Although none of these questions will be found in the exam itself, they will allow you to review your knowledge of the subject.

Select the correct answers, and then check the completeness of your answers in the next section. Remember that on the exam, you must select all correct answers—and only correct answers—to receive credit for the question.

1. What must a purchase order item for direct consumption include? (There are three correct answers.)

☐ **A.** A plant

☐ **B.** An account assignment category

☐ **C.** A material number

☐ **D.** An account assignment object

2. True or false: If required, a stock material can be ordered directly for a cost center or an order without being stored beforehand.

☐ **A.** True

☐ **B.** False

3. Which of the following statements apply to the account assignment category? (There are two correct answers.)

☐ **A.** Account assignment categories cannot be configured. They are defined by SAP. You can only change the description.

☐ **B.** The account assignment category is a factor that contributes to determining whether a field in the purchase order item is mandatory for input or not.

☐ **C.** Depending on the account assignment category, you must or can post a non-valuated goods receipt to the purchase order item.

☐ **D.** The account assignment category controls the number assignment of a purchase order.

4. True or false: You can only change the account assignment data relating to a purchase order item during invoice entry if no goods receipt may be posted for it.

☐ **A.** True

☐ **B.** False

5. True or false: When you create a purchase order item assigned to an account, you can always decide whether a goods receipt is to be posted and, if so, valuated.

☐ **A.** True

☐ **B.** False

6. When you create a purchase order item with account assignment in your system, the general ledger account and sometimes the account assignment objects (cost center, order, and so on) are only guide values. The accounts payable clerk should be able to decide in the last instance to which account the invoice is to be posted. Under what conditions can this requirement be met? (There are two correct answers.)

☐ **A.** The **AA Chgable at IR** checkbox must be set in Customizing of the account assignment category, and no goods receipt is posted.

☐ **B.** The **AA Chgable at IR** checkbox must be set in Customizing of the account assignment category, and the goods receipt is posted valuated.

☐ **C.** The **AA Chgable at IR** checkbox must be set in Customizing of the account assignment category, and the goods receipt is posted non-valuated.

☐ **D.** The **AA Chgable at IR** checkbox must be set and the **Account Modification** field must be maintained in Customizing of the account assignment category.

7. You want to order the same office chairs for different cost centers. How do you proceed?

☐ **A.** You must create a purchase order item for each cost center.

☐ **B.** You create one item and use the multiple account assignment option.

☐ **C.** You create one item and use the delivery schedule lines tab.

8. True or false: With self-service procurement, SAP S/4HANA offers user-friendly applications for ordering products and services for your own needs. The objects and process steps in the background (purchase requisition, purchase order, goods receipt, invoice receipt) are the same as in the classic procurement process.

☐ **A.** True

☐ **B.** False

9. Which field in a material master record controls that you must order this material with account assignment?

☐ **A.** Purchasing group

☐ **B.** Material group

☐ **C.** Material type

☐ **D.** Valuation class

10. True or false: If you manually create a purchase requisition item with account assignment, you must always enter a valuation price.

☐ **A.** True

☐ **B.** False

11. In a purchase order item assigned to an account, the **Goods Receipt** and **Invoice Receipt** checkboxes are selected. The **GR Non-Valuated** checkbox is not selected. Now you post the goods receipt for this purchase order item. Which of the following general ledger accounts are posted to at goods receipt? (There are two correct answers.)

☐ **A.** GR/IR clearing account

☐ **B.** Price difference account

☐ **C.** Cost account from purchase order item

☐ **D.** Liability account

12. True or false: You can prevent limit items from being created with account assignment category U (unknown).

☐ **A.** True

☐ **B.** False

13. Which of the following features are features of a blanket purchase order item? (There are three correct answers.)

☐ **A.** No purchase order quantity

☐ **B.** Account assignment category mandatory

☐ **C.** Goods receipt mandatory

☐ **D.** Expected value required

14. When you create a purchase order item with account assignment, the system can propose a general ledger account. Which of the following key in Customizing of the account assignment category can you use to influence account determination?

☐ **A.** The **Consumption Posting** indicator

☐ **B.** The **Derive Account Assignment** checkbox

☐ **C.** The **Account Modification** field

☐ **D.** The **Distribution** indicator

15. True or false: At a valuated goods receipt, the account assignment data from the purchase order can be changed.

☐ **A.** True

☐ **B.** False

16. True or false: The GR/IR clearing account is not applicable in a procurement process without a goods receipt or with a non-valuated goods receipt.

☐ **A.** True

☐ **B.** False

17. You create a purchase order item assigned to several cost centers. You expect partial deliveries for this item and want each partial delivery to be allocated proportionally to the individual cost centers. What must you do?

☐ **A.** Use the **Distribution** indicator.

☐ **B.** Use the item category B.

☐ **C.** Use the **Partial Invoice** indicator.

☐ **D.** Use the **Limits** tab.

Practice Question Answers and Explanations

1. Correct answers: **A, B, D**

   Like all purchase order items, a purchase order item with account assignment must contain a plant. You create a purchase order item with account assignment by entering an account assignment category. You must enter a general ledger account and, depending on the account assignment category, one or more account assignment objects (e.g., a cost center, an order, etc.). You can place purchase orders for consumption with or without a material master record.

2. Correct answer: **A**

   True. You can indeed order a stock material with account assignment, if required.

3. Correct answers: **B, C**

   Together with other factors, the account assignment category determines the field selection. You can customize the account assignment category. Among other things, you can specify whether a goods receipt must be posted non-valuated or whether it is only an option. Number assignment is controlled by the purchase order type and not by the account assignment category.

4. Correct answer: **B**

   False. Remember that the goods receipt for a purchase order item assigned to an account can be skipped or posted without valuation. In these cases, the accounts in financial accounting are updated when the invoice is posted. In both cases, and provided this is allowed in Customizing for the account assignment category (the **AA Chgable at IR** checkbox), you can change the account assignment data (general ledger account, cost center, and so on) during invoice entry.

5. Correct answer: **B**

   False. When you create a purchase order with account assignment, you cannot always decide whether a goods receipt should be posted and, if so, valuated. You can only do this if the account assignment category allows this.

6. Correct answers: **A, C**

   To enable you to adjust the account assignment data at invoice entry, the **AA Chgable at IR** checkbox must be selected for the account assignment category in Customizing, and the goods receipt should be either non-valuated or not posted at all.

   Despite the name, account modification has nothing to do with adjusting the account during invoice entry. It is a key for determining a suitable account for a purchase order item with account assignment, depending on the account assignment category.

   If the goods receipt is valuated, the cost components are debited at goods receipt so that you can no longer change them when you enter the invoice.

7. Correct answer: **B**

   Of course, you can create a purchase order item with account assignment for each cost center, but you do not have to. You can create a single item and use the multiple account assignment option.

   The schedule lines do not allow differentiated account assignments, only different delivery dates.

8. Correct answer: **A**

   True. With self-service procurement, SAP S/4HANA offers user-friendly applications for ordering products and services for your own needs. The objects and process steps in the background (purchase requisition, purchase order, goods receipt, invoice receipt) are the same as in the classic procurement process with account assignment.

9. Correct answer: **C**

   The material type in a material master record controls whether a material is managed in stock and, if so, whether it is valuated. If a material is either not managed in stock (material type NLAG) or only managed on a quantity basis (material type UNBW), then the material can only be ordered with account assignment.

10. Correct answer: **B**

    False. When you create a purchase requisition item with account assignment for a valuated material, the valuation price is taken from the material master record. You do not have to enter it manually.

11. Correct answers: **A, C**

    If the **Goods Receipt** and **Invoice Receipt** checkboxes are selected and the **GR Non-Valuated** checkbox is not selected in a purchase order item assigned to an account, the goods receipt will be posted valuated: the cost account in the purchase order item will be debited and the offsetting entry made to the GR/IR clearing account.

12. Correct answer: **A**

    True. You can define in Customizing which account assignment categories—including account assignment category U—you allow in combination with the limit item categories.

13. Correct answers: **A, B, D**

    A blanket purchase order item is characterized by a corresponding item category (B or E), it contains value limits, it is assigned to an account, and a goods receipt is not foreseen.

14. Correct answer: **C**

    The account modification is one of the influencing factors for account determination.

15. Correct answer: **B**

    False. You cannot adjust the account assignment data at goods receipt. When you enter an invoice, this may be possible only if you do not post a goods receipt or if you post a non-valuated goods receipt and Customizing of the account assignment category allows this.

16. Correct answer: **A**

    True. For a procurement process without a goods receipt or with a non-valuated goods receipt not applicable, the account updates only take place when the invoice is posted. There is no need for clearing.

17. Correct answer: **C**

    The **Partial Invoice** indicator controls the assignment of both a partial delivery and a partial invoice to account assignment objects in the case of a purchase order item with multiple account assignment. It enables proportional distribution or assignment to a cost element according to the fill-up principle.

    The **Distribution** indicator provides an entry aid for the user when distributing quantities and values of an order item to several account assignments.

    The item category controls the procurement process.

    The **Limits** tab is available for blanket orders, for example, to set value limits.

Takeaway

You should now understand the procurement process with account assignment. You should be able to explain how you can configure an account assignment category in Customizing, how you can assign account assignment categories to procurement processes (item categories), and how you can create purchase order items with account assignment. You should understand the effects of a valuated or non-valuated goods receipt posting. You should also be able to explain the postings that result from a goods receipt and from entering a vendor invoice. You should be familiar with the process of blanket purchase orders and be able to describe the advantages of self-service procurement.

Summary

In this chapter, the aim was to understand the characteristics and special features of a procurement process with account assignment: What (i.e., which materials) can you order with account assignment, what does a purchase order item with account assignment look like, what happens if we post a goods receipt valuated or non-valuated with reference to a purchase order item with account assignment, what does the value flow look like, and under which conditions can we change the account assignment data when we post an invoice?

Now you should know the main differences between the procurement of stock material and the procurement of consumables.

This basic knowledge, which we have refreshed in this section, especially the configuration options for account assignment categories, is a prerequisite for making the procurement process for consumable materials as efficient and simplified as possible.

Chapter 5
Sources of Supply

5

Techniques You'll Master

- Name the various sources of supply in SAP S/4HANA.
- Create and describe a purchasing info record.
- Create and describe a contract.
- List the differences between quantity contract and value contract.
- Create and describe a scheduling agreement.
- Name characteristics of a scheduling agreement with release documentation.
- Specify characteristics of a scheduling agreement without release documentation.

In this chapter, you will learn about the various sources of supply that SAP S/4HANA offers. We will explain the special features of the individual objects (purchasing info record, contract, and scheduling agreement) and examine the possible variants. We will discuss the advantages and disadvantages of the various options and how they can be deployed.

Real-World Scenario

The so-called sources of supply are primarily there to record the results of the purchasing negotiations, especially prices and other conditions. Depending on what is agreed and at what level, a purchasing info record, a contract, or a scheduling agreement is more suitable. If, for example, strategic purchasing makes a price agreement valid for one year for a raw material used worldwide, a central value contract may be the appropriate instrument. If it is only a matter of entering valid prices in the system together with some other purchasing parameters (e.g., the planned delivery time), then the purchasing info record may be the most suitable tool. Sources of supply are also used to store special agreements or control parameters for a material-vendor combination: In the info record, for example, you record whether the use of the evaluated receipt settlement (ERS) procedure has been agreed with the supplier. You also record the dunning rules that apply in the event of late delivery. As a consultant, you must be familiar with the special features of the respective sources of supply in order to be able to recommend the use of suitable objects to your customers. Knowledge of the organizational structure is essential, since the sources of supply are created along this structure.

Objectives of This Portion of the Test

In this part of the certification exam, your knowledge of the available sources of supply in SAP S/4HANA and their possible uses will be tested. For the certification exam, you must be able to:

- Describe the most important information and control parameters that can be stored in a purchasing info record, a contract, and a scheduling agreement.
- Describe the organizational levels at which the respective sources of supply can be created.
- Explain important elements of pricing, such as the condition and the calculation schema, as well as the difference between time-dependent and time-independent conditions.
- Describe the main differences between scheduling agreements and contracts and then between the different categories of contracts and the different categories of scheduling agreements.

- Explain the principles of a scheduling agreement and the use of the various associated tools, such as delivery schedule lines, forecast delivery schedules, just-in-time delivery schedules, and release creation profile.

Key Concept Refresher

In this section, we go through the three different sources of supply in detail:

- Purchasing info record
- Contract
- Scheduling agreement

Let's start with the purchasing info record.

Purchasing Info Record

The purchasing information record (or info record, for short) is part of the *master data* of purchasing. It contains information about a material (or material group) and a vendor who supplies the material (or items assigned to the material group). The data stored in the info record (especially prices) is proposed as default values for the procurement process (as in purchase orders, for example).

Structure of the Purchasing Info Record

A purchasing info record applies either to an entire purchasing organization or to a specific plant within a purchasing organization.

A purchasing info record is valid for a procurement process. For example, if a material is ordered from a vendor once as standard and once on consignment, then you create a purchasing info record for each procurement process.

A purchasing info record is created for the following:

- A combination of a material or material group and a supplier
- A procurement process (i.e., the info record category stands for the process)

It contains the following data:

- General data at the client level
- Purchasing data and conditions at the purchasing organization level and possibly at the purchasing organization and plant level

With both the Manage Purchasing Info app records and Transaction ME11, you must specify the following when you create an info record:

- A supplier
- A material (or a material group—we will come back to this special type of info record later)

- A category corresponding to a procurement process
- A purchasing organization in which the info record is valid
- Optionally, a plant that is managed by the purchasing organization; if you specify a plant, you reduce the validity area of the info record: instead of being valid for all plants in the purchasing organization's area of responsibility, the purchasing info record is only valid for the one plant

Figure 5.1 shows how to start entering an info record and its required data with the Manage Purchasing Info Record app.

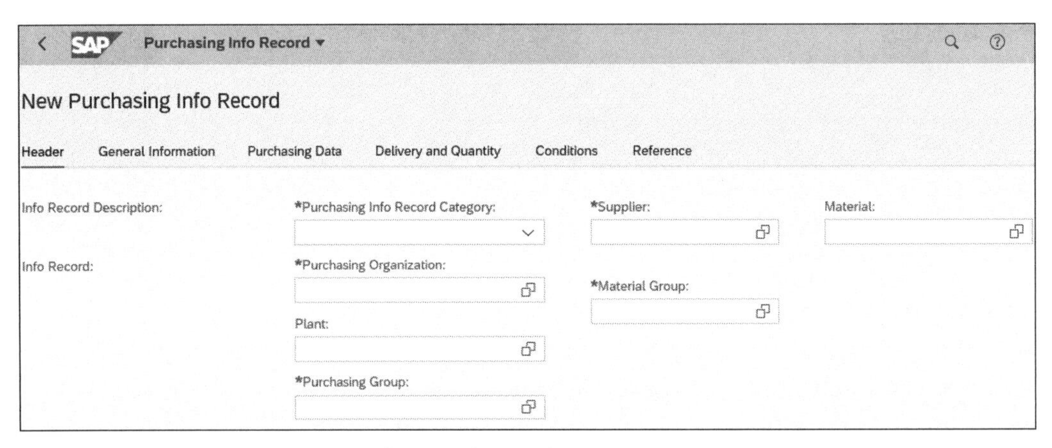

Figure 5.1 The Manage Purchasing Info Record App

Figure 5.2 shows how to start creating an info record using the Create Purchasing Info Record app (Transaction ME11). The initial screen clearly shows the structure of the info record with material, vendor, category, and organizational units.

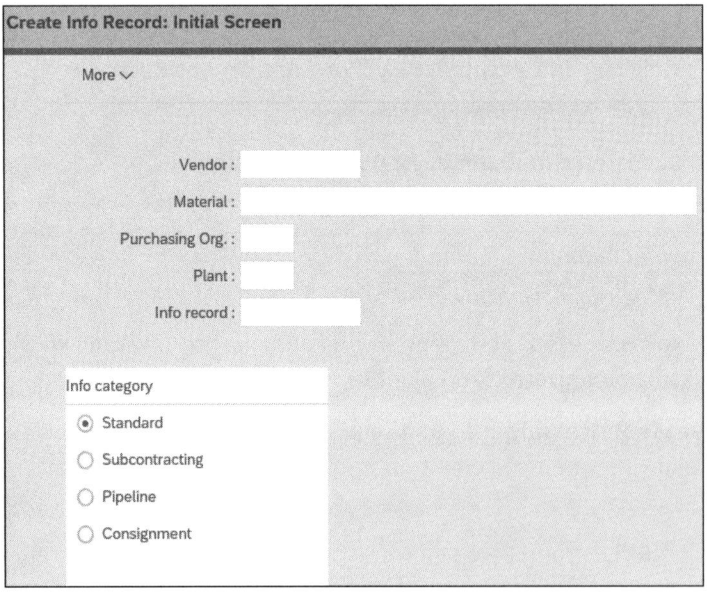

Create Info Record: Initial Screen
More
Vendor :
Material :
Purchasing Org. :
Plant :
Info record :
Info category
Standard
Subcontracting
Pipeline
Consignment

Figure 5.2 The Create Purchasing Info Record App (Transaction ME11)

Note

Conditions are maintained at the purchasing organization level. The possibility of maintaining conditions at a lower level, namely at plant level within a purchasing organization, is defined in Customizing.

Figure 5.3 shows the step in Customizing of **Material Management • Purchasing**, and Figure 5.4 shows the possible settings:

- Both plant and purchasing organization conditions allowed
- Only plant conditions allowed
- Only conditions at purchasing organization level allowed (no plant conditions)

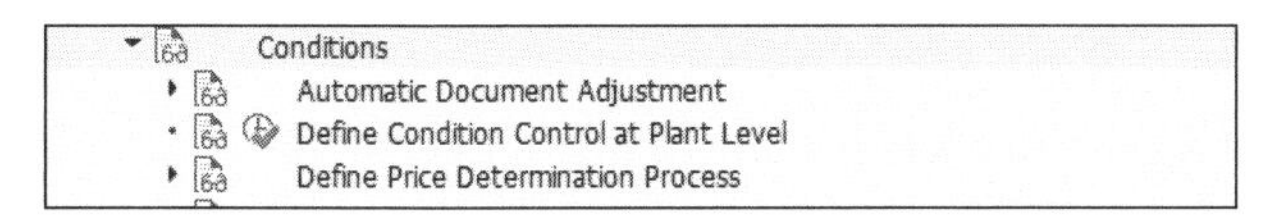

Figure 5.3 Setting the Condition Level in Customizing

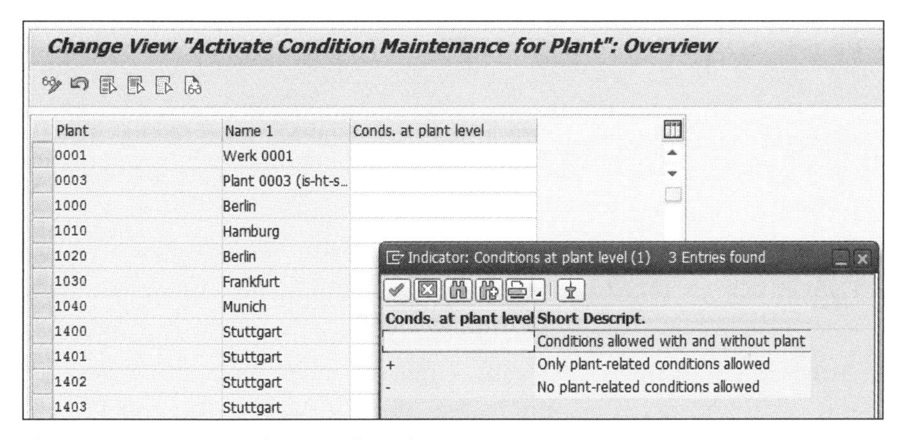

Figure 5.4 Setting the Condition Level per Plant in Customizing

Information in the Purchasing Info Record

The purchasing info record contains the following information:

- General information, as shown in Figure 5.5, that is valid for the entire client—that is, information that cannot be defined specifically for each purchasing organization. This includes the following:
  - The supplier material number (i.e., the number that the material has at the supplier)
  - The responsible salesperson at the supplier
  - Supply options (available to/from)
  - Number of days before and after the scheduled delivery date that should trigger a reminder or a dunning letter

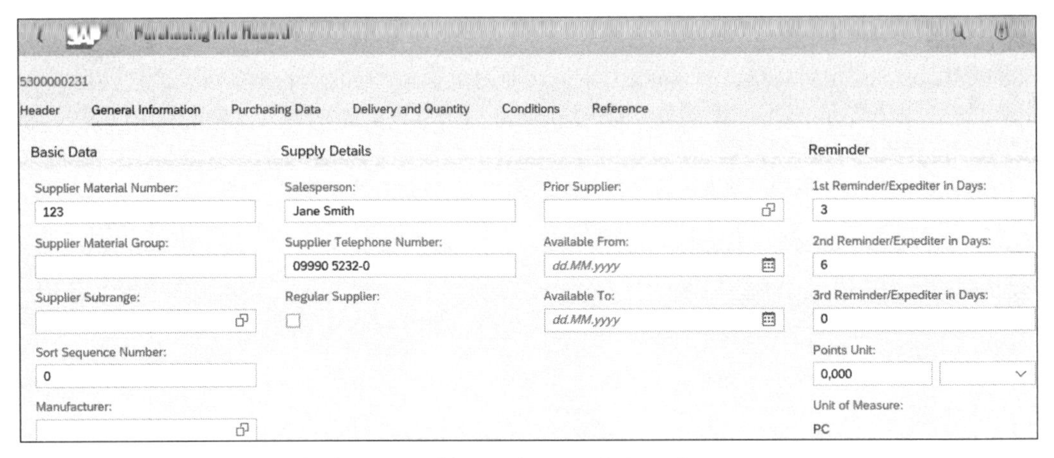

Figure 5.5 General Information of the Purchasing Info Record

- Data that are valid for a purchasing organization or for a plant within a purchasing organization. This includes the following:
  - Current and future prices and conditions (e.g., gross price and discounts)
  - Incoterms
  - Price determination date control; this field indicates which date (e.g., purchase order date or goods receipt date) should be used for determining the relevant price for a purchasing process
  - Tolerance limits for over-deliveries or under-deliveries
  - Planned delivery time for the material
  - References like the last purchase order number or the quotation number
  - Texts like the info memo that is an internal comment and the purchase order text that is transferred to the purchase order item and can be printed
  - The **Automatic Sourcing** indicator, which means that the info record is considered as a possible source of supply during the material requirements planning (MRP) run

Creating an Info Record

In addition to the possibility of migrating info records from a legacy system when implementing SAP S/4HANA, info records can be created in a running system either manually or automatically.

Creating Info Records Manually

Purchasing info records can be created on the SAP Fiori launchpad with the Manage Purchasing Info Records app or with the Create Purchasing Info Record app (Transaction ME11). You can also use Transaction ME11 (Create Info Record), Transaction ME12 (Change Info Record), or Transaction ME13 (Display Info Record) on the SAP GUI in the backend system.

Note

The Manage Purchasing Info Records app presents an important limitation: You cannot create texts with it. If you must create texts, use the Create Purchasing Info Record app (Transaction ME11).

Creating Info Records Automatically

Purchasing info records can also be created or updated automatically by setting the **Info Record Update** indicator when maintaining a purchasing document (purchase order, outline agreement, or quotation).

One option is to create or update info records automatically from a purchase order or a scheduling agreement. When you create a purchase order or a scheduling agreement, you can set the **Info Record Update** indicator at the item level. This has the following effect:

- If an info record already exists for the vendor-material combination, it is updated, meaning that the reference to the last purchasing document and item number is actualized. However, the price of the info record stored in the **Conditions** section is not overwritten by the purchase order price.

Note

If both purchasing organization data and plant-specific data exist for the material-vendor combination, the plant-specific data is updated.

- If no info record exists for the vendor-material combination, the **Info Record Update** indicator ensures that one is created, but without conditions. In this case, too, only the reference to the purchase order number/purchase order item is added (or scheduling agreement number/scheduling agreement item).

The following figures show the **Info Record Update** indicator in the two apps: Manage Purchase Orders is shown in Figure 5.6 and Create Purchase Order—Advanced (Transaction ME21N) is shown in Figure 5.7 and Figure 5.8.

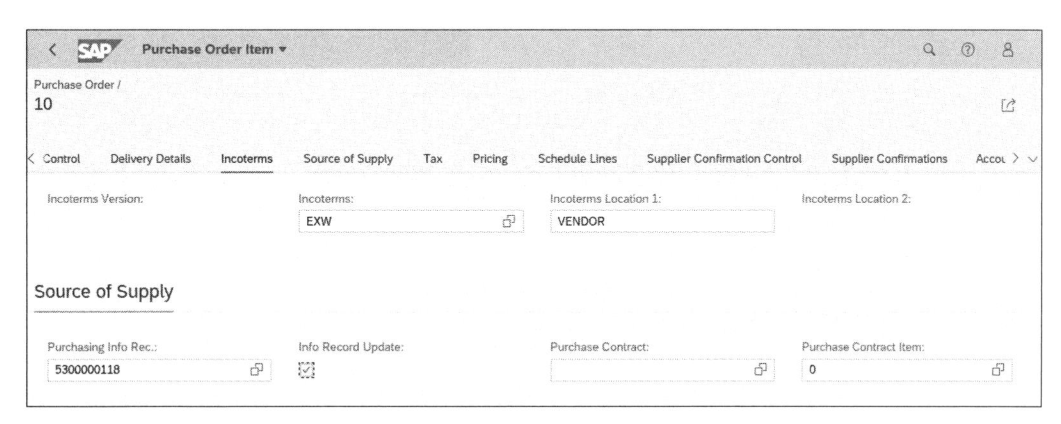

Figure 5.6 The Info Record Update Indicator in the Manage Purchase Orders App

Note

The **Info Record Update** indicator is also called the **Info Update** indicator and appears as such on the Create Purchase Order—Extended app (Transaction ME21N).

Figure 5.7 and Figure 5.8, respectively, show a purchase order item and the corresponding item details with the **Info Update** indicator set in the **Material Data** tab.

Standard PO | 4500000047 | Vendor | 10300001 Walter Stahl AG | Doc. Date | 08.06.2020

Header

S..	Itm	A	I	Material	Short Text	PO Quantity	O...	C	Deliv. Date	Net Price	Curr...	Per	O...	Plnt	Stor. L
	10			TG0011	CityFit20	10	PC	D	09.06.2020	30,00	EUR	1	PC	Hamburg	

Figure 5.7 Purchase Order Item in the Create Purchase Order—Advanced App (Transaction ME21N)

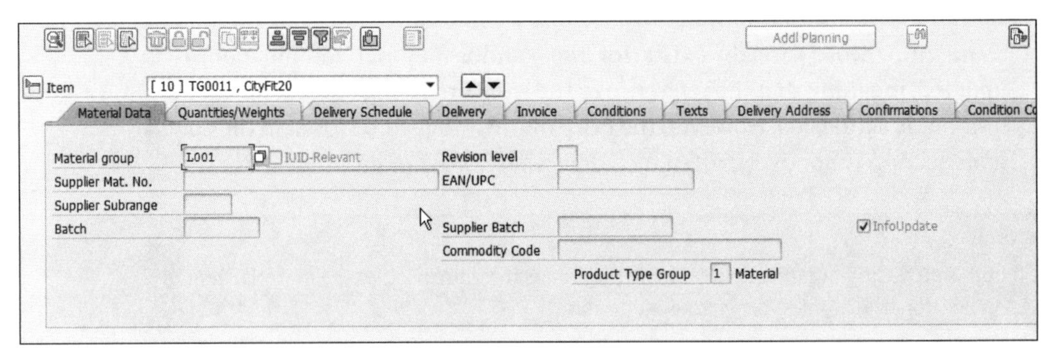

Figure 5.8 Item Details in the Create Purchase Order—Advanced App (Transaction ME21N)

Figure 5.9 shows the purchasing info record. Here the last purchase order number has been updated and is displayed in the **Reference** tab. Remember that the conditions are not affected by updates from a purchase order.

Another possibility is to create/update info records automatically from a contract. When you create a contract, you can also use the **Info Update** (or **Info Record Update**) indicator to create an info record automatically. However, it depends on whether an info record already exists for the vendor-material combination:

- If you create or change a contract and no info record exists for the combination of vendor, material, and organizational level, then the system creates an info record with the information from the contract (including conditions).
- If an info record already exists, the info record is not updated.

Last, you can create/update info records automatically from a quotation. When you create a quotation, the **Info Update** indicator not only ensures that the document and item numbers are updated in the info record but also the conditions.

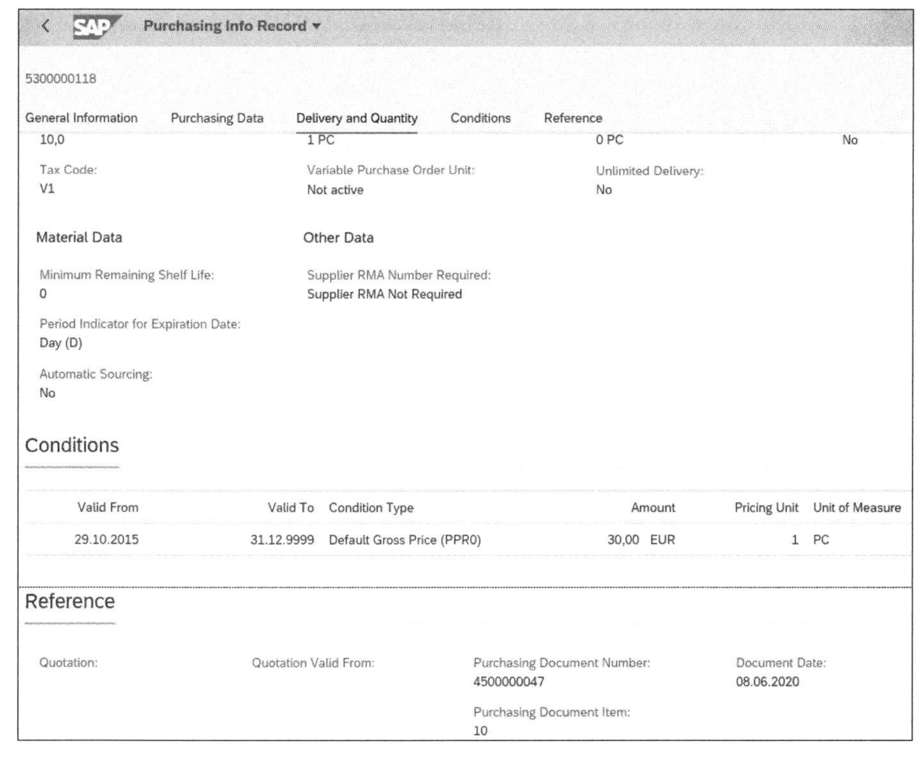

Figure 5.9 The Latest Purchase Order Number in the Purchasing Info Record

Conditions in the Purchasing Info Record

Conditions represent prices agreed with suppliers, discounts and surcharges, freight costs, and so forth. Conditions can be stored in info records or outline agreements and are then proposed in the purchase orders.

Note

You can also maintain general conditions at the vendor level. The system then also applies these conditions for price determination in purchasing documents.

You can manually enter further conditions in the purchase order itself.

Now let's start with some general terms related to the conditions. (Although detailed pricing and the condition technique are not part of the exam, you must be familiar with the terms *condition type* and *calculation schema*.)

Depending on whether you want to enter a price, a discount, a surcharge, or freight costs, choose a corresponding condition type. The condition type is a key that represents the type of pricing element and controls how the pricing element is calculated.

Condition types are configured in Customizing. The standard condition type PB00, for example, stands for a gross price; standard condition type RA01 stands for a discount that is to be entered as a percentage of the net price.

The condition types allowed for a transaction are combined in a calculation schema in Customizing. The calculation schema also determines the sequence in which the condition types are considered and therefore included in the calculation of the net or effective price. Figure 5.10 shows the condition schema RM0000 that is often used for purchase orders.

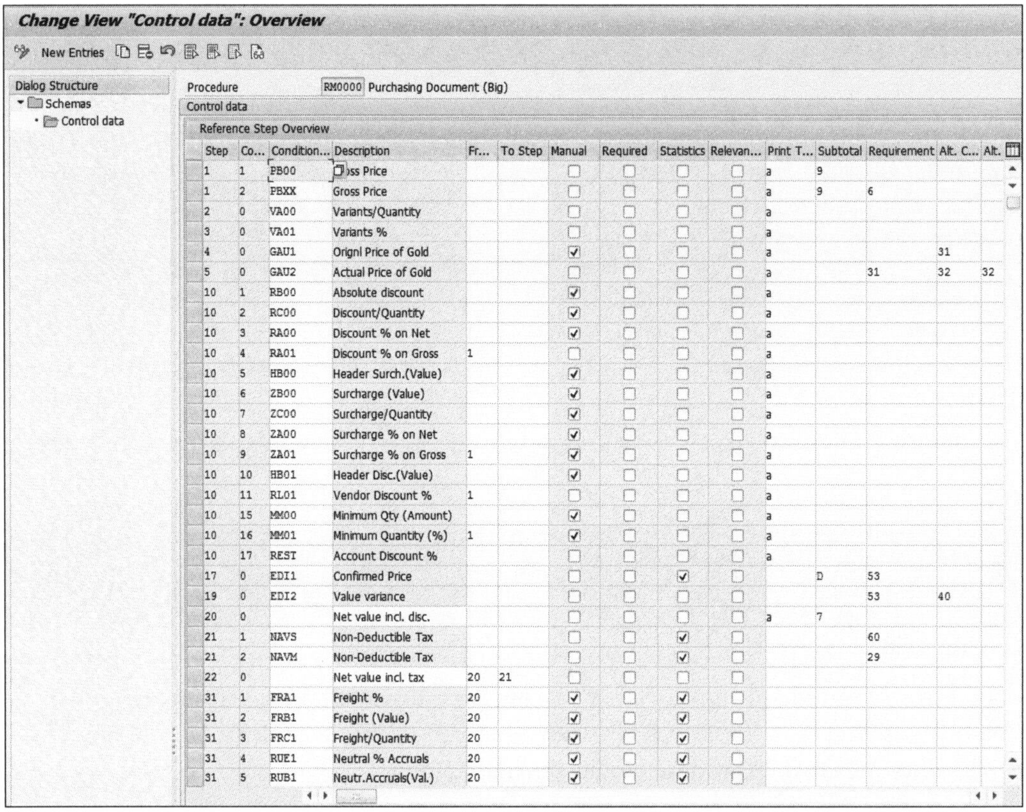

Change View "Control data": Overview

New Entries

Dialog Structure
- Schemas
  - Control data

Procedure: RM0000 Purchasing Document (Big)

Control data

Reference Step Overview

Step	Co...	Condition...	Description	Fr...	To Step	Manual	Required	Statistics	Relevan...	Print T...	Subtotal	Requirement	Alt. C...	Alt.
1	1	PB00	Gross Price			☐	☐	☐	☐	a	9			
1	2	PBXX	Gross Price			☐	☐	☐	☐	a	9	6		
2	0	VA00	Variants/Quantity			☐	☐	☐	☐	a				
3	0	VA01	Variants %			☐	☐	☐	☐	a				
4	0	GAU1	Orignl Price of Gold			☑	☐	☐	☐	a			31	
5	0	GAU2	Actual Price of Gold			☐	☐	☐	☐	a		31	32	32
10	1	RB00	Absolute discount			☑	☐	☐	☐	a				
10	2	RC00	Discount/Quantity			☑	☐	☐	☐	a				
10	3	RA00	Discount % on Net			☑	☐	☐	☐	a				
10	4	RA01	Discount % on Gross	1		☐	☐	☐	☐	a				
10	5	HB00	Header Surch.(Value)			☑	☐	☐	☐	a				
10	6	ZB00	Surcharge (Value)			☑	☐	☐	☐	a				
10	7	ZC00	Surcharge/Quantity			☑	☐	☐	☐	a				
10	8	ZA00	Surcharge % on Net			☑	☐	☐	☐	a				
10	9	ZA01	Surcharge % on Gross	1		☑	☐	☐	☐	a				
10	10	HB01	Header Disc.(Value)			☑	☐	☐	☐	a				
10	11	RL01	Vendor Discount %	1		☐	☐	☐	☐	a				
10	15	MM00	Minimum Qty (Amount)			☑	☐	☐	☐	a				
10	16	MM01	Minimum Quantity (%)	1		☑	☐	☐	☐	a				
10	17	REST	Account Discount %			☐	☐	☐	☐	a				
17	0	EDI1	Confirmed Price			☐	☐	☑	☐		D	53		
19	0	EDI2	Value variance			☐	☐	☐	☐			53	40	
20	0		Net value incl. disc.			☐	☐	☐	☐	a	7			
21	1	NAVS	Non-Deductible Tax			☐	☐	☑	☐			60		
21	2	NAVM	Non-Deductible Tax			☐	☐	☑	☐			29		
22	0		Net value incl. tax	20	21	☐	☐	☐	☐					
31	1	FRA1	Freight %	20		☑	☐	☑	☐					
31	2	FRB1	Freight (Value)	20		☑	☐	☑	☐					
31	3	FRC1	Freight/Quantity	20		☑	☐	☑	☐					
31	4	RUE1	Neutral % Accruals	20		☑	☐	☑	☐					
31	5	RUB1	Neutr.Accruals(Val.)	20		☑	☐	☑	☐					

Figure 5.10 Example of a Condition Schema

The conditions in the purchasing info record are *time-dependent* conditions. This means that you can create several validity periods with different prices.

Depending on the condition type, you can also specify scales in the purchasing info record—for example, if a discount of 3% is granted from a purchasing quantity of 100 pieces and 5% from 200 pieces onward.

Data Proposal from the Purchasing Info Record

The idea behind the info record is to store data in a way that you do not have to search for and enter it again and again if, for example, you repeatedly procure certain materials from the same suppliers under the same conditions. The system should propose this data—especially the price—for each procurement process, when a purchase order is created.

The proposal logic from the purchasing info record appears as follows when creating a purchase order item:

- If valid conditions exist, these are proposed.
- If no valid conditions exist, the system proposes the price from the last purchase order or scheduling agreement that was updated as a reference in the purchasing info record.
- If neither conditions nor references exist, no price is proposed.
- If conditions exist at both the purchasing organization level and plant level, the system proposes the appropriate plant price.

We will use some examples to illustrate how the system proposes the conditions from the info record. For the following examples, we consider a purchasing organization 1010, which is assigned to plants 1010, 1020, and 1030.

Example

You agree on a price of EUR 10 with vendor L1 for the procurement of material M1 in all plants. Accordingly, you manually create an info record with the following data: material M1, vendor L1, purchasing organization 1010, price 10 EUR per piece.

When you order M1 from L1 for plant 1010, 1020, or 1030, the system proposes a price of 10 EUR.

Example

Now you agree with supplier L1 on a lower price for the purchase of M1 for plant 1030, since this plant requires larger quantities and is also closer to the location of the supplier, so that transport costs are lower.

You enhance the info record from the previous example with plant data for plant 1030: material M1, vendor L1, purchasing organization 1010, plant 1030, price 9 EUR per piece. When you order M1 from L1 for plant 1010 or 1020, the system still proposes the price of EUR 10. However, if you order for plant 1030, then the system proposes EUR 9.

Example

Since you suddenly require a larger quantity of M1 for plant 1010 at short notice, you must include a second vendor. You order this material for the first time from vendor L2 at a price of EUR 11, which you requested by telephone. You set the **Info Update** indicator in the purchase order item, and an info record with a reference to the purchase order is created in the background.

A few days later, you order material M1 from vendor L2 again. The system proposes the price of 11 EUR: it is the price from the last purchase order recorded as a reference in the purchasing info record.

Since it looks as if you now regularly buy from L2, you negotiate a price of 10.5 EUR for the next six months. You adjust the info record accordingly: you enter the newly agreed price in the conditions. The next time you order, the system proposes the price of EUR 10.5 (from the conditions—no longer the price from the last purchase order).

Having examined the special features and characteristics of the info record, we now turn to another source of supply: the contract.

Contracts

While the info record is a master record that contains information on a material-vendor combination (or material-group-vendor combination), the contract is a purchasing document with the corresponding document structure: header and items. In a contract, you can record (price) agreements that affect several materials (items).

We distinguish between *quantity contracts* and *value contracts*. When you create a contract, you determine whether it is a quantity or value contract by using the contract type. In standard SAP S/4HANA, the contract type MK is available for quantity contracts and WK for value contracts.

Let's begin our discussion of contracts by focusing on quantity contracts.

Quantity Contracts

In a quantity contract, you specify that you will procure a certain quantity of defined materials or services from a vendor under agreed conditions within a fixed period.

You make the following mandatory entries in the contract header:

- The contract type (in the standard system, the contract type for quantity contract is MK)
- A supplier
- A purchasing organization
- A purchasing group
- A validity period

You can optionally enter an overall value, but this only has an informative character.

You then create items with the following data:

- **Item category (optional entry)**
  You can create contract items for different procurement processes like standard, subcontracting, and consignment. You can also choose special item categories, which we will discuss in the section on special contracts that follows.
- **Account assignment category (optional entry)**
  You can create contract items for the warehouse or for consumption. If you create a contract item for the warehouse, leave the account assignment category blank; otherwise, select the appropriate indicator.

  For a contract item, you can choose the account assignment category U (unknown), which indicates that the account assignment data is not yet known. This is determined with the contract release order.

- **A material or a material group**
  The general rules for purchasing documents, which we have already described for purchase orders, apply here. For a stock item, you must enter a material number; for an item assigned to an account, you can do without the material number and enter a material group and a description.
- **A target quantity that you want to purchase in the period defined in the header**
  This entry is mandatory.
- **The conditions (price) agreed with the vendor at which you pay for the material or service**
  This entry is mandatory.

A quantity contract is fulfilled when the defined target quantity per contract item is reached by issuing contract release orders.

A contract release order is a purchase order that refers to a contract item. In a contract release order, the reference to the contract is contained at the item level. *Contract release documentation* is updated with each contract release order and enables you to monitor contract fulfillment. The release order documentation contains the following data on each release order assigned to the contract item:

- Number and associated item number of the release order
- Order date of the release order
- Release quantity
- Value of the release

In addition, the release order documentation for the item of a quantity contract contains the following elements:

- The total quantity already released
- The target quantity compared to the outstanding quantity

Figure 5.11 shows a quantity contract using the Manage Purchase Contract app. This contract consists of two items, as shown in Figure 5.12. Two release orders have already been issued for the first contract item; they are displayed in Figure 5.13.

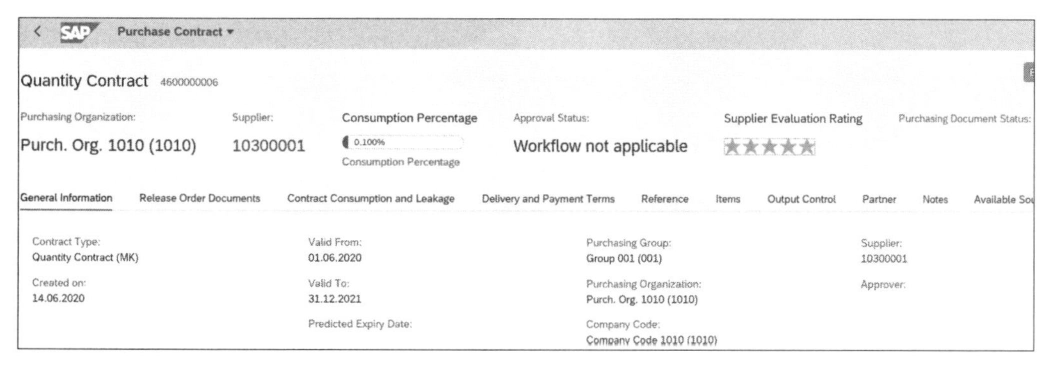

Figure 5.11 Header Data of a Quantity Contract in the Manage Purchase Contract App

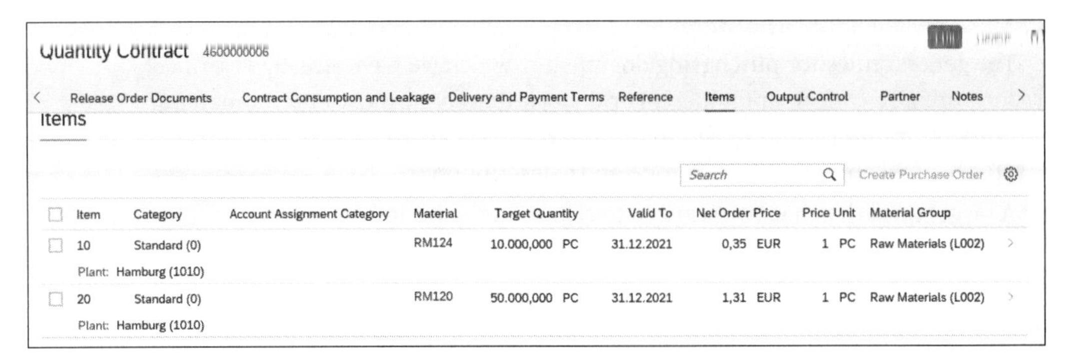

Quantity Contract 4600000006

Release Order Documents | Contract Consumption and Leakage | Delivery and Payment Terms | Reference | Items | Output Control | Partner | Notes

Items

Search | Create Purchase Order

Item	Category	Account Assignment Category	Material	Target Quantity	Valid To	Net Order Price	Price Unit	Material Group
10	Standard (0)		RM124	10.000,000 PC	31.12.2021	0,35 EUR	1 PC	Raw Materials (L002)
Plant: Hamburg (1010)								
20	Standard (0)		RM120	50.000,000 PC	31.12.2021	1,31 EUR	1 PC	Raw Materials (L002)
Plant: Hamburg (1010)								

Figure 5.12 Contract Items in the Manage Purchase Contract App

Quantity Contract 4600000006

Edit | Delete

Release Order Documents | Contract Consumption and Leakage | Delivery and Payment Terms | Reference | Items | Output Control | Partner | Notes | Avai

Release Orders (2)

Purchasing Document	Purchasing Document Item	PO Date	Order Quantity	Net Order Value	Company Code	Plant	Purchasing Organization	Changed On
4500000046	00010	14.06.2020	100,000 PC	35,00 EUR	Company Code 1010 (1010)	Hamburg (1010)	Purch. Org. 1010 (1010)	14.06.2020
4500000047	00010	14.06.2020	20,000 PC	7,00 EUR	Company Code 1010 (1010)	Hamburg (1010)	Purch. Org. 1010 (1010)	14.06.2020

Figure 5.13 Release Order Documents in the Manage Purchase Contract App

Figure 5.14 shows the consumption and the residual value in the purchase contract.

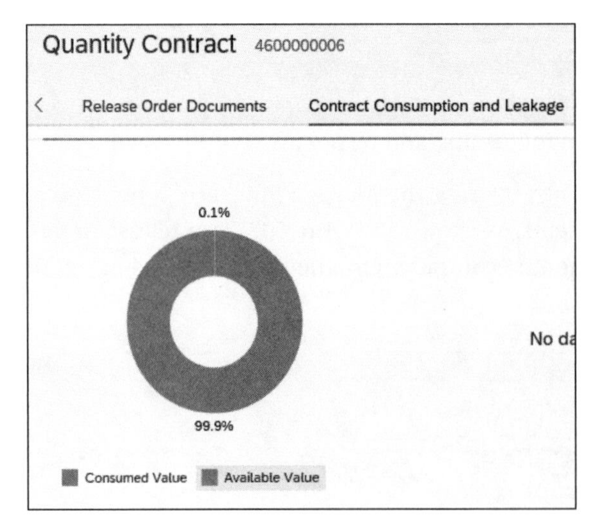

Figure 5.14 Contract Consumption and Leakage in the Manage Purchase Contract App

Figure 5.15 shows the release order documentation for the same contract using Transaction ME33K (Display Contract) on the SAP GUI.

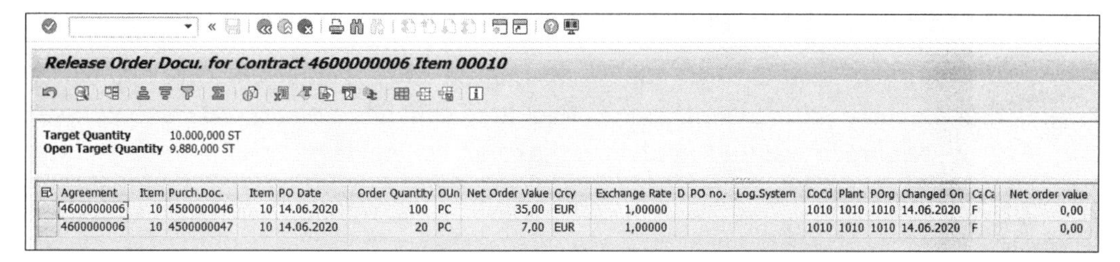

Figure 5.15 Transaction ME33K (Display Contract)

Value Contracts

In a value contract, you specify that you will purchase a certain value from a vendor within a fixed period. The value contract contains the list of materials or services affected by this agreement and the agreed prices.

You make the following mandatory entries in the contract header:

- The contract type (in the standard system, the contract type for quantity contract is WK)
- A supplier
- A purchasing organization
- A purchasing group
- A validity period
- The agreed target value

You then create items with the following data:

- Item category (optional entry—the same rules apply as for quantity contracts)
- Account assignment category (optional entry—the same rules apply as for quantity contracts)
- A material or a material group (the same rules apply as for quantity contracts)
- The conditions (price) agreed with the vendor at which you pay for the material or service

You can optionally enter a target quantity, but this is for information purposes only.

A value contract is fulfilled when the target value stipulated in the contract header is reached by issuing contract release orders.

> **Note**
>
> The target quantity, target value, and validity period are checked when you create a release order against a contract. If the quantity or value is exceeded, or in the case of release orders outside the validity period, the system issues warning messages. You can set these as error messages in Customizing to prevent a release order from being issued.

Centrally Agreed Contracts

When you create a contract item, you may or may not enter a plant as opposed to a purchase order item. If the contract item contains a plant, you can only issue release orders against this plant. If the contract item does not contain a plant, you can issue release orders for all plants assigned to the purchasing organization in the contract header.

A contract item without a plant is called a *centrally agreed contract*. The term is somewhat confusing, as it suggests that the entire contract is negotiated centrally but it is only for items. A contract can have both plant-specific and plant-independent items.

Note

In the case of centrally agreed contract items, however, you can define conditions for individual plants. These can result from different transportation costs for different distances, for example.

Warning

Be careful not to confuse the centrally agreed contract items just discussed with the central contracts that are created in an SAP S/4HANA hub system in the context of Central Procurement. Central Procurement enables the integration of an SAP S/4HANA hub system with SAP ERP, SAP S/4HANA, and SAP S/4HANA Cloud backend systems. Contract management, purchasing, and requisitioning processes can be centralized using the hub system. The central purchase contracts are created in SAP S/4HANA, which acts as the hub system, distributed to the connected systems.

Special Contracts

In this section, we look at special contracts. Two contract-specific item categories allow you to create contract items for material groups without specifying the material number:

- **Item category M (material unknown)**
  You use the item category M if you have an outline agreement for several similar materials at the same price. For example, you conclude a contract with your supplier for different types of letter paper, such as lined, checkered, unlined, and pre-punched paper. The different types of paper have the same weight and quality and cost the same price. When you create a contract item of category M, you enter a short description (like letter paper), a corresponding material group, a target quantity, a unit of measure, and a price, but no material number. You enter the exact description (ruled, checkered, and so on) or, if available, a material number for the material group from the contract item in the release order.

  You can use this item category in both quantity and value contracts.
- **Item category W (material group)**
  This item category is also used for a group of materials, but these materials cost

different prices. For example, you conclude a contract with your vendor for different types of electrical equipment. The contract is to cover all the materials included in the vendor's price list. The exact item is specified in the release order.

Instead of having a contract item for each article in the vendor list, you can use the W item category and the corresponding material group—for example, *electrical equipment*. The short text makes it clear that the contract item covers all electrical equipment supplied by the supplier.

In the release orders relating to this contract, you enter the exact information on the article—for example, a package of cable ties or a power outlet or, if available, a material number assigned to the material group from the contract item.

You also specify the price and quantity. If an info record exists, the conditions maintained in the info record are proposed in the release order. Items without a material number must be assigned to an account.

You can only use this item category in value contracts.

Note

The Manage Purchase Contracts app only supports item category W. Use the Create Purchase Contract app (Transaction ME31K) to create items with item category M.

Scheduling Agreements

Like the contract, the scheduling agreement is an outline agreement belonging to the purchasing documents.

To create a scheduling agreement, you can use both the Manage Scheduling Agreements app and the Create Scheduling Agreement app (Transaction ME31L). You enter a supplier, an agreement type, a validity period, and organizational units in the scheduling agreement header, as shown in Figure 5.16.

Figure 5.16 Example of a Scheduling Agreement Header in the Manage Scheduling Agreements App

In the items shown in Figure 5.17, you specify the materials that are part of the agreement together with the respective planned purchase quantities in the scheduling agreement period and the agreed prices.

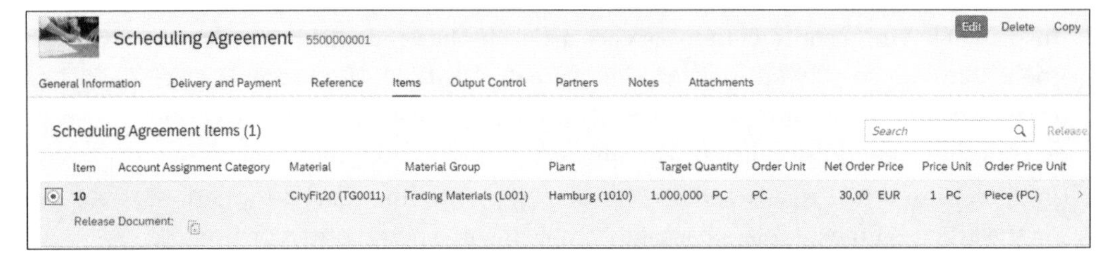

Figure 5.17 Example of a Scheduling Agreement in the Manage Scheduling Agreements App

Scheduling agreement items can be created for various procurement processes, such as standard, subcontracting, vendor consignment. The item category is available for this purpose.

Note

Item categories M (material unknown) and W (material group) are not permitted in scheduling agreements.

It is common for scheduling agreements to be used for the supply of raw materials or components that are required on a very regular basis and are placed in storage first. Nevertheless, it is possible to create scheduling agreement items with account assignment. The account assignment category is available for this purpose.

Note

Account assignment categories U (unknown) is not permitted in scheduling agreements.

As for a purchase order item, you must always specify the plant in a scheduling agreement item. There are no central scheduling agreements!

As mentioned above, a delivery schedule is a framework agreement. Concrete requirements (e.g., quantity and dates) must be reported separately.

Schedule Lines

Unlike the contract, scheduling agreements do not require release orders in the form of additional documents. The actual delivery quantities and dates are defined in *schedule lines*. Schedule lines are not separate purchasing documents, unlike release orders, which are purchase orders. Schedule lines are part of the scheduling agreement.

Schedule lines can be created manually with the Manage Scheduling Agreements app shown in Figure 5.18 or with Transaction ME38 (Maintain Delivery Schedule), as shown in Figure 5.19.

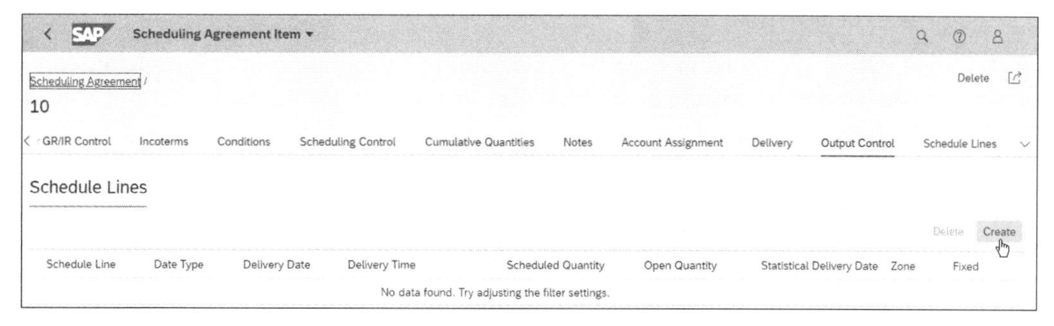

Figure 5.18 Creating Schedule Lines for an Item in the Manage Scheduling Agreements App

Maintain Sch. Agmt Schedule : Delivery Schedule for Item 00010

Agreement	5500000001	Quantity	700 PC
Material	TG0011 CityFit20		
Last Transm.	00:00:00	Next Transmission Number	1
Cum. Rec. Qty	0	Old Qty	0

C	Delivery D...	Scheduled Quantity	Time	F	C...	St.DelDate	Purcha...	Item	Cum. ...	Prev. CQ	Sc...	Previous Qty	GR qty	N...	Fi...	Open Quantity
D	17.06.2020	100			R	17.06.2020			100		1				☑	100
D	18.06.2020	200			R	18.06.2020			300		2				☑	200
D	19.06.2020	50			R	19.06.2020			350		3				☑	50
D	22.06.2020	70			R	22.06.2020			420		4				☑	70
D	10.07.2020	90			R	10.07.2020			510		5				☑	90
D	12.07.2020	100			R	12.07.2020			610		6				☑	100
D	15.07.2020	90			R	15.07.2020			700		7				☑	90

Figure 5.19 Example of Schedule Lines for a Scheduling Agreement Item in Transaction ME38 (Maintain Delivery Schedule)

In fact, scheduling agreements are typically used to automate and accelerate the procurement process. Therefore, scheduling agreement schedule lines are often preferably created automatically by the MRP run. If a scheduling agreement exists for a material, the system can create schedule lines instead of purchase requisitions during the MRP run. Of course, this only applies to stock materials that are planned. The automatic creation of schedule lines by the MRP run is described in more detail in Chapter 8.

The schedule lines can be transmitted directly to the vendor (scheduling agreement without releaase documenettion) or be prepared first (with release documentation). We examine these two options in the next sections.

Scheduling Agreement Types without Release Documentation

Here we are discussing scheduling agreements without release documentation when the schedule lines for a scheduling agreement item are transmitted to the vendor as they are.

Whether the schedule lines for a scheduling agreement item are transmitted to the vendor as they are or whether they are prepared beforehand is controlled by the scheduling agreement type.

In Customizing for the scheduling agreement type, the indicator called **Release Documentation** controls this. Figure 5.20 shows the path to Customizing for scheduling agreement types. Standard SAP S/4HANA provides the document types LP (without release documentation) and LPA (with release documentation).

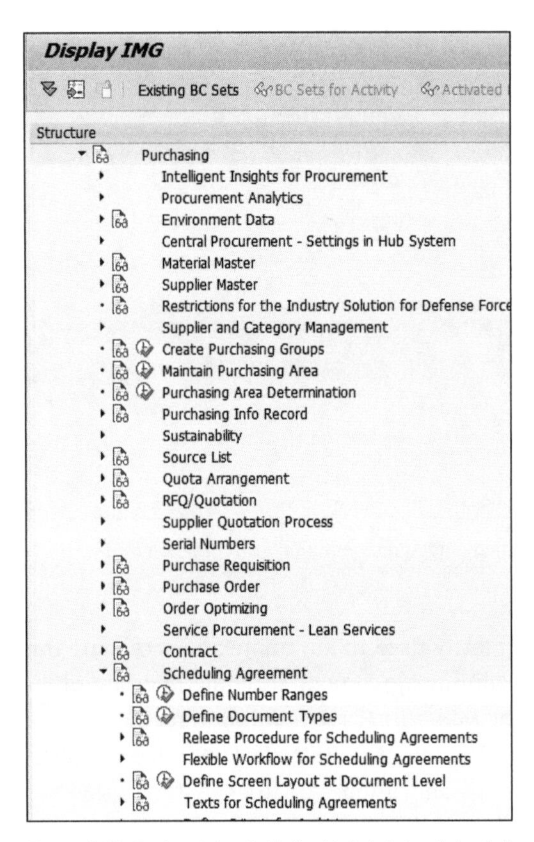

Figure 5.20 Customizing Path for Maintaining Scheduling Agreement Types

Figure 5.21 shows standard scheduling agreement types. You can see the previously mentioned scheduling agreement types LP and LPA. Look at the last customizing column, named **Release...**, stands for **Release Documentation**, and is checked for type LPA.

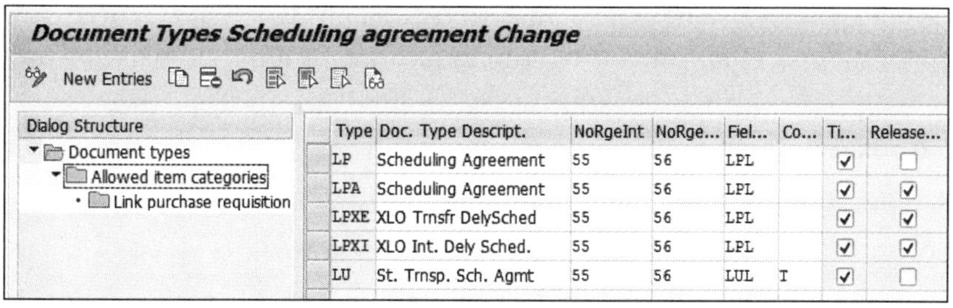

Document Types Scheduling agreement Change

New Entries

Dialog Structure
- Document types
  - Allowed item categories
    - Link purchase requisition

Type	Doc. Type Descript.	NoRgeInt	NoRge...	Fiel...	Co...	Ti...	Release...
LP	Scheduling Agreement	55	56	LPL		☑	☐
LPA	Scheduling Agreement	55	56	LPL		☑	☑
LPXE	XLO Trnsfr DelySched	55	56	LPL		☑	☑
LPXI	XLO Int. Dely Sched.	55	56	LPL		☑	☑
LU	St. Trnsp. Sch. Agmt	55	56	LUL	T	☑	☐

Figure 5.21 Customizing of the Scheduling Agreement Types

The scheduling agreement schedule lines are transmitted to the vendor as they are stored in the system. The system does not document these transmissions in detail. All open schedule lines are communicated with each transmission. This type of document, which is a simple form of scheduling agreement, is therefore more suitable if the number of schedule lines is small and these schedule lines are not changed frequently.

Scheduling Agreement Types with Release Documentation

If you use scheduling agreements with release documentation, the schedule lines are not transmitted directly to the vendor. Initially, they are for internal information purposes only and correspond to actual requirements.

In this type of scheduling agreement, an additional element is introduced: the *delivery schedule*. The schedule lines have an internal character only. The schedule lines are not transmitted to the vendor until you explicitly generate a release—that is, either a forecast delivery schedule or a *just-in-time* (JIT) delivery schedule. Release documentation is created in the process; using it, you can display the releases transmitted to a vendor at any time within a certain period and thus trace exactly when you sent which information to the vendor.

Delivery schedules are simply prepared schedule lines: the schedule lines can be aggregated, for example, on a weekly basis if the vendor only delivers once a week. Delivery schedules are generated at regular intervals based on the schedule lines or are only generated if the schedule line quantities are changed importantly.

The overall process flow for a scheduling agreement with release documentation could, for example, appear as follows:

- A requirement planning run is carried out at regular intervals for the materials concerned. This automatically generates schedule lines for scheduling agreements.
- Based on these schedule lines, delivery schedules are created at certain intervals and transmitted to the vendor:
  - The MRP controller or purchaser can first check the schedule lines and then create a release manually.

- The releases can also be generated automatically by the system. There is an app for this. For example, this can run by job directly after the planning run.
- The delivery schedules can either be transmitted directly when they are created or later.

You can choose whether you want to work with both forecast and JIT delivery schedules or only with forecast delivery schedules.

If you work with both forecast and JIT delivery schedules, you have the advantage that the vendor can receive detailed, binding delivery data from you at shorter intervals in JIT delivery schedules, and at longer intervals, he can also receive a preview (forecast schedule).

You can only generate both forecast and JIT delivery schedules for a scheduling agreement with release documentation if you have first set the **JIT Delivery Schedule** indicator in the material master record (the Purchasing or MRP 2 view; see Figure 5.22 for material TG0011) and then in the scheduling agreement item.

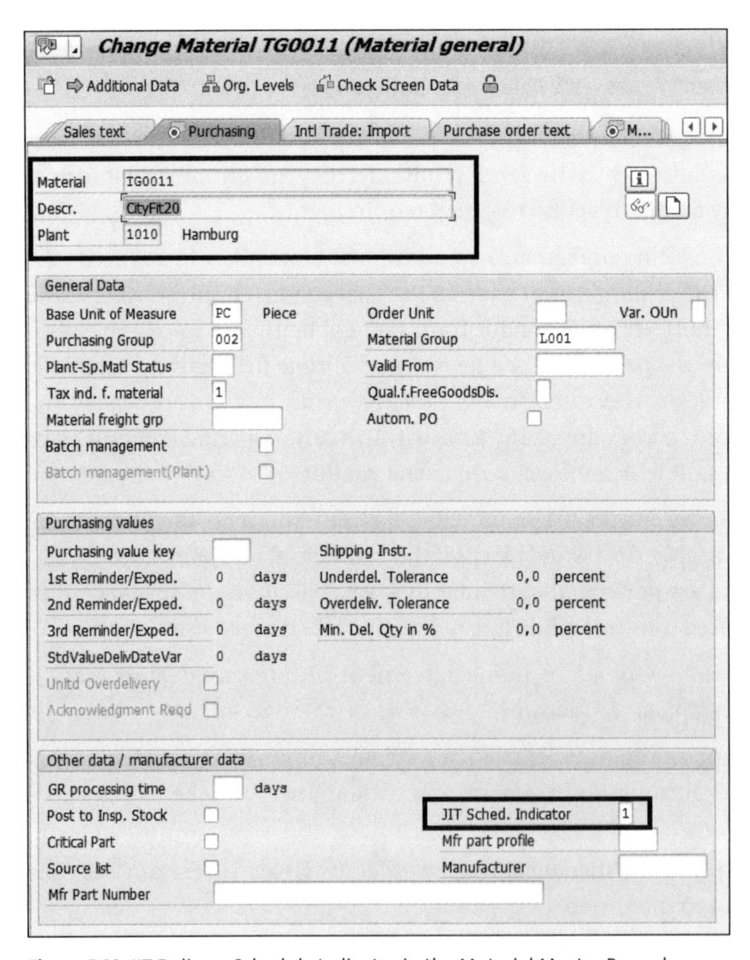

Figure 5.22 JIT Delivery Schedule Indicator in the Material Master Record

Figure 5.23 shows a scheduling agreement with release documentation (scheduling agreement type LPA). It consists of one item (see Figure 5.24) with material TG001, for which the JIT indicator has been set.

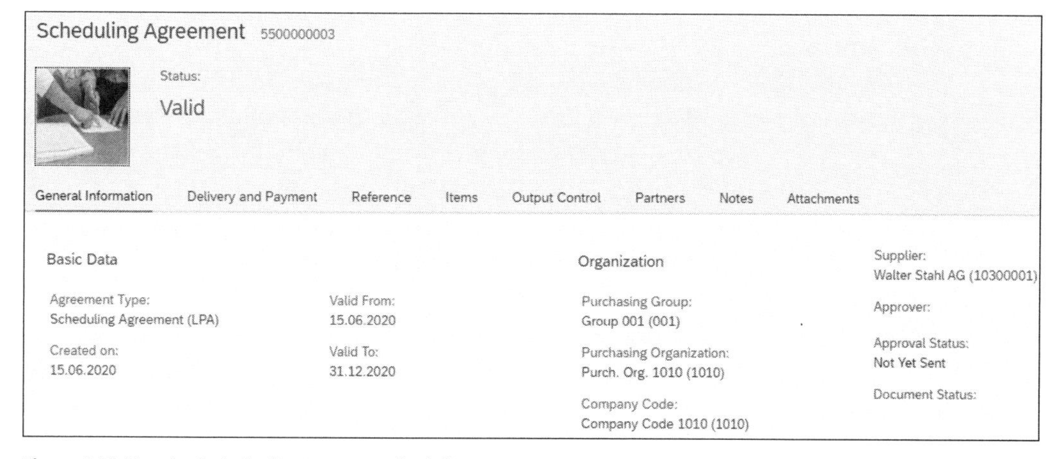

Figure 5.23 Header Data in the Manage Scheduling Agreements App

Scheduling Agreement 5500000003

Scheduling Agreement Items (1)

Item	Account Assignment Category	Material	Material Group	Plant	Target Quantity	Order Unit	Net Order Price	Price Unit	Order Price Unit	Release Document
10		CityFit20 (TG0011)	Trading Materials (L001)	Hamburg (1010)	50.000,000 PC	PC	30,00 EUR	1 PC	Piece (PC)	

Figure 5.24 Item Data in the Manage Scheduling Agreements App

In the **Output Control** tab in the details of a scheduling agreement item, you can see the **JIT Indicator** data in the center of Figure 5.25. Since it has already been set in the material master record, it is proposed in the scheduling agreement item, where it can either be kept or removed.

Scheduling Agreement /
10

Output Control

Creation Profile:	JIT Indicator:	Next FRC Schedule:	Next JIT Schedule:
Profile Purchasing (Z001)	1		

Figure 5.25 The JIT Delivery Schedule Indicator in the Scheduling Agreement Item in the Manage Scheduling Agreements App

Besides the JIT indicator, you can specify other important control parameters in a scheduling agreement item. In the **Scheduling Control** tab, in the details of a scheduling agreement item (see Figure 5.26), the following important parameters are significant:

- **Firm zone**
  The first firming period defines the point in time (calculated from today's date) at which the production release period ends. The production release period begins on the current date. The production release period should be defined in consultation with the vendor and usually covers the time required by the vendor to produce the scheduled quantities. You can agree with the vendor that the schedule lines in this period are to be regarded as fixed.
- **Trade-off zone**
  The second firming period defines the point in time (calculated from today's date) at which the material release period ends. The material release period begins with and includes the end of the production release period. Schedule lines that fall within this firming period can have a lower commitment level than schedule lines in the first firming period. The material release period is usually based on the delivery or production time of the components that the vendor requires to produce the ordered materials.
- **Binding for MRP**
  You can use this indicator to control requirements planning in such a way that either only the schedule lines within the first period are automatically firmed (indicator value 1) or even the schedule lines within both periods are automatically firmed (indicator value 2).

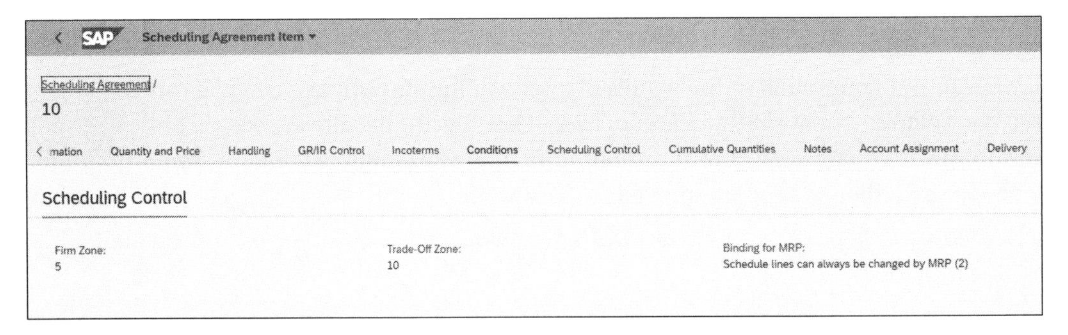

Figure 5.26 Scheduling Control in the Manage Scheduling Agreements App

On the **Output Control** tab in the details of a scheduling agreement item, you find the **Creation Profile** parameter, shown in Figure 5.27. The creation profile for delivery schedules is an instrument that controls how delivery schedules are to be generated for a scheduling agreement. It includes the following information:

- **Rhythm of delivery schedule creation (e.g., daily, weekly, every Tuesday...)**
  You can define different time intervals for forecast and JIT delivery schedules. For example, you can set daily generation for the JIT delivery schedules and weekly generation for the forecast delivery schedules, as shown in Figure 5.28.
- **Prerequisites for the delivery schedule creation**
  The profile in Figure 5.29 controls that both JIT and forecast releases should be generated, either when the next periodic date is reached or when the schedule line quantities have changed.

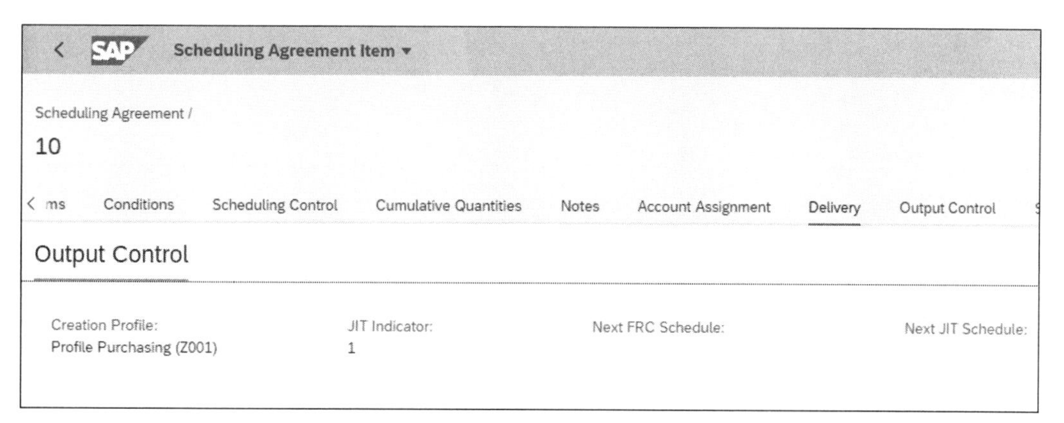

Figure 5.27 Output Control in the Manage Scheduling Agreements App

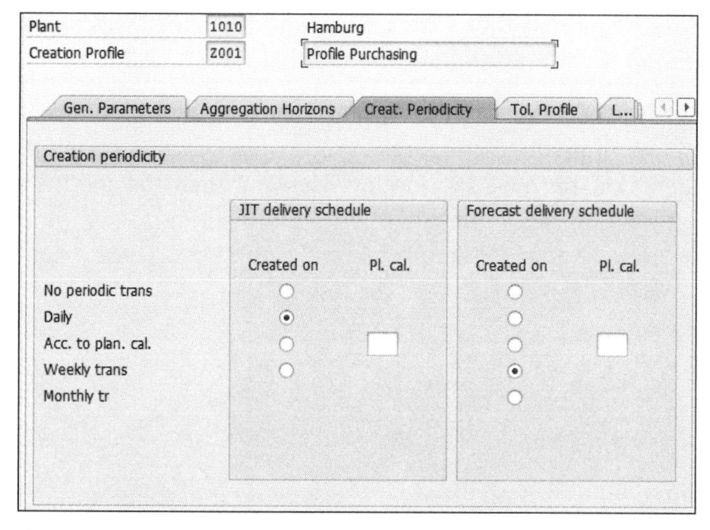

Figure 5.28 Customizing: Creation Profile—Periodicity

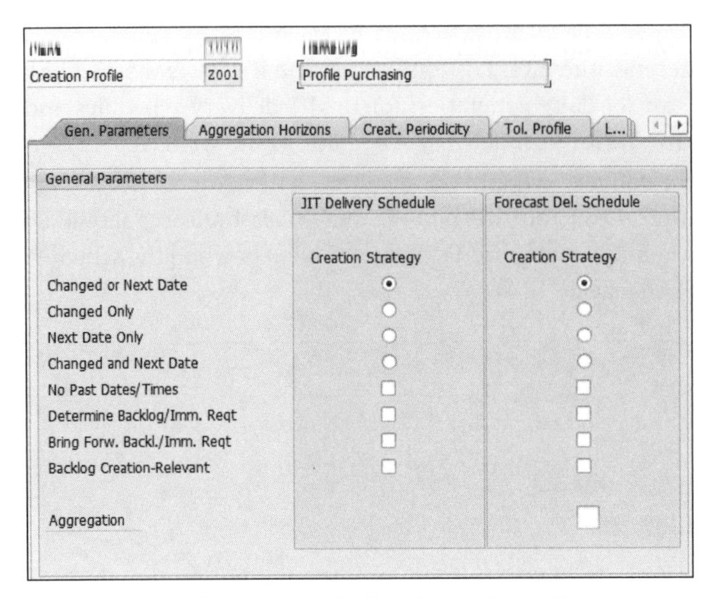

Figure 5.29 Customizing: Creation Profile—General Parameters

- **Type of aggregation of schedule line quantities**
  You can summarize the schedule line quantities on a daily or weekly or monthly basis.
- **The delivery schedule horizon in the release creation profile**
  This enables you to transmit only those dates and quantities to the vendor that they need for planning.

In the Figure 5.30 for profile Z001, JIT releases represent the daily requirements of the next 10 working days, while the forecast releases provide information about the weekly requirements of the next 20 working days and then the monthly requirements up to 60 working days.

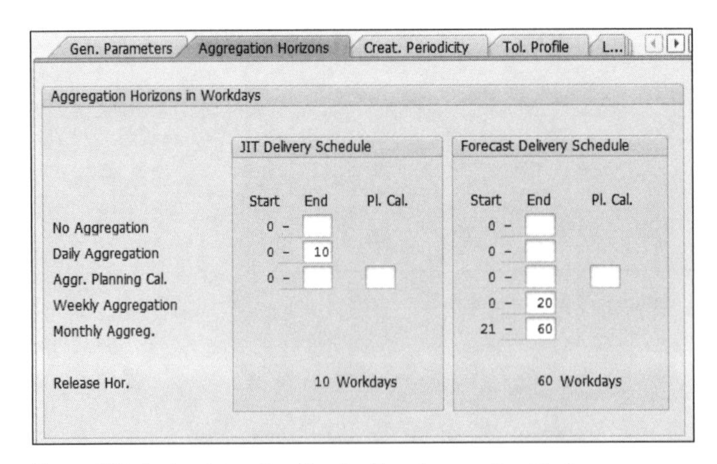

Figure 5.30 Customizing: Creation Profile—Aggregation Horizons

Release creation profiles are created per plant in Customizing using the menu path **Materials Management • Purchasing • Scheduling Agreement**.

Important Terminology

The following important terminology was used in this chapter:

- **Purchasing info record**
  A purchasing info record belongs to the master data of purchasing and contains information about a supplier-material combination, such as price, planned delivery time, and so forth.
- **Contract**
  A contract is a purchasing document. We distinguish between quantity and value contracts. In a quantity contract, it is agreed to purchase a defined quantity of a specific material (or service) at a defined price from a specific supplier within a certain period. In a value contract, it is agreed to purchase a specified value from a supplier within a given period. The relevant materials or services are recorded in the contract items.
- **Scheduling agreement**
  Like the contract, the scheduling agreement is an outline agreement in which purchase quantities are specified within a period. However, in contrast to the contract, no extra release orders are created for the scheduling agreement. Instead, the exact requirement quantities and dates are generated in the form of delivery schedules.
- **Info Record Update indicator**
  The **Info Record Update** indicator can be set in a purchase order item, an outline agreement item, or a quotation item and causes a purchasing info record to be automatically created or updated. Note that the conditions in the purchasing info record are never automatically created or overwritten with the purchase order price.
- **Condition type**
  A condition type is a key, is defined and configured in Customizing, and represents a pricing element (e.g., a price, a discount, or a surcharge). The condition type controls how this price element is calculated (e.g., for a discount, as a percentage, or per quantity).
- **Calculation schema**
  The calculation schema, also called condition schema or calculation procedure, is the set of rules for determining the net and effective price in a purchasing document. Among other things, it contains all permissible condition types and the sequence in which they are included in the calculation.
- **Contract release documentation**
  If you work with contracts, you must issue the exact quantity requirements in the form of release orders. These release orders are generated with reference to

the respective contract item and then updated in the contract item as release documentation. The contract release documentation provides information on the extent to which the contract item has already been consumed.

- **Scheduling agreement schedule line**
  Schedule lines belong to a scheduling agreement item and contain exact quantities and dates. These can be generated manually or automatically by the MRP run. In case of a delivery schedule without release documentation, the schedule lines are transmitted as is to the supplier. Otherwise, they must first be prepared in the form of delivery schedules. Be careful not to confuse the two terms!
- **Delivery schedule**
  Delivery schedules are ready-made schedule lines. Schedule lines can, for example, be combined, prepared on certain weekdays, and sent to the supplier, but only if the required quantity changes by a specified percentage. Delivery schedules can only be created for scheduling agreements with release documentation. Delivery schedules are generated manually or in the background. You can use the delivery schedule creation profile to configure a set of rules for the automatic generation of delivery schedules. We distinguish between forecast and JIT delivery schedules.
- **Creation profile**
  The delivery schedule creation profile represents an option to automatically create forecast and JIT delivery schedules. It is defined and configured in Customizing and stored in the scheduling agreement item (of course, only for scheduling agreements with release documentation).

 Practice Questions

These questions will help you evaluate your understanding of the topics covered in this chapter. They are similar in nature to those on the certification examination. Although none of these questions will be found in the exam itself, they will allow you to review your knowledge of the subject.

Select the correct answers, and then check the completeness of your answers in the next section. Remember that on the exam, you must select all correct answers—and only correct answers—to receive credit for the question.

1. Which of the following objects are sources of supply? (There are three correct answers.)

☐ **A.** Purchasing group

☐ **B.** Value contract

☐ **C.** Scheduling agreement

☐ **D.** Purchase order

☐ **E.** Purchasing info record

2. Where can you determine whether conditions can be maintained at the plant level?

☐ **A.** In the supplier master per plant

☐ **B.** In Customizing per plant

☐ **C.** In the material master per plant

☐ **D.** In Customizing per purchasing organization

3. Under which conditions is the most recent purchase order number by which a material was ordered from a vendor updated in the purchasing info record?

☐ **A.** Always, provided there is already an info record for the supplier-material combination

☐ **B.** If the **Info Record Update** indicator is set and on the purchase order item and under the condition that an info record already exists for the vendor-material combination

☐ **C.** If the **Info Record Update** indicator is set in the purchase order item; where no info record exists for the vendor-material combination, one is created.

4. True or false: You can use the **Info Record Update** indicator to have the price from a purchase order automatically transferred to an info record.

☐ **A.** True

☐ **B.** False

5. You have agreed to a price of EUR 4 for the procurement of a material from a vendor for the next six months. You have created a purchasing info record with the appropriate conditions for this. However, within this period, you order at a price of EUR 5 because you need the material urgently and the vendor must send it by express delivery. By default, the **Info Record Update** indicator is set in the purchase order item so that this purchase order is updated in the info record. The next time you order the material from this vendor, which price is defaulted by the system?

☐ **A.** 4 EUR

☐ **B.** 5 EUR

☐ **C.** No price

6. Which of the following types of purchasing info records can you create? (There are three correct answers.)

☐ **A.** Subcontracting

☐ **B.** Third-party

☐ **C.** Standard

☐ **D.** Consignment

7. True or False: A purchasing info record can be valid for a single plant of a purchasing organization.

☐ **A.** True

☐ **B.** False

8. Via which field can you decide whether you create a quantity contract or a value contract?

☐ **A.** The item category

☐ **B.** The contract type

☐ **C.** The account assignment category

☐ **D.** The material group

9. The contract is a framework agreement. How do you order the materials or services contained in a contract?

☐ **A.** You need a specific purchasing document with a specific order type.

☐ **B.** You must create schedule lines or have them generated by the MRP run.

☐ **C.** You do not need an additional document. The contract contains the concrete delivery dates.

☐ **D.** You just need a normal purchase order that contains a reference to the contract.

10. Which of the following statements apply to a centrally agreed contract? (There are three correct answers.)

☐ **A.** A centrally agreed contract can only be created by a central purchasing organization.

☐ **B.** In a centrally agreed contract, no plant is specified at the item level. You must specify the plant when you create a release order.

☐ **C.** All plants assigned to the purchasing organization can release against a centrally agreed contract.

☐ **D.** A centrally agreed contract item does not contain a plant but can contain plant-specific conditions.

11. How can you prevent a user from exceeding the agreed quantity in a contract when creating a release order?

☐ **A.** Users never can exceed the quantity because the system issues error messages by default when the agreed quantity is reached. No action is needed.

☐ **B.** It is always permitted to exceed the agreed quantity. The system only issues an information message. You cannot prevent it.

☐ **C.** The system issues a warning message by default when the agreed quantity is reached. You can change this message to an error message in Customizing to prevent the quantity from being exceeded.

☐ **D.** You can implement a BAdI to generate an error message when the agreed quantity is reached.

12. True or false: If you do not enter a plant in a scheduling agreement item, you create a central scheduling agreement item. You specify the plant when creating the schedule lines.

☐ **A.** True

☐ **B.** False

13. True or false: Scheduling agreements schedule lines contain the concrete quantities required on specific dates and are independent documents with their own document type and number range.

☐ **A.** True

☐ **B.** False

14. You are creating a scheduling agreement for a material. Both forecast and JIT delivery schedules are to be generated. How can you achieve this? (There are three correct answers.)

☐ **A.** You select the scheduling agreement type LP.

☐ **B.** You select the scheduling agreement type LPA.

☐ **C.** You set the **JIT Delivery Schedule** indicator in the material master.

☐ **D.** You set the **JIT Delivery Schedule** indicator in the supplier master.

☐ **E.** You set the **JIT Delivery Schedule** indicator in the scheduling agreement item.

15. Which of the following does the creation profile in a scheduling agreement item control? (There are two correct answers.)

☐ **A.** How schedule line quantities are to be aggregated

☐ **B.** Whether schedule lines can be generated by the MRP run

☐ **C.** In which rhythm JIT delivery schedules are to be generated

☐ **D.** Whether JIT delivery schedules may be generated in addition to forecast delivery schedules

16. Your vendor needs a certain degree of planning reliability and must not adjust production quantities at the last minute. How can you ensure that MRP does not change schedule line quantities with a short-term requirement date?

☐ **A.** Enter a suitable creation profile in the scheduling agreement item.

☐ **B.** Enter some days in the **Firm Zone** field in the scheduling agreement item and set the **Binding for MRP** indicator to 1.

☐ **C.** Set the **JIT Delivery Schedule** indicator in the material master, which is copied into the scheduling agreement item.

17. True or false: You can create a scheduling agreement item with account assignment to a production order.

☐ **A.** True

☐ **B.** False

18. True or false: You can create a contract item or a scheduling agreement item for subcontracting.

☐ **A.** True

☐ **B.** False

Practice Question Answers and Explanations

1. Correct answers: **B, C, D**

   The possible sources of supply are purchasing info records, contracts (quantities and value contracts), and scheduling agreements. They contain agreements and prices that are valid for longer periods of time, which you can refer to in procurement transactions. A purchasing group refers to one or more responsible purchasers, and a purchase order contains data that is only valid for the purchasing process in question.

2. Correct answer: **B**

   For each plant, you specify in Customizing whether conditions can be created only per plant or only per purchasing organization or whether both levels are permitted for condition maintenance.

3. Correct answer: **C**

   The last purchase order number and purchase order item are updated in the purchasing info record if you have set the **Info Record Update** indicator in the purchase order item. If an info record already exists for the vendor-material combination, it is updated. If no info record exists, one is created but without conditions. Only the reference to the purchase order is updated. In the case of a new purchase order, the price from the last purchase order will be proposed until conditions are maintained in the purchasing info record. These then take precedence.

4. Correct answer: **B**
   False. Using the **Info Record Update** indicator in the purchase order item, you can only have the purchase order number and the purchase order item updated in the info record as a reference. The price in the info record (in the **Conditions** section) is not overwritten by a purchase order price.

5. Correct answer: **A**
   Valid conditions have priority to the price from the last order. Therefore, the price of 4 EUR is proposed.

6. Correct answers: **A, C, D**
   You can create info records for the procurement processes standard, consignment, pipeline, and subcontracting.

7. Correct answer: **A**
   True. A purchasing info record can be valid for a single plant of a purchasing organization if you create it specifically for a combination of purchasing organization and plant. If you enter only one purchasing organization when creating a purchasing info record, the purchasing info record is valid for all plants of the purchasing organization.

8. Correct answer: **B**
   You use the contract type to decide whether you create a quantity contract or a value contract. SAP S/4HANA offers the contract type MK for quantity contracts and WK for value contracts in the standard system. If you require other contract types, for example, with another number range or another field selection, you can create these in Customizing. You can copy the existing contract types as a reference.

9. Correct answer: **D**
   The contract does not contain concrete delivery dates and quantities. It is indeed an outline agreement that belongs to the sources of supply. You order the materials or services in a contract via a normal purchase order item that refers to a contract item. This reference to a contract item can already be set in the purchase requisition. It is then adopted when the purchase requisition is converted into a purchase order. Schedule lines are only relevant for scheduling agreements.

10. Correct answers: **B, C, D**
    A centrally agreed contract item is characterized by the fact that it contains no plant. You can issue release orders against the contract for any plant assigned to the purchasing organization in the contract header. Optionally, you can store separate conditions for individual plants in the contract item. The purchasing organization in the contract header must not necessarily be a central purchasing organization; it can be any purchasing organization responsible for several plants.

11. Correct answer: **C**

    The system issues a warning message by default when the agreed quantity is reached. You can change this message to an error message in Customizing to prevent ordering. You can even define several message versions and assign these versions to different users, so that different users can have different types of messages: some have an error message, whereas others only get a warning message.

12. Correct answer: **B**

    False. The plant is mandatory in a delivery schedule item.

13. Correct answer: **B**

    False. Schedule lines contain the concrete quantities required on specific dates but are not independent documents with their own document type and number range. They are part of the scheduling agreement. For this reason, the use of scheduling agreements helps reduce the volume of documents.

14. Correct answers: **B, C, E**

    Selecting a scheduling agreement type with the release documentation indicator set in Customizing—like scheduling agreement type LPA—is a prerequisite for being able to generate delivery schedules at all. If you then wish to generate both forecast and JIT delivery schedules, you must first set the **JIT Delivery Schedule** indicator in the material master record and then also in the scheduling agreement item.

15. Correct answers: **A, C**

    Whether JIT delivery schedules may be generated in addition to scheduling agreement releases is controlled by the **JIT Delivery Schedule** indicator in the material master, which is then set in the scheduling agreement item. Schedule lines can always be generated by the MRP run, provided the material is planned: always with MRP Live always, but only if desired with classic MRP.

    Among other things, the creation indicator controls the frequency with which schedule lines are processed in delivery schedules and the way in which schedule line quantities are aggregated.

16. Correct answer: **B**

    You can define two periods in a delivery schedule: the first is the *firm zone*; the second is the *trade-off zone*. You can set up MRP so that either only the schedule lines within the first period are automatically firmed—that is, they cannot be changed (**Binding for MRP** indicator 1) or that even the schedule lines within both periods are automatically firmed (**Binding for MRP** indicator 2).

17. Correct answer: **A**

    True. A scheduling agreement item can basically have an account assignment.

18. Correct answer: **A**

    True. You can choose item category L for subcontracting when you create a scheduling agreement item or a contract item.

Takeaway

You should now understand the term *source of supply* and be able to list the possible sources of supply in SAP S/4HANA. At this point, you should be so familiar with the special features and capabilities of the respective sources of supply that you could recommend the most suitable object to be used in a customer situation.

Summary

In this chapter, we examined the three sources of supply: info record, contract, and scheduling agreement.

We analyzed the structure of the info record, noting that we can create info records for different procurement processes, and we explored some important fields in the info record. We looked at the creation of the info record and discussed the role of the info record update indicator in detail.

We also analyzed the different types of contracts, focusing on quantity, value, and centrally agreed contracts. We reviewed how contracts are released.

The scheduling agreement and its various characteristics were also explored in this chapter. This is a powerful tool for simplifying and automating the procurement process. Schedule lines can be generated automatically by the MRP run, and if necessary, forecast and JIT delivery schedules can also be issued automatically.

In the next chapter, we will look further at automation and simplification and learn how the system can automatically determine existing supply sources.

Chapter 6
Purchasing Optimization

6

Techniques You'll Master

- Create a purchase requisition with source determination.
- Describe the logic of the source determination.
- Create and describe a source list.
- Understand the functions of the source list.
- Create and describe quota arrangements.
- Explain the possible uses of a quota arrangement.
- Convert purchase requisitions to purchase orders.
- Describe the prerequisites for the automatic conversion of requisitions into purchase orders.

The aim of this chapter is to investigate how you can optimize and automate the procurement process. Material requirements planning (MRP) makes a significant contribution to the automation of the procurement process and is dealt with in Chapter 8. Here we concentrate on the automatic determination of sources of supply when creating a purchase requisition manually and on the possibilities to convert purchase requisitions into purchase orders automatically.

Real-World Scenario

In many companies, certain goods are purchased very regularly from a few fixed suppliers, such as raw materials and supplies needed for production. Often, companies maintain long and trusting relationships with their suppliers, so that nothing stands in the way of automating the procurement process with them.

The first step on the path to automation is the creation of purchase requisitions and the determination of sources of supply by materials planning. The second step is the automatic conversion of purchase requisitions into purchase orders. However, the system can also provide strong support in determining the sources of supply in the case of manual requisitions. As an internal or external consultant, you should always look for ways to help your customer reduce manual, repetitive activities that do not actually require expertise and replace them with automation. To do this, you need to know the options that SAP S/4HANA offers for sourcing and automatic conversion so that you can propose them for use.

Objectives of This Portion of the Test

In this part of the certification exam, your knowledge of how to optimize the procurement process in SAP S/4HANA will be tested. For the certification exam, you must be able to do the following:

- Create a purchase requisition and describe its main features.
- Explain the logic the system uses to search for a suitable source of supply.
- Create a source list.
- Describe the possible uses of a source list.
- Create a quota arrangement.
- Explain the functions and logic of a quota arrangement.
- Describe how to convert a purchase requisition into a purchase order.
- Enumerate the prerequisites for automatic conversion.

Key Concept Refresher

In this chapter, we will create purchase requisitions manually and thereby explore the automatic source of supply determination. We will also cover the creation and use of source lists and quota arrangements.

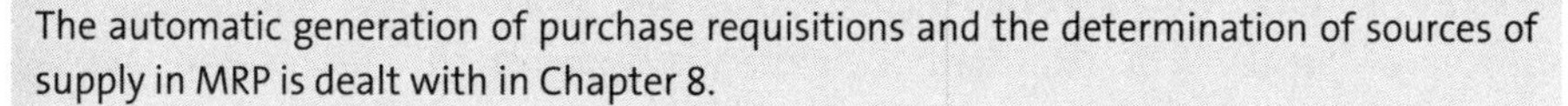

Note

The automatic generation of purchase requisitions and the determination of sources of supply in MRP is dealt with in Chapter 8.

Purchase Requisition with Source Determination

Purchase requisitions are internal documents used for requesting the purchasing department to procure a certain quantity of a material or service for a certain date.

A purchase requisition can be created manually or automatically. The person who creates a purchase requisition *manually* determines which material or service is ordered, the quantity, and the date. A purchase requisition can also be created *automatically* from another component, such as through MRP, via maintenance orders, via production orders, and via networks.

A purchase requisition can be intended for the warehouse or for direct consumption. This is determined by the account assignment category. If a purchase requisition is intended for consumption, but you do not yet exactly know how precise the account assignment will be, then you can choose account assignment category U (unknown). The account assignment category and account assignment details are specified during conversion into a purchase order.

Figure 6.1 shows one way of creating purchase requisitions manually using the Manage Purchase Requisition—Professional app.

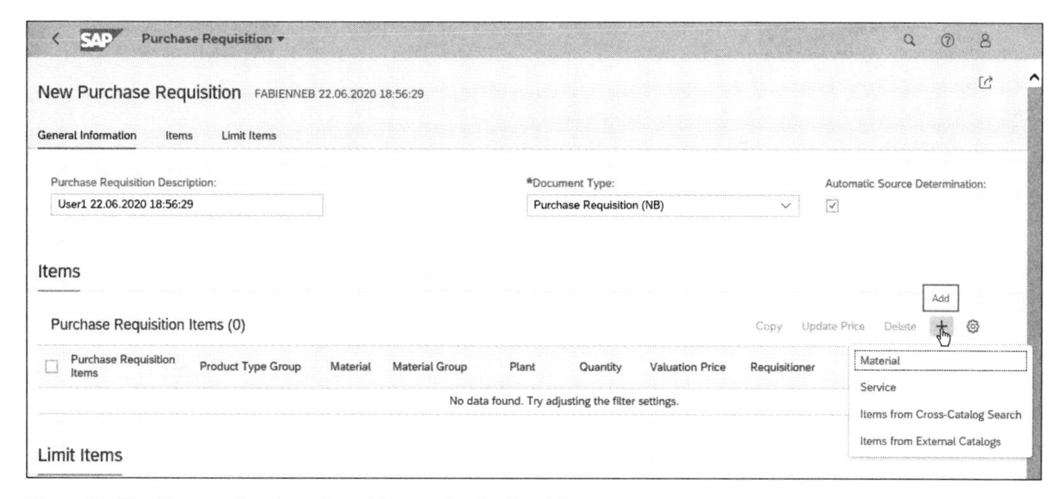

Figure 6.1 The Manage Purchase Requisition—Professional App

If you create a purchase requisition manually, you can set the **Automatic Source Determination** indicator. In this way, the system automatically attempts to determine a suitable source in the background for the materials or services contained in the requisition.

If the system determines several valid sources of supply for an item, and none of them can be identified as a preferred source of supply (we will see later how a source of supply can be defined as preferred), then it will display a list of possible sources of supply. The user can then manually select the appropriate supply source.

Figure 6.2 shows a purchase requisition item for which the system could not determine a unique source of supply. The user can select **Assign Source of Supply** to obtain a list of possible sources of supply; this list is shown in Figure 6.3.

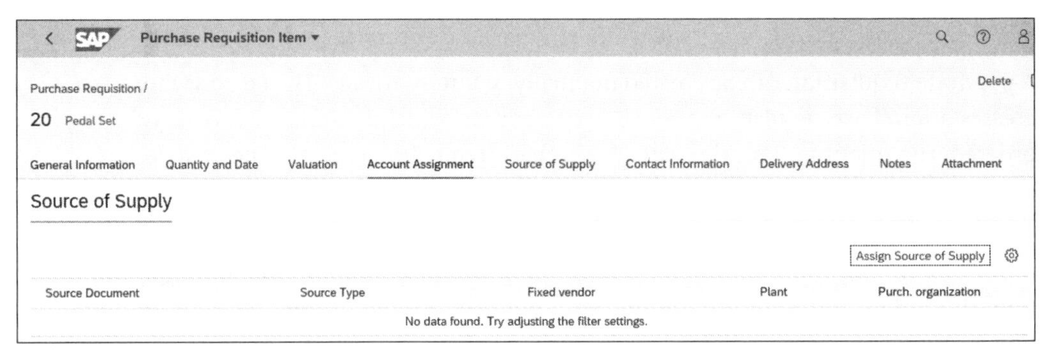

Figure 6.2 No Unique Source of Supply in the Manage Purchase Requisition—Professional App

Select Source Of Supply

PROPOSED SoS | INFO RECORD | CENTRAL CONTRACT | AGREEMENT | FIXED SUPPLIER

Line Items (2)

Source Document	Source Type	Vendor	Plant	Purch. organization	Net Order Price
5300000123	Info Record	Walter Stahl AG (10300001)	Hamburg (1010)	Purch. Org. 1010 (1010)	1,31 EUR
5300000124	Info Record	Johan Kraft GmbH Inlandslieferant D (10300002)	Hamburg (1010)	Purch. Org. 1010 (1010)	1,37 EUR

Cancel

Figure 6.3 Possible Sources of Supply in the Manage Purchase Requisition—Professional App

Figure 6.4 shows another alternative for creating a purchase requisition. This is the Create Purchase Requisition—Advanced app. The **Source Determination** indicator is set. It corresponds to the **Automatic Source Determination** indicator, which we are already familiar with from the SAP Fiori app, and causes the system to automatically attempt to determine a suitable source for the materials or services contained in the purchase requisition in the background. If several are found, a list is displayed to the user in the **Source of Supply** tab.

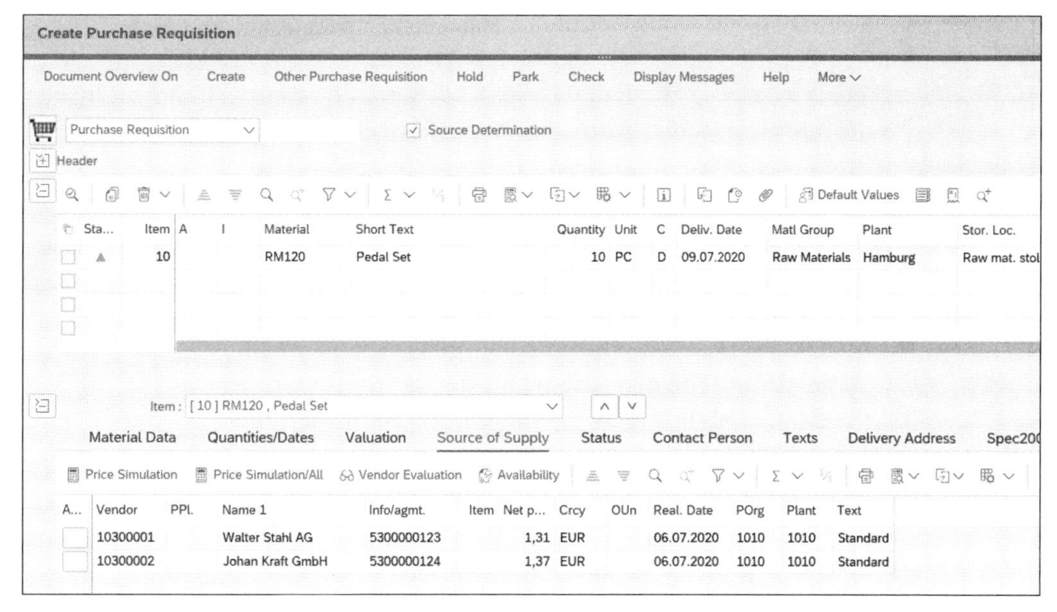

Figure 6.4 Choosing One of the Proposed Sources of Supply in the Create Purchase Requisition—Advanced App

When will the system then be able to determine a unique source of supply? If several valid sources of supply exist for the item in the system, the system uses the following logic to try to find a unique source:

- The system first checks whether a *quota arrangement* exists for the material whose validity period covers the delivery date of the purchase requisition. If a quota arrangement exists, the system assigns the source of supply with the lowest quota rating. How the quota arrangement works is described in a further section of this chapter.
- If no valid quota arrangement exists, the system checks whether an entry exists in the *source list* for the material in the relevant plant. The source list is a register of sources of supply for a material in a certain plant and for certain periods of time. We will discuss this in detail in the next paragraph. The validity period of the source list entry must cover the delivery date in the purchase requisition. The entry can either be an info record or an outline agreement (contract or scheduling agreement). If the source list contains only one item, the system assigns this source to the purchase requisition. If the source list contains several valid entries, the system assigns the source with the **Fixed** indicator. If both an outline agreement item and an info record are flagged as fixed sources, the outline agreement item has higher priority and is assigned.
- If neither a quota arrangement nor a source list exists, or if none of the entries in the source list is marked as fixed, the system searches for a valid *outline agreement* item. If only one valid outline agreement item exists, the system can assign it. If more than one valid outline agreement item exists, users must select manually from a list.

- If neither a quota arrangement nor a source list with a unique or fixed entry nor an outline agreement item exists, the system searches for valid *info records*. If there is only one valid info record, the system can assign it. If more than one valid info record exists, users can choose from a list.

Note

This logic applies to purchase requisitions created outside MRP with source determination. Please refer to Chapter 8 for details on automatic source determination during MRP.

The benefit of source determination and, especially, automatic source determination at the time of the purchase requisition is that the subsequent process flow (i.e., the creation of the purchase order) can be automated and accelerated. A purchase requisition with an assigned source of supply can easily be converted into a purchase order—under certain circumstances, even automatically.

Supply source determination requires that sources of supply have been stored in the system. Automatic supply source determination even requires additional master records such as the source list and/or quota arrangement if several valid sources of supply exist for an article simultaneously.

Source Lists

If you create a purchase requisition manually and the system determines several valid equivalent sources of supply, you must select the appropriate source manually from a list. If the system is to make this selection on its own, you can additionally use the *source list*.

The source list is a register of sources of supply for a material in a certain plant. Each source list record contains a validity period, a source of supply, and parameters that control the source of supply during this period. One of these parameters is used to define a source as your preferred source of supply. The source list also performs other functions that we will examine in this section.

Figure 6.5 and Figure 6.6 show the structure of a source list—once with the Manage Source Lists app and once with the SAP GUI transaction, Transaction ME01. For each material and plant, the source list contains a list of sources of supply that have certain characteristics within a certain validity period.

A source of supply is identified in the source list as follows:

- **By the vendor number only**
  In this case, the source of supply is an info record. Although the number of the info record is not visible in the source list, it is unique based on the material number and the vendor number.
- **By the vendor number, outline agreement number, and item**
  In this case, the source of supply is a scheduling agreement or a contract.

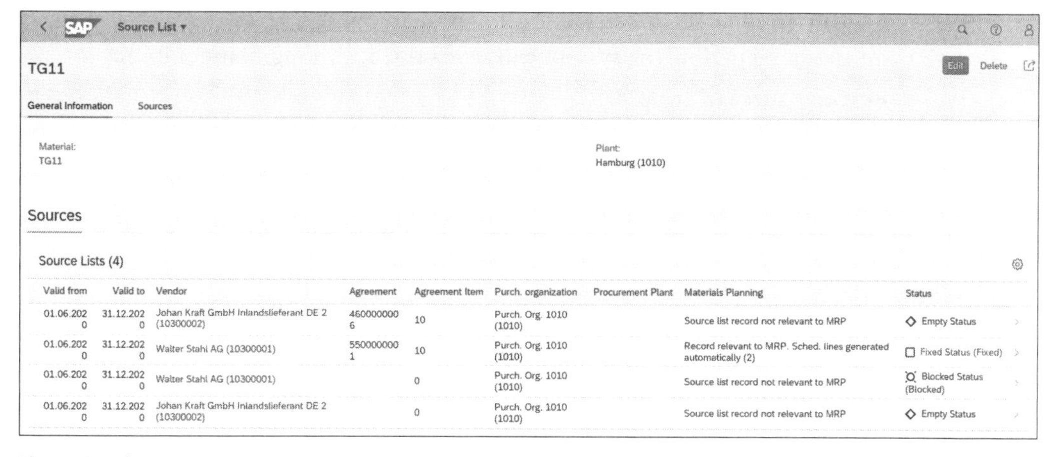

Figure 6.5 The Manage Source Lists App

Maintain Source List: Overview Screen

Material TG11 Helmet
Plant 1010 Hamburg

Source List Records

Valid from	Valid to	Vendor	POrg	PPl	OUn	Agmt	Item	Fix	Blk	MRP	MRP Area
01.06.2020	31.12.2020	10300002	1010		PC	4600000006	10	☐	☐		1010
01.06.2020	31.12.2020	10300001	1010		PC	5500000001	10	☑	☐	2	1010
01.06.2020	31.12.2020	10300001	1010		PC			☐	☑		
01.06.2020	31.12.2020	10300002	1010		PC			☐	☐		

Figure 6.6 Transaction ME01 (Maintain Source List)

Note

If another plant is a possible source of supply (stock transfer), enter the plant number in the **PPl** field of the SAP GUI transaction or in the **Procurement Plant** field of the SAP Fiori app and do not enter a vendor number or outline agreement item number. The plant is only determined as the source of supply for a purchase requisition item with item type U (stock transfer). For purchase requisitions that are created by the planning run, the materials require a corresponding special procurement type in the MRP data of the material master record.

Here are the options that the source list offers for managing sources of supply:

- **Fix a source of supply for source determination**
  You can use the **Fixed** status in the SAP Fiori app or the **Fix** checkbox in the SAP GUI transaction to define a source of supply as the preferred source of supply for automatic sourcing for a certain period. This fixed source of supply is then always selected automatically, unless you have maintained a quota arrangement for the material (see the source determination logic described previously) and except during the MRP run. At any given time, only one source of the same category (e.g., an info record or an outline agreement item) can be marked as

fixed. If you mark both an info record and an outline agreement item as fixed for the same period, the system will select the outline agreement item.

- **Block a source of supply**
  The **Blocked** status in the SAP Fiori app or the **Blk** checkbox in the SAP GUI transaction allows you to block a source of supply for a certain period. If you block a source of supply, you cannot use the info record or outline agreement item in source determination, and you can create neither a purchase order nor a requisition using this supply source in this period.
- **Set a source of supply to relevant for MRP or not**
  You can use the **Materials Planning** field in the SAP Fiori app or the **MRP** checkbox in the SAP GUI transaction to override the source determination logic during the planning run (see Chapter 8). If you set the indicator to 1 for a source of supply, this source of supply is selected during the MRP run and assigned to a purchase requisition. If a source of supply is a scheduling agreement item, you can set the indicator to 2, then the MRP run will create schedule lines for this scheduling agreement.

Note

In the Manage Source Lists app, indicators or checkboxes are replaced by two dropdown fields:

- You can use the **Status** field to lock a source of supply or define it as fixed.
- You use the **Materials Planning** field to specify whether the source of supply is not relevant to MRP, whether it is to be assigned to purchase requisitions generated via MRP (corresponds to indicator 1), or whether schedule lines are to be generated via MRP (corresponds to indicator 2 and is only useful for scheduling agreements).

Figure 6.7 shows the two fields and their possible values.

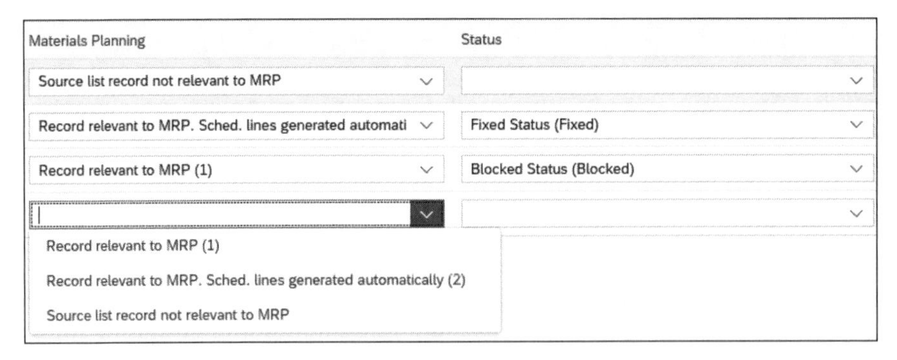

Figure 6.7 The Status and Materials Planning Fields in the Manage Source Lists App

In summary, we can say that the source list is a management tool for sources of supply. A source list contains part or all the sources of supply of a material with properties (fixed, blocked, relevant to MRP) valid for a certain period.

The use of source lists can be optional or mandatory. The mandatory use of source lists can be defined per material or even at the plant level. Figure 6.8 and Figure 6.9

show the plant-specific Purchasing view of a material. You can set the **Source List Requirement** indicator there.

Figure 6.8 Purchasing Data in the Change Material App (Transaction MM02)

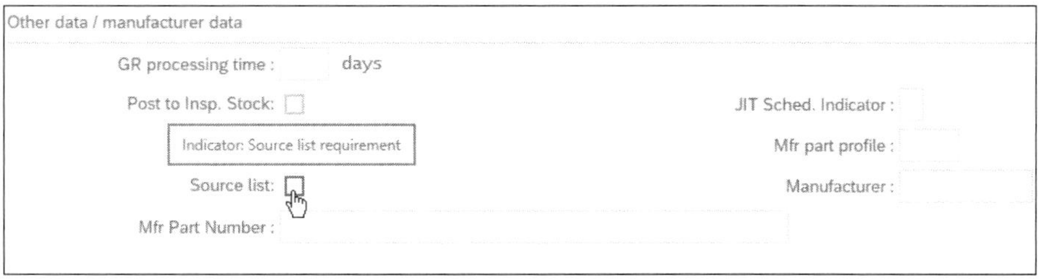

Figure 6.9 Source List Requirement in the Material Master Record using the Change Material App (Transaction MM02)

Figure 6.10 shows the customizing path for setting the source list requirement at the plant level.

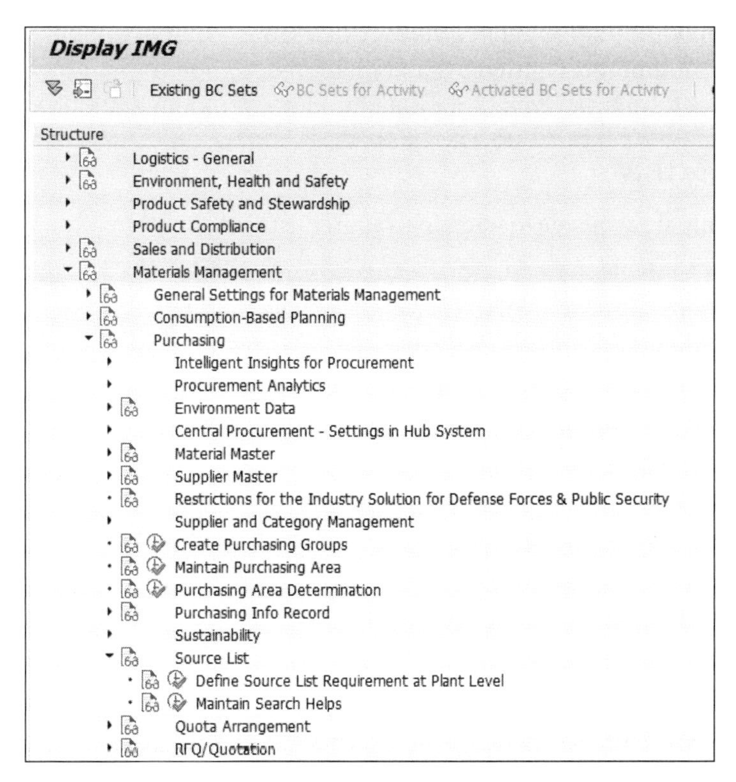

Figure 6.10 Customizing Path: Source List Requirement at the Plant Level

Figure 6.11 shows you the settings. In this example, no plant is subject to the source list requirement in the system.

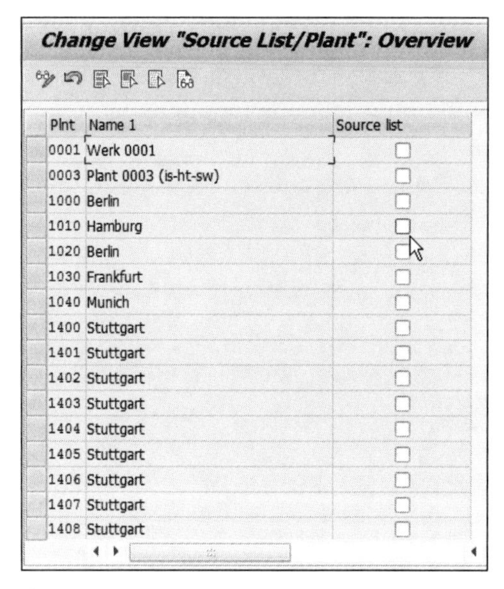
Change View "Source List/Plant": Overview

Plnt	Name 1	Source list
0001	Werk 0001	☐
0003	Plant 0003 (is-ht-sw)	☐
1000	Berlin	☐
1010	Hamburg	☐
1020	Berlin	☐
1030	Frankfurt	☐
1040	Munich	☐
1400	Stuttgart	☐
1401	Stuttgart	☐
1402	Stuttgart	☐
1403	Stuttgart	☐
1404	Stuttgart	☐
1405	Stuttgart	☐
1406	Stuttgart	☐
1407	Stuttgart	☐
1408	Stuttgart	☐

Figure 6.11 Customizing: Source List Requirement at the Plant Level

If you declare the source lists as mandatory, a source list then assumes the role of a list of approved sources. You can only order a material from a certain supplier if an info record or an outline agreement exists for this vendor-material combination and this source of supply is registered in the source list.

Maintaining Source Lists

The following options are available for creating a source list:

- Manual creation of a single source list for a material-plant combination with the SAP GUI transaction Transaction ME01 or with the Manage Source Lists app
- Creation when entering a contract or a scheduling agreement; from the SAP GUI transactions for creating an outline agreement, you can branch to the source list maintenance transaction so that you can directly add the new outline agreement to the source list
- Creation when creating a purchasing info record; from the SAP GUI transactions for creating an info record, you can branch to the source list maintenance screen so that you can directly add the new info record to the source list
- Automatic creation using the Generate Source List app (Transaction ME05; see Figure 6.12)

You can use this last procedure to generate source list entries for several materials and plants. But you must decide whether you wish to generate source list entries for info records only, for outline agreements only, or for all sources of supply of the selected materials and plants. The **Relevant to MRP** indicator can be set either for all records or no records. Subsequent maintenance thus may be necessary.

Generate Source List

Save as Variant... More

Material: to:

Plant: to:

Generate

All Records for Material/Plant

Excl. Outline Agreements

Outline Agreements Only

Control

Valid from: 01.06.2020

Valid To: 31.12.2999

MRP Indicator:

Add Planned Delivery Time:

Existing Records

Do Not Change

Delete

Become Invalid

Test Run:

Figure 6.12 Selection Screen of the Generate Source List App

Note

You may have outline agreement items that refer to a material group and not to a single specific material—that is, an outline agreement item with item category M (material unknown) or W (material group). You can create material-specific source list records for these contract items and by doing so decide whether you will exclude or include materials of the relevant material group.

Example

You create a contract for highlighters. It is not necessary to include all possible highlighters as separate items in the contract. The contract only needs to have a single item for all highlighters with item category M and the corresponding material group. However, the pink test markers should be excluded from this contract. You thus decide to use the source list for the material group item as an exclusion list: In the exclusion list, you enter the material numbers that correspond to the pink markers. Accordingly, the system creates blocked source list entries for them, so that you cannot release these materials from the contract item.

The example just described is illustrated by the following three figures. Figure 6.13 shows a value contract item with item category M. Select the menu item on the top menu line and choose **Maintain Source List**, as shown in Figure 6.14.

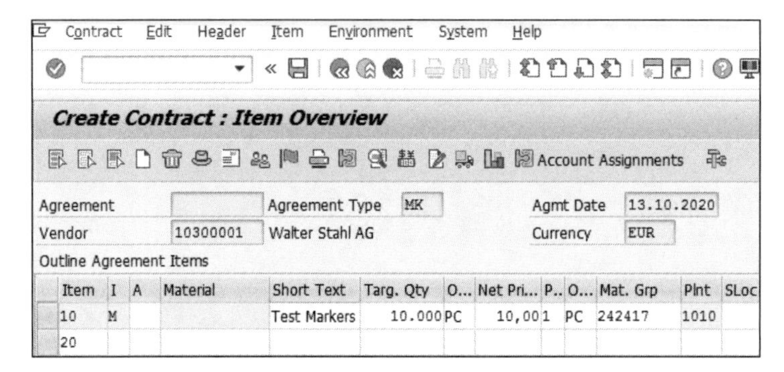

Figure 6.13 Item Category M in Transaction ME31K

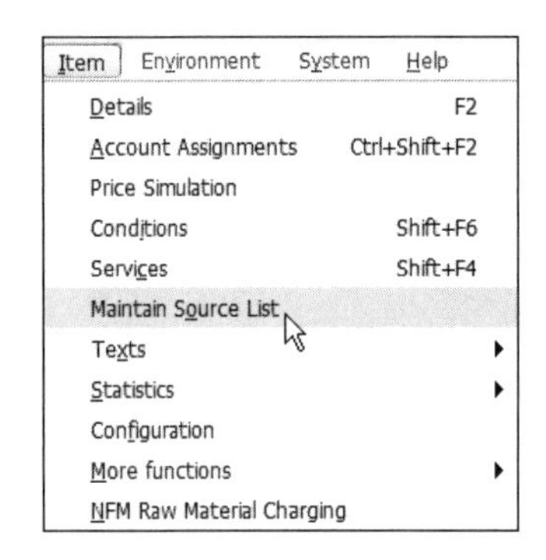

Figure 6.14 Maintaining a Source List in Transaction ME31K (I)

On the screen for maintaining the source list, select the **Exclusion** indicator shown in Figure 6.15 and enter the materials that are not part of the contract.

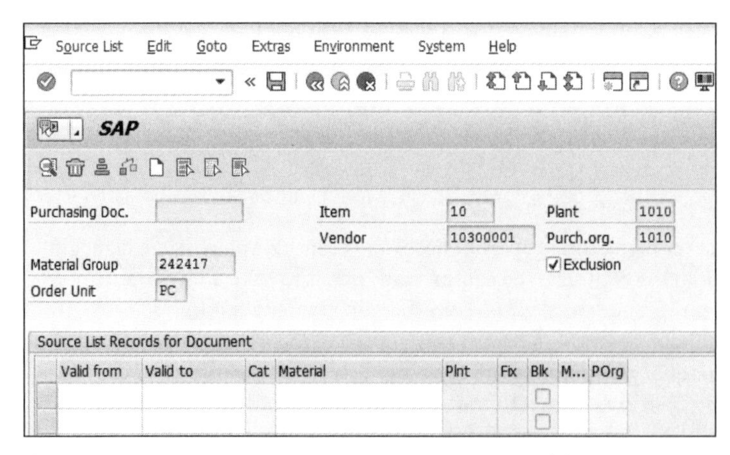

Figure 6.15 Maintaining a Source List in Transaction ME31K (II)

In addition to the source list, the quota arrangement offers a further option for managing sources of supply and controlling automatic source determination.

Quota Arrangements

Use quota arrangements if you want to procure a material from several vendors within a certain period. This will determine which proportion of the requirements for this material is to be allocated to which vendor within the period.

If a valid quota arrangement exists for a material, it will be considered first in the source determination process. A quota arrangement is defined for a certain period, and a quota arrangement item is created for every possible vendor within this period.

Note

If a quota arrangement exits for a material, this does not mean that the quantity of a single manually created purchase requisition is automatically distributed among various suppliers. The total requested quantity of a purchase requisition item is allocated in full to the vendor who is next in line according to the quota arrangement. The distribution according to the quota arrangement must be adhered to over the entire requirements of the defined period.

An individual requirement can only be split up among several vendors according to the quota arrangement during the planning run.

Figure 6.16 shows an example of a simple quota arrangement for the standard external procurement process: vendor 10300001 and vendor 10300002, respectively, are each to supply 70% and 30% of the requirements of material TG11 in the period from June 1 to December 31.

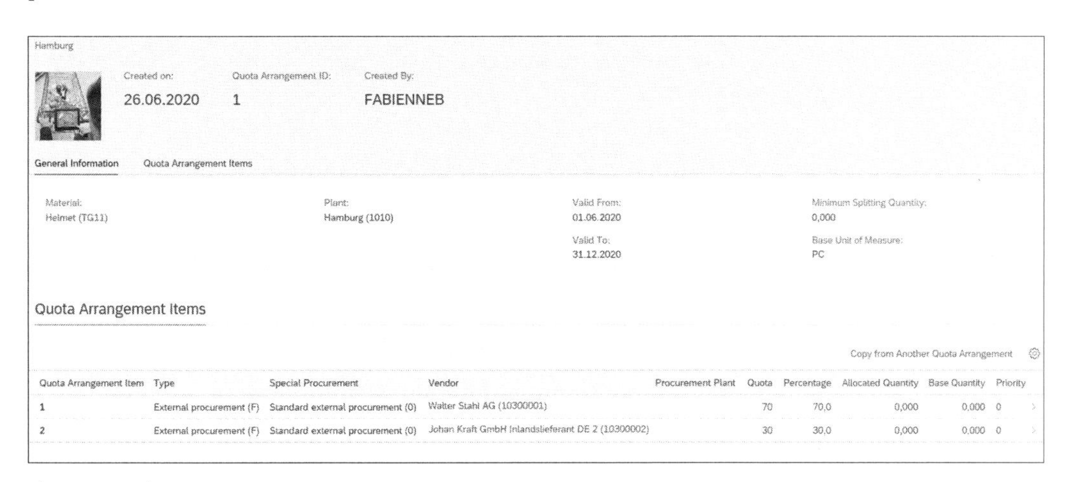

Figure 6.16 The Manage Quota Arrangement App

A quota arrangement contains general information and items. When you create a quota arrangement, enter the following general information (header data):

- Material
- Plant
- Validity period

You also have the option to enter the minimum splitting quantity, which defines the smallest quantity that a lot must have in order to be divided among several vendors during the planning run.

Then you create items with the following data:

- Type of procurement, either external procurement or internal procurement (in-house production)
- Special procurement
  - Standard, consignment, subcontracting, and stock transfer for external procurement
  - Standard in-house production or production in another plant for internal procurement
- Supplier or procurement plant, depending on the procurement type
- Quota you want to assign to this supplier or plant

We will explore further optional information in the following discussion.

The system then calculates the percentage distribution of the requirements based on the quotas. The allocated quantity is the total quantity that has been allocated to a source of supply since the beginning of the quota arrangement. The allocated quantity is updated automatically if, for example, you create a purchase requisition with this source assigned or if you order directly from that source. Updating also takes place if you delete or change objects (e.g., purchase requisition, purchase order) after allocation.

Now let's look at how the system proceeds to allocate a source of supply using a quota arrangement.

Source Determination Using Quota Arrangements

When you create a purchase requisition with source determination, the system first searches for a valid quota arrangement. If it finds one, it calculates a so-called quota rate. The vendor (source) with the lowest quota rating is then selected. The quota rating is calculated using the following formula:

Quota Rating = (Allocated Quantity + Base Quantity) / Quota

Let's consider an example to illustrate this approach. Figure 6.17 shows a quota arrangement for material TG11: vendor 10300001 is to receive 70% of the requirements within the validity period, and vendor 10300002 is to receive 30%. At this point in time, 10300002 has already been allocated 330 pieces, while 10300002 has only obtained 200 pieces.

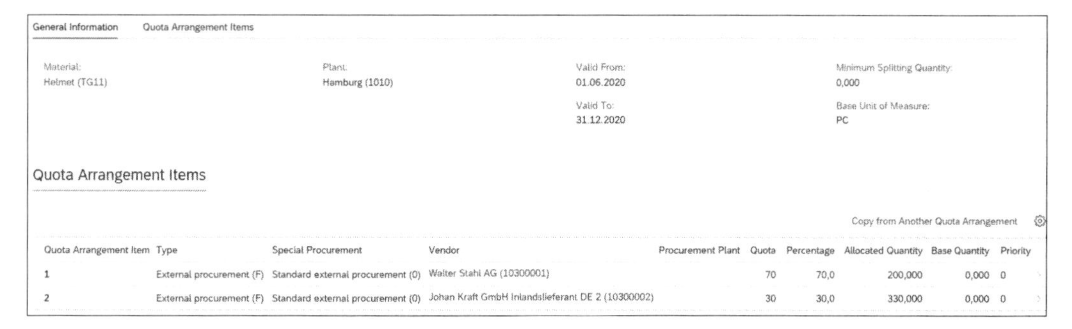

Figure 6.17 Example of a Quota Arrangement in the Manage Quota Arrangements App

Now we create a purchase requisition for material TG11, and the system should determine the source of supply.

The system calculates the quota rating for each quota arrangement item:

- Item 1: *Quota rating 1 = (200 + 0) / 70 = 2.86*
- Item 2: *Quota rating 2 = (330 + 0) / 30 = 11*

Item 1 (vendor 10300001) has the smaller quota rate and is selected accordingly, as shown in Figure 6.18.

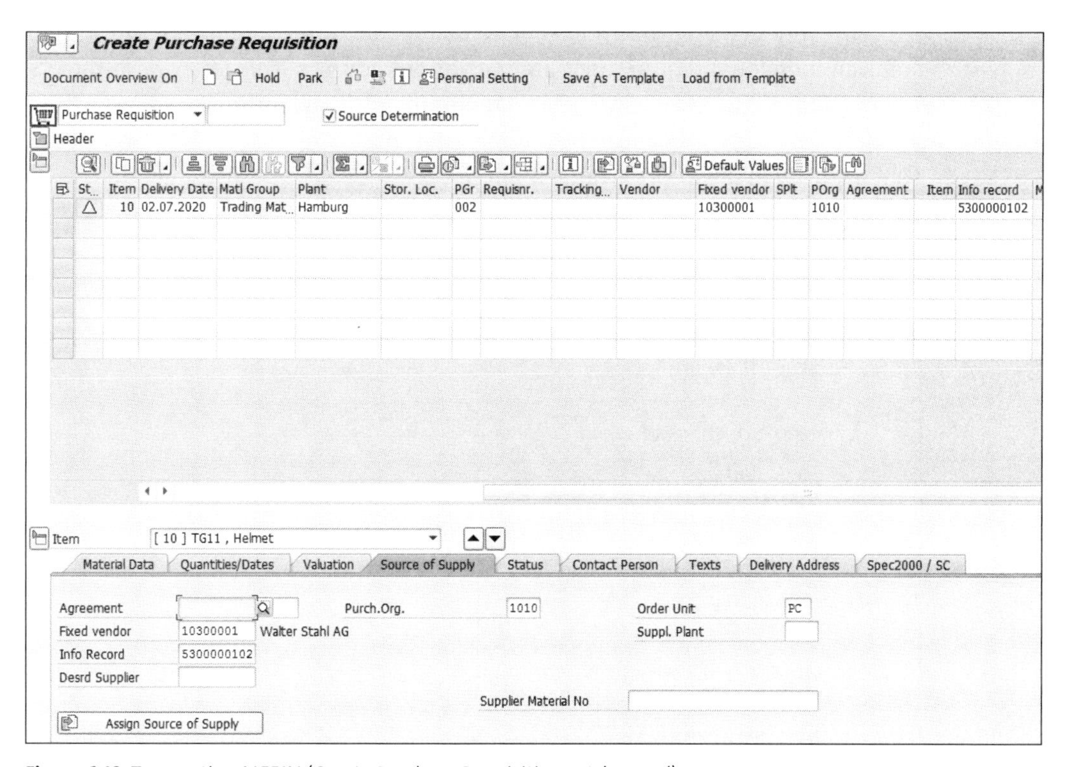

Figure 6.18 Transaction ME51N (Create Purchase Requisition—Advanced)

After the purchase requisition has been saved, the allocated quantity is updated on the quota arrangement item for vendor 10300001, as shown in Figure 6.19.

Material: Helmet (TG11)
Plant: Hamburg (1010)
Valid From: 01.06.2020
Valid To: 31.12.2020
Minimum Splitting Quantity: 0,000
Base Unit of Measure: PC

Quota Arrangement Items

Copy from Another Quota Arrangement

Quota Arrangement Item	Type	Special Procurement	Vendor	Procurement Plant	Quota	Percentage	Allocated Quantity	Base Quantity	Priority
1	External procurement (F)	Standard external procurement (0)	Walter Stahl AG (10300001)		70	70,0	290,000	0,000	0
2	External procurement (F)	Standard external procurement (0)	Johan Kraft GmbH Inlandslieferant DE 2 (10300002)		30	30,0	330,000	0,000	0

Figure 6.19 Allocated Quantity Updated in the Manage Quota Arrangements App

Now what is this base quantity that appears in the formula? The base quantity is maintained manually and is used to influence the allocation. You typically use the quota base quantity when a new source of supply is added to an existing quota arrangement or to correct an imbalance between quota and quota-allocated quantity.

Quota Arrangements in MRP

In the planning run, source determination with quota arrangement either runs in the same way as when you create a purchase requisition manually, or the system splits an individual requirement among various sources. This latter procedure is called *splitting*. You control whether a requirement is split or not by means of the assigned lot-sizing procedure in the material master record, and the minimum split quantity in the quota arrangement header.

You must set the indicator for splitting in Customizing for the lot-sizing procedure. Figure 6.20 shows the customizing path.

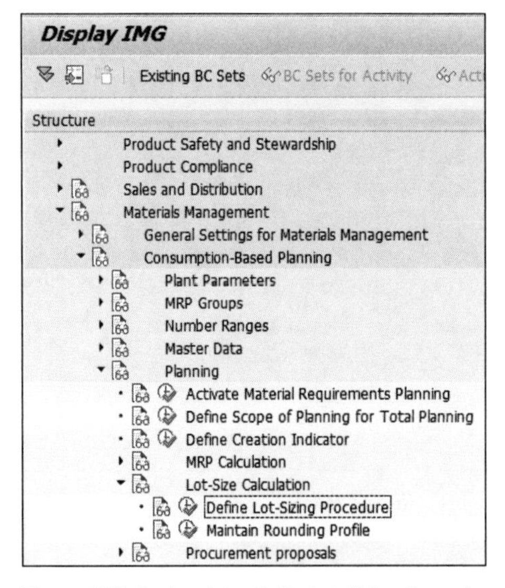

Figure 6.20 Customizing Path: Lot-Sizing Procedures

In the standard SAP S/4HANA system, there is the procedure ES (lot-sizing procedure with splitting) where the splitting indicator is set. This procedure is shown in Figure 6.21 and Figure 6.22.

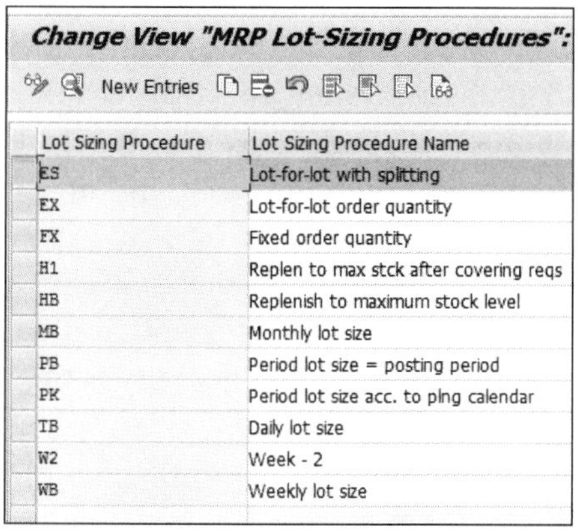

Change View "MRP Lot-Sizing Procedures":

New Entries

Lot Sizing Procedure	Lot Sizing Procedure Name
ES	Lot-for-lot with splitting
EX	Lot-for-lot order quantity
FX	Fixed order quantity
H1	Replen to max stck after covering reqs
HB	Replenish to maximum stock level
MB	Monthly lot size
PB	Period lot size = posting period
PK	Period lot size acc. to plng calendar
TB	Daily lot size
W2	Week - 2
WB	Weekly lot size

Figure 6.21 Customizing: Lot-Sizing Procedure ES

Change View "MRP Lot-Sizing Procedures": Details

New Entries

Lot Sizing Proc ES Lot-for-lot with splitting
☐ Last lot exact

Horizon lot-for-lot ord qty

Lot Size in Short-Term Horizon

Lot-sizing procedure	S Static lot-sizing procedure	No. of Periods	
Lot-size indicator	E Lot-for-lot order quantity		
Max. Stock Level			
Scheduling	Requirements Date = Availability Date		
Date Interpretation			
Overlapping		☑ Splitting quota	

End of Short-Term/Start of Long-Term Horizon

PerInd: LT Lot Size	Initial value	Number of periods	

Lot Size in Long-Term Horizon

Long-Term LSP		No. of periods	
Lot-Size Indicator			
Scheduling	Requirements Date = Availability Date	☐ Check Min. Loz Size	
Date Interpretation		☐ Check Max. Lot Size	
Overlapping		☐ Splitting quota	

Lot Size: MtO ☐ Lot-for-lot order quantity ☐ Underdel. Tol.

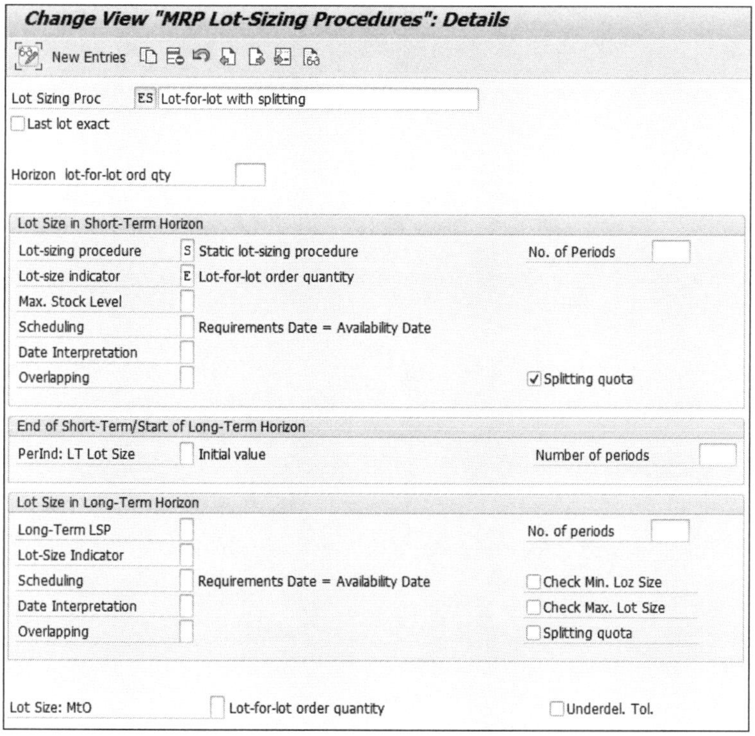

Figure 6.22 Customizing: Splitting Indicator for Lot-Sizing Procedure ES

To prevent the system from splitting a requirement that is too small, you can specify a minimum quantity for the split in the quota header (see Figure 6.16). The *minimum splitting quantity* defines the lowest quantity that a lot must have in order to be split between several sources during the planning run.

As shown in Figure 6.23, a quota arrangement contains also the following additional fields whose values are only of relevance for MRP:

- The **Maximum Lot Size** value is the maximum quantity that a procurement proposal assigned to this source of supply may contain. If the requirements quantity that should be allocated to this source of supply according to the quota arrangement exceeds the maximum lot size, the system splits the requirements quantity over several procurement proposals, each with a quantity up to the maximum lot size. Accordingly, several purchase requisitions can be assigned to a vendor, each with a quantity up to the maximum lot size. To prevent this, you can select the **Only Once** checkbox. If you set this indicator, a source of supply cannot be assigned to more than one procurement proposal with a quantity up to the maximum lot size. The remaining quantity is then assigned to the remaining sources of supply.
- The **Minimum Lot Size** value is the minimum quantity that a procurement proposal assigned to the source of supply may contain. If the actual allocated requirement quantity is smaller than the minimum lot size, the procurement proposal is increased to the minimum lot size.

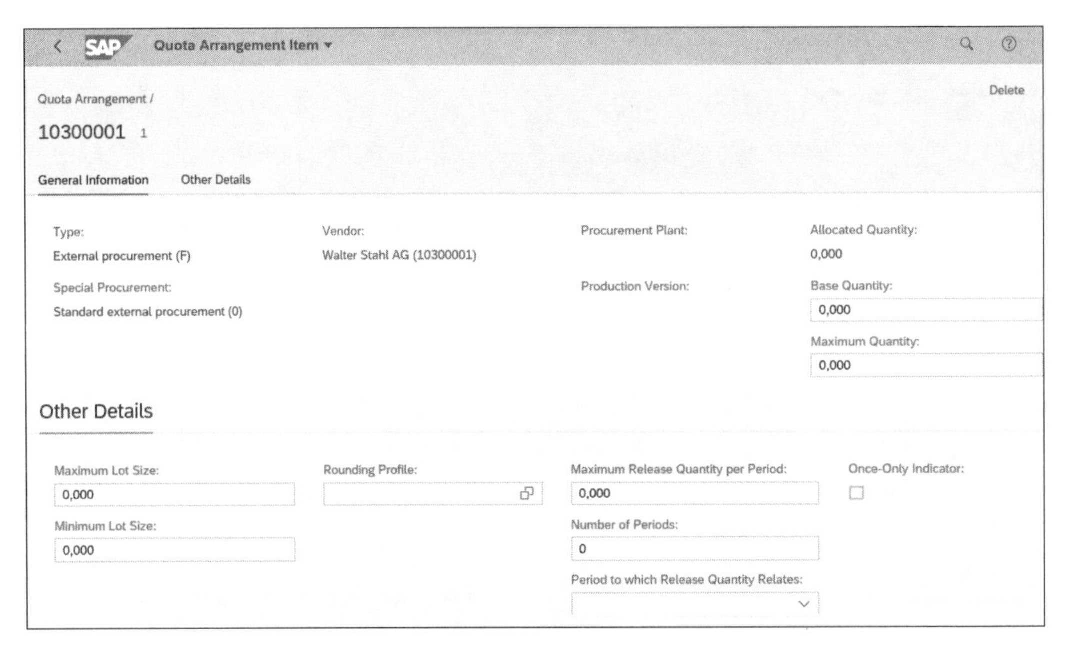

Figure 6.23 Quota Item in the Manage Quota Arrangements App

Further Features of the Quota Arrangement

A quota arrangement contains the following other important features:

- The **Maximum Quantity** value is considered both when manually creating purchase requisitions and when automatically generating procurement proposals in MRP. The maximum quantity serves as an upper limit for the quantity assigned to a source of supply in the validity period of the quota arrangement.

A quota arrangement item is no longer considered in source determination if the quantity already assigned is equal to the maximum quantity or would become higher as a result of the new allocation.

- You can also define a **Maximum Allocation Quantities per Period** value. The maximum allocation quantity per period is often used in conjunction with a priority. For example, you prefer supplier 10300001 and primarily want to order from them. However, their maximum delivery capacity is 1,000 pieces per month. You therefore would like to order from them until their maximum quantity per month is reached, then order from vendor 10300002. To be supported by the system, you can create a quota arrangement as follows:
  - You create two quota arrangement items for each material concerned in the respective plant: one for vendor 10300001, one for vendor 10300002, as shown in Figure 6.24.
  - Instead of assigning quotas, you assign priorities: 1 for vendor 10300001, 2 for vendor 10300002.
  - You define a maximum quantity per month for vendor 10300001, as shown in Figure 6.25.

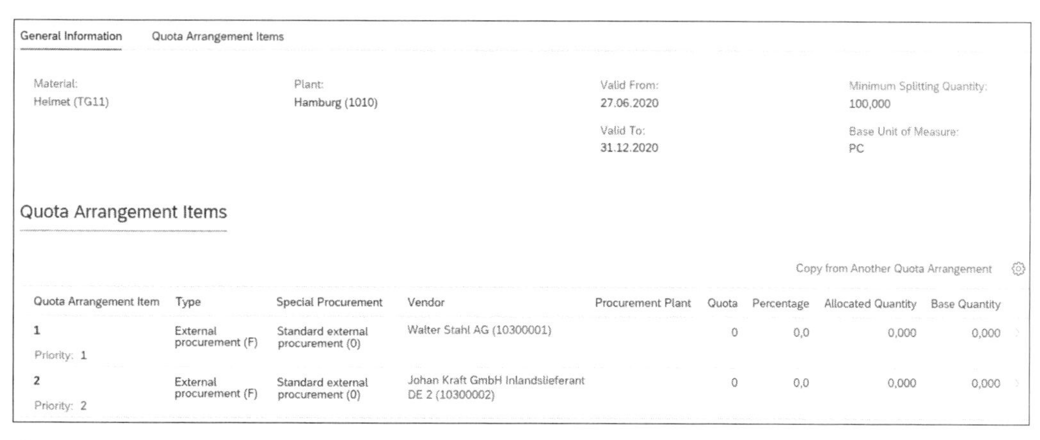

Figure 6.24 Example of Priorities in the Manage Quota Arrangements App

Figure 6.25 Example of Priorities and Maximum Quantity per Month in the Manage Quota Arrangements App

So far, we have focused on how the system can support source assignment when creating a purchase requisition (manually or via MRP). However, the assignment of purchase requisitions to sources of supply can take place at a later point in time.

Collective Assigning of Purchase Requisitions

As consultants, we would probably advise our customers to maintain their master data, including source lists and quota arrangements, so that the assignment of purchase requisitions to sources of supply takes place directly when they are created. This saves time, avoids additional manual steps, and is a good prerequisite for the subsequent automatic conversion of requisitions into purchase orders.

Nevertheless, it can happen that the assignment of purchase requisitions to sources of supply is to be carried out in a second separate step. Now let's look at how purchasers can carry out this task in the most efficient way.

You can process requisition items that were not assigned to a source of supply at the time they were created collectively and assign them to sources.

Use the Assign and Process Purchase Requisitions app (Transaction ME57) to list, process, assign, group, and then generate purchase orders. To assign non-assigned requisitions in collective processing, select them and choose the **Assign Source of Supply** button shown in Figure 6.26.

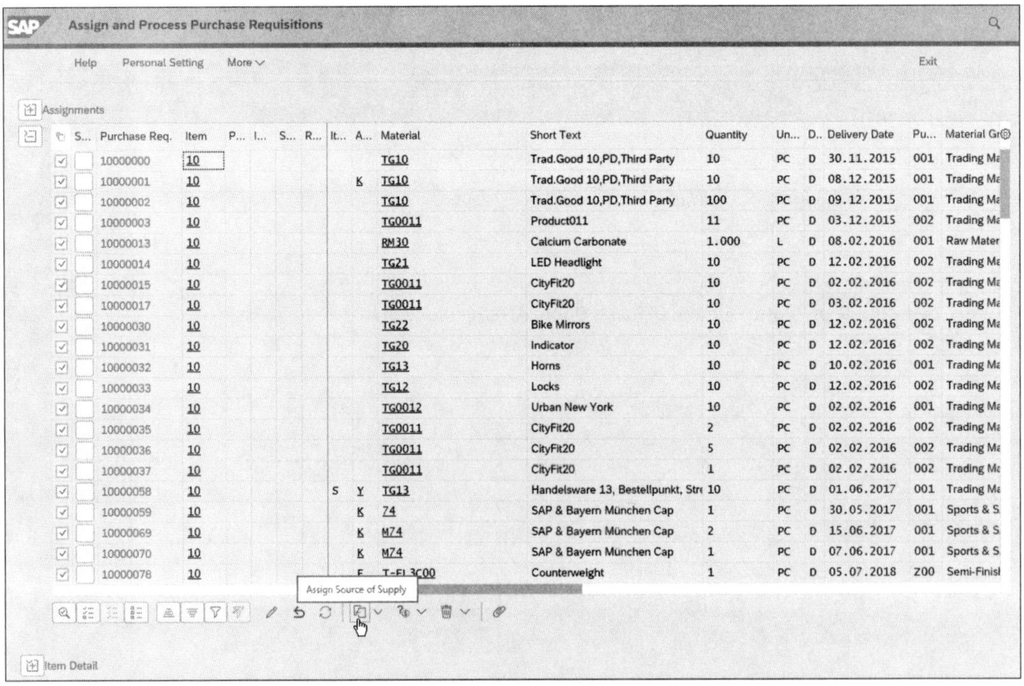

S...	Purchase Req.	Item	P...	I...	S...	R...	It...	A...	Material	Short Text	Quantity	Un...	D..	Delivery Date	Pu...	Material Gr
	10000000	10							TG10	Trad.Good 10,PD,Third Party	10	PC	D	30.11.2015	001	Trading Ma
	10000001	10						K	TG10	Trad.Good 10,PD,Third Party	10	PC	D	08.12.2015	001	Trading Ma
	10000002	10							TG10	Trad.Good 10,PD,Third Party	100	PC	D	09.12.2015	001	Trading Ma
	10000003	10							TG0011	Product011	11	PC	D	03.12.2015	002	Trading Ma
	10000013	10							RM30	Calcium Carbonate	1.000	L	D	08.02.2016	001	Raw Mater
	10000014	10							TG21	LED Headlight	10	PC	D	12.02.2016	002	Trading Ma
	10000015	10							TG0011	CityFit20	10	PC	D	02.02.2016	002	Trading Ma
	10000017	10							TG0011	CityFit20	10	PC	D	03.02.2016	002	Trading Ma
	10000030	10							TG22	Bike Mirrors	10	PC	D	12.02.2016	002	Trading Ma
	10000031	10							TG20	Indicator	10	PC	D	12.02.2016	002	Trading Ma
	10000032	10							TG13	Horns	10	PC	D	10.02.2016	001	Trading Ma
	10000033	10							TG12	Locks	10	PC	D	12.02.2016	002	Trading Ma
	10000034	10							TG0012	Urban New York	10	PC	D	02.02.2016	001	Trading Ma
	10000035	10							TG0011	CityFit20	2	PC	D	02.02.2016	002	Trading Ma
	10000036	10							TG0011	CityFit20	5	PC	D	02.02.2016	002	Trading Ma
	10000037	10							TG0011	CityFit20	1	PC	D	02.02.2016	002	Trading Ma
	10000058	10					S	Y	TG13	Handelsware 13, Bestellpunkt, Str	10	PC	D	01.06.2017	001	Trading Ma
	10000059	10						K	74	SAP & Bayern München Cap	1	PC	D	30.05.2017	001	Sports & S
	10000069	10						K	M74	SAP & Bayern München Cap	2	PC	D	15.06.2017	001	Sports & S
	10000070	10						K	M74	SAP & Bayern München Cap	1	PC	D	07.06.2017	001	Sports & S
	10000078	10						F	T-FL3C00	Counterweight	1	PC	D	05.07.2018	Z00	Semi-Finisl

Figure 6.26 The Assign and Process Purchase Requisitions App (Transaction ME57)

The system searches for suitable sources of supply according to the logic previously learned for source determination outside MRP. If the system cannot determine a unique source from several existing sources, it offers the user a selection option. You can manually assign requisitions for which no source was found to a vendor.

The assigned and saved purchase requisitions are separated into bundles for each source of supply, as shown in Figure 6.27. Each bundle of requisitions sorted by source of supply can then be selected for generating purchase orders.

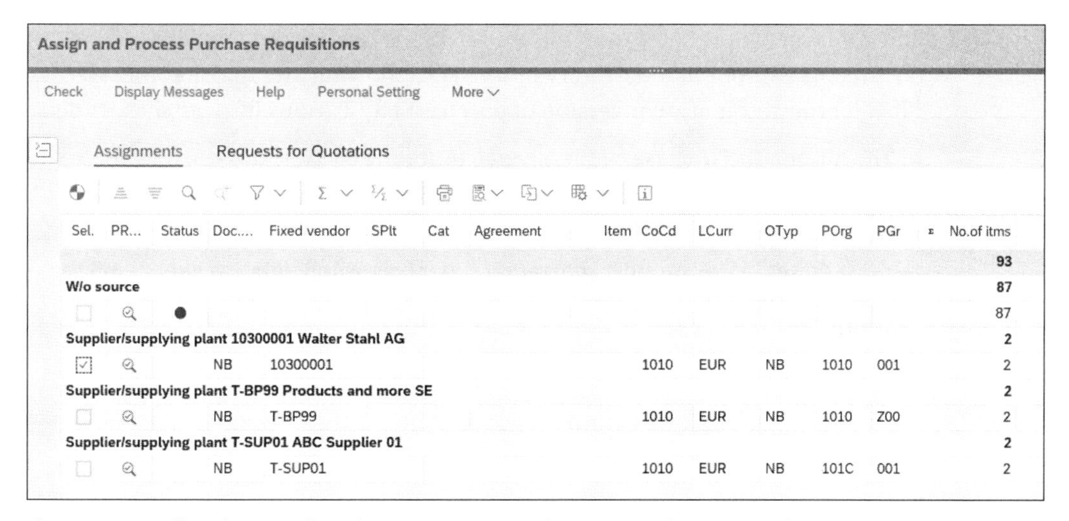

Figure 6.27 Bundles of Assigned Purchase Requisitions in the Assign and Process Purchase Requisitions App (Transaction ME57)

Note

You can call the Assign and Process Purchase Requisitions app on the SAP Fiori launchpad or choose Transaction ME57 in SAP GUI. The Assign Source of Supply to Purchase Requisitions app (Transaction ME56) is also available but only supports assignment, not conversion. There is currently no SAP Fiori app for collective automatic assignment.

Converting Purchase Requisitions into Purchase Orders

Purchase requisitions must be converted into purchase orders. In the process, the data from the purchase requisition is copied into the purchase order: material, quantity, delivery date, and the like, along with source of supply, if available. Technically speaking, an assigned purchase requisition can be converted automatically into a purchase order without manual intervention, provided it contains all the necessary data. The system allows this automatic conversion. We will now turn to examine the prerequisites for automatic conversion, and we will also look at other possible options for converting purchase requisitions into purchase orders.

Mass Conversion of Purchase Requisitions

To simplify and streamline the procurement process, the automatic purchase order generation from purchase requisitions is available. As an alternative to the automatic conversion of purchase requisitions, you can first list and group the purchase requisitions and then convert them in bundles into purchase orders.

Automatic Purchase Order Generation from Purchase Requisitions

Automatic processing is recommended if you have well-maintained master data and if it is highly probable that most requisitions can be converted into follow-on documents without manual intervention. The following prerequisites must be fulfilled for the automatic conversion of purchase requisitions into purchase orders:

- The **Automatic Purchase Order Allowed (Autom. PO)** indicator must be set in the material master record (plant-specific purchasing data), as shown in Figure 6.28.
- The **Automatic Generation of Purchase Order Allowed (Automatic PO)** indicator must be set in the supplier master record (purchasing data), as shown in Figure 6.29.
- A valid source of supply must be assigned to the purchase requisition item.
- During conversion, no additional entries may be necessary, such as the tax code for automatic goods receipt settlement.

To create purchase orders automatically, the Automatic Creation of Purchase Orders from Requisitions app (Transaction ME59N, shown in Figure 6.30) is available. You can start it online or as a background job. For example, you can schedule it as a job to start immediately after the MRP run.

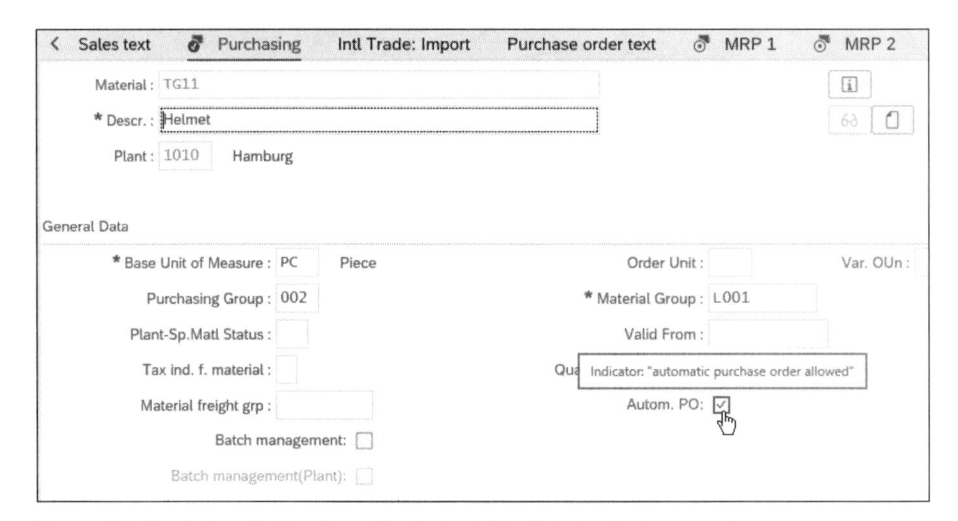

Figure 6.28 The Automatic PO Allowed Indicator in the Material Master Record

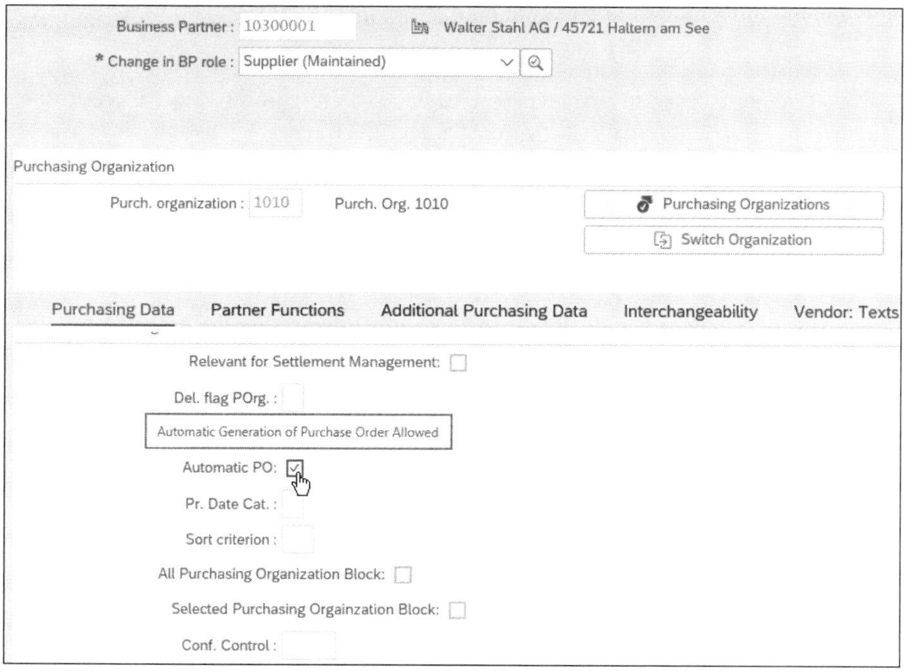

Figure 6.29 The Automatic Generation of PO Allowed Indicator in the Business Partner (Supplier) Master Record

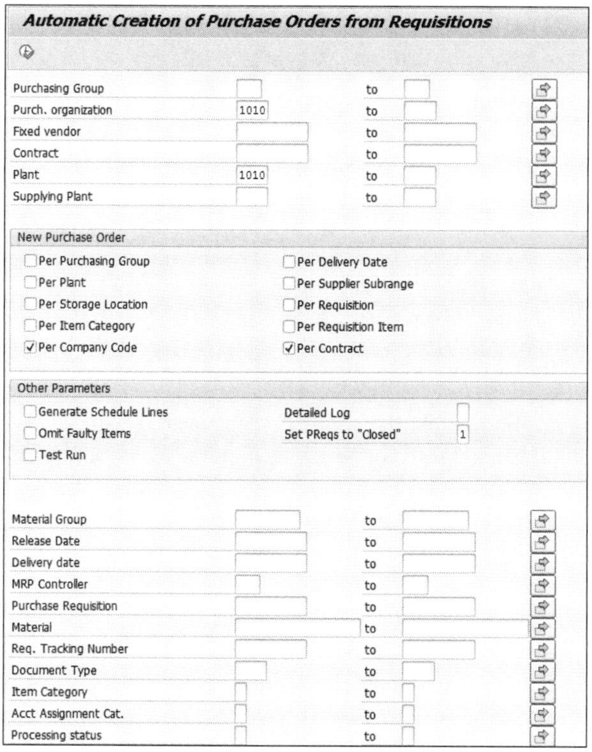

Figure 6.30 Selection Screen for the Automatic Creation of Purchase Orders from Requisitions App (Transaction ME59N)

Note

The Automatic Creation of Purchase Orders from Purchase Requisitions app (Transaction ME59N) cannot be used to create delivery schedules from purchase requisitions. These can only be created automatically by the MRP run. The **Generate Schedule Lines** checkbox can be confusing. It merely refers to the possibility of creating delivery schedule lines within one purchase order item instead of separate purchase order items in the case of purchase requisitions for the same material with different delivery dates.

Conversion of Purchase Requisitions into Purchase Orders Using the Assignment List

Before you can perform a mass conversion of purchase requisitions to purchase orders or delivery schedules with the Create Purchase Order via Purchase Requisition Assignment List app (Transaction ME58) or the Assign and Process Purchase Requisitions app (Transaction ME57), you must already have assigned supply sources to the requisition items to be converted.

The system groups together assigned requisitions for each source and displays a list showing the number of requisition items assigned to each vendor (assignment list). You can convert all requisition items assigned to a vendor without reference to a scheduling agreement into purchase orders in one step. Items for which a scheduling agreement has been assigned form a separate bundle and must be converted separately.

Note

Instead of converting requisitions via the assignment list, you can configure the document overview of the Create Purchase Order—Advanced app (Transaction ME21N) so that requisition items are displayed as a worklist. The document overview represents an alternative to the assignment list if you select released, already-assigned, and still-open requisition items for your area of responsibility (e.g., your purchasing group).

Note

As an alternative to the Assign and Process Purchase Requisitions app (Transaction ME57), the Manage Purchase Requisitions app (Transaction F1048) is also suitable for changing assigning and converting purchase requisitions.

First, you select the purchase requisition items that you want to convert. For this, you enter the relevant values in the filter bar and start the selection. In the search results, you can see whether sources for an item are available. After you assign a source of supply, select the purchase requisition items and choose **Create Purchase Order**. It is possible to convert multiple items in one step. Then, the system bundles requisition items with the same assigned supplier to create a purchase order with several items. If no source of supply is available, you can manually assign a supplier (for this, choose **Edit**) or start a request for quotation.

Note that you can only assign requisitions to one source individually, not in collective processing.

Individual Conversion of Purchase Requisitions

While the automatic or at least the mass conversion of purchase requisitions into purchase orders is the most effective way of creating purchase orders, there is also the possibility of individual conversion.

While checking the planning results, the purchaser/planner can directly convert individual purchase requisitions into purchase orders, for example with the Monitor Material Coverage—Net and Individual Segments app (Transaction F2101).

Important Terminology

The following important terminology was used in this chapter:

- **Source determination**
  In automatic source determination, the system searches for a suitable source of supply according to a certain logic, which we have described in this chapter.
- **Source list**
  A source list contains a list of possible sources of supply for a certain material and plant. The source list has various functions; it is used for source determination but can also be used to block a specific source temporarily. If the source list is set to mandatory, then it can be considered a form of supply source release.
- **Quota arrangement**
  Quota arrangements are used if you want to divide your requirements between several suppliers over a period of time according to fixed shares. Quota arrangements are used in automatic source determination for both manual transactions and MRP. With MRP, it is even possible to split a single requirement between several suppliers.
- **Splitting indicator**
  If you work with quota arrangements and want MRP to split individual requirements between several suppliers, use a lot-sizing procedure with the splitting indicator for the affected materials.
- **Assigned purchase requisitions**
  An assigned purchase requisition is a purchase requisition that contains a source of supply. This assignment can be the result of a manual or automatic process and is a prerequisite for the automatic conversion of purchase requisitions into purchase orders.

Practice Questions

These questions will help you evaluate your understanding of the topics covered in this chapter. They are similar in nature to those on the certification examination. Although none of these questions will be found in the exam itself, they will allow you to review your knowledge of the subject.

Select the correct answers, and then check the completeness of your answers in the next section. Remember that on the exam, you must select all correct answers—and only correct answers—to receive credit for the question.

1. You create a purchase requisition manually using source determination. In which of the following situations can the system determine a unique source of supply? (There are three correct answers.)

☐ **A.** There is only one valid source of supply for the requisition item. Neither quotation nor source list are in use.

☐ **B.** There are several valid sources of supply for the requisition item, one of which is a scheduling agreement. Neither quotation nor source list are in use.

☐ **C.** There are several valid sources of supply for the requisition item, one of which is marked in the source list as a fixed source. Quotation is not in use.

☐ **D.** There are several valid info records for the requisition item, one of which has the **Automatic Sourcing** indicator activated. Neither the quotation nor the source list is in use.

2. True or false: The source list is used exclusively for automatic source determination.

☐ **A.** True

☐ **B.** False

3. True or false: The system is only able to determine a source of supply—both in MRP and when creating a purchase requisition manually—if a quota arrangement or source list exists for the requested material.

☐ **A.** True

☐ **B.** False

4. True or false: In the source list of a material, an info record is defined for a certain period as the single source relevant to materials planning. Consequently, this info record is assigned to every MRP-generated purchase requisition whose requirement date lies within this period, even if a scheduling agreement also exists for this material. We assume that no quota arrangement exists.

☐ **A.** True

☐ **B.** False

5. In the system, you store purchase prices from various possible vendors for your purchasing parts. However, sources of supply are subject to an approval process, and purchasers may only procure from authorized sources. Which tool can you use to manage authorized sources of supply?

☐ **A.** Quota arrangement

☐ **B.** Release creation profile

☐ **C.** Source list

☐ **D.** Calculation schema

6. Which of the following features belong to the possible source list functions? (There are four correct answers.)

☐ **A.** Block a source of supply

☐ **B.** Override source determination during MRP

☐ **C.** Allow sources of supply

☐ **D.** Exclude materials from a contract item for a material group

☐ **E.** Override quota arrangement

7. You manually create a purchase requisition with source determination. In what logical sequence does the system search for a source of supply?

☐ **A.** 1. Quota arrangement; 2. Info records; 3. Source list; 4. Outline agreements items

☐ **B.** 1. Quota arrangement; 2. Source list; 3. Outline agreement items; 3. Info records

☐ **C.** 1. Source list; 2. Info records; 3. Outline agreement items; 3. Quota arrangement

☐ **D.** 1. Info records; 2. Quota arrangement; 3. Outline agreement items; 3. Source list

8. True or false: If you have created a quota arrangement for a certain material in a plant, each requirement for this material is distributed among the various sources of supply according to the quota arrangement.

☐ **A.** True

☐ **B.** False

9. You use quota arrangements for some materials. Which field in the material master record controls whether an individual requirement is to be split up among the various sources of quota arrangement?

☐ **A.** The MRP type

☐ **B.** The special procurement indicator

☐ **C.** The lot-sizing procedure

☐ **D.** The purchasing value key

10. You use quota arrangements and splitting for the determination of sources of supply during MRP. How can you prevent the smallest requirement quantities from being split between the different vendors in the quota arrangement?

☐ **A.** Use a rounding profile in the quota arrangement

☐ **B.** Use the minimum lot size in the quota arrangement

☐ **C.** Use the once-only indicator in the quota arrangement

☐ **D.** Use the minimum splitting quantity in the quota arrangement

11. You can produce 10,000 pieces of a material per month yourself. You want to procure the requirements that exceed the quantity of 10,000 pieces within one month from an external vendor. Can a quota arrangement be used to support you in this scenario?

☐ **A.** Yes: Create a quota arrangement with one item for in-house production and one item for the external supplier. Enter a maximum lot size of 10,000 and a priority of 1 for the in-house production item.

☐ **B.** Yes: Create a quota arrangement with one item for in-house production and one item for the external supplier. Enter a maximum release quantity of 10,000 per month and a priority of 1 for the in-house production item.

☐ **C.** Yes: Create a quota arrangement with one item for in-house production and one item for the external supplier. Enter a base quantity of 10,000 and a priority of 1 for the in-house production item.

☐ **D.** No: A quota arrangement with one item for in-house production and one item for the external supplier is not allowed.

12. True or false: Using the quota arrangement, you can, for example, assign half of the requirement quantities of a material externally and procure half by means of stock transfers from another plant.

☐ **A.** True

☐ **B.** False

13. Which of the following fields do you use to manually control the quota arrangement, when, for example, a new vendor is included in the quota arrangement in the middle of the arrangement period?

☐ **A.** Base quantity

☐ **B.** Allocated quantity

☐ **C.** Quoting rate

☐ **D.** Priority

14. What requirements must be met for a purchase requisition to be automatically converted into a purchase order? (There are three correct answers.)

- ☐ **A.** The indicator for automatic purchase orders must be set in the material master.
- ☐ **B.** The indicator for automatic purchase orders must be set in the source list.
- ☐ **C.** The purchase requisition must be assigned to a source of supply.
- ☐ **D.** The purchase requisition must be assigned to a contract item.
- ☐ **E.** The indicator for automatic purchase orders must be set in the supplier master.

15. Which of the following options does the Automatic Creation of Purchase Orders from Purchase Requisitions app (Transaction ME59N) offer? (There are two correct answers.)

- ☐ **A.** You can schedule it as a background job.
- ☐ **B.** You can use it to create scheduling agreement schedule lines.
- ☐ **C.** You can use it to create contract release orders.
- ☐ **D.** You can use it to assign sources to purchase requisitions.

16. True or false: By means of the assignment list, you can convert several requisition items from the same vendor into one purchase order.

- ☐ **A.** True
- ☐ **B.** False

17. True or false: Both an SAP Fiori app and an SAP GUI transaction are available for simultaneously converting several purchase requisitions into one purchase order.

- ☐ **A.** True
- ☐ **B.** False

18. You want to block a vendor for the procurement of certain materials for a certain period. Which of the following options are available to you?

- ☐ **A.** Set the blocking indicator in the supplier master record.
- ☐ **B.** Set the blocking indicator in the source list entry for the relevant materials.
- ☐ **C.** Use the material status in the master records of the relevant materials.
- ☐ **D.** Set the deletion flag in the info records of the relevant materials.

Practice Question Answers and Explanations

1. Correct answers: **A, B, C**

   You manually create a purchase requisition item and set the Automatic Source Determination indicator for the system to assign a source. If only one source of supply exists, the system can clearly assign this source. If several sources of supply exist, including a single outline agreement, the system assigns the outline agreement item. Otherwise, one source must be specified in the source list as a fixed source of supply so that the system can assign it as a unique source. If a valid quota arrangement exists, this has priority.

   The **Automatic Sourcing** indicator is only intended for source determination during MRP.

2. Correct answer: **B**

   False. The source list is used for automatic sourcing by declaring a source of supply as fixed or relevant for MRP, but the source list can also be used to block a source of supply for a certain period.

   In addition, the source list can be used to manage approved sources of supply. If the source list is made mandatory at the plant level or for certain materials, the permitted sources of supply must be entered in the source list.

3. Correct answer: **B**

   False. MRP considers valid sources of supply in the following sequence: scheduling agreement, contract, and info records with the **Automatic Sourcing** indicator. You can overrule this source determination logic by using source lists or quota arrangements. Quota arrangements have higher priority than source lists. Read Chapter 8 for more details about source determination during MRP.

   When you create a purchase requisition for a material manually, the system can automatically assign a source if there is only this one source or if this source is the only outline agreement for the material.

4. Correct answer: **A**

   True. Indeed, the source list can be used to override the standard logic of source determination during an MRP run, at least for a certain period.

5. Correct answer: **C**

   This is one of the source list's functions: if you make the use of source lists obligatory (per material or even at plant level), the source list corresponds to the list of authorized sources.

   As a reminder: the calculation schema controls the price structure, the release creation profile controls the creation of scheduling agreement releases, and the quota arrangement determines how the requirements of a material in a period are to be distributed among various vendors.

6. Correct answers: **A, B, C, D**

   Quota arrangement has the highest priority in source determination, both for MRP and outside MRP. The source list cannot therefore override it. All other items are possible functions of the source list.

7. Correct answer: **B**

   If you manually create a purchase requisition with source determination, the system first searches for a quota arrangement, then for a source list, then for outline agreement items, and finally for info records.

8. Correct answer: **B**

   False. The requirements can only be split up under certain conditions during the planning run. Otherwise, the requirement is allocated to the source that is next in line according to a certain formula. During the period of validity of the quota, the distribution of the total quantity among the various sources must be made in accordance with the quota arrangement.

9. Correct answer: **C**

   The lot-sizing procedure controls whether a requirement generated during MRP is split between the different sources of a quota arrangement. If you wish this splitting to take place, you must select a lot-sizing procedure where the splitting quota indicator is set in Customizing.

10. Correct answer: **D**

    You use the minimum splitting quantity to prevent requirement quantities that are smaller than this value from being split between the different vendors in the quota arrangement.

11. Correct answer: **B**

    A quota arrangement with one item for in-house production and one item for the external vendor is allowed and is suitable for this scenario. A maximum release quantity per month and priority 1 for the in-house production item would correspond to the scenario.

12. Correct answer: **A**

    True. Create a quota arrangement with items for standard external procurement and the corresponding external vendors, and an item for stock transfer and the corresponding supplying plant. Depending on the number of external vendors, you allocate quotas so that half of the requirement will be allocated to the other plant.

    Figure 6.31 shows an example where half of the requirements for material TG11 in plant 1010 are to be covered by stock transfer from plant 1020. The other half is shared equally between two vendors.

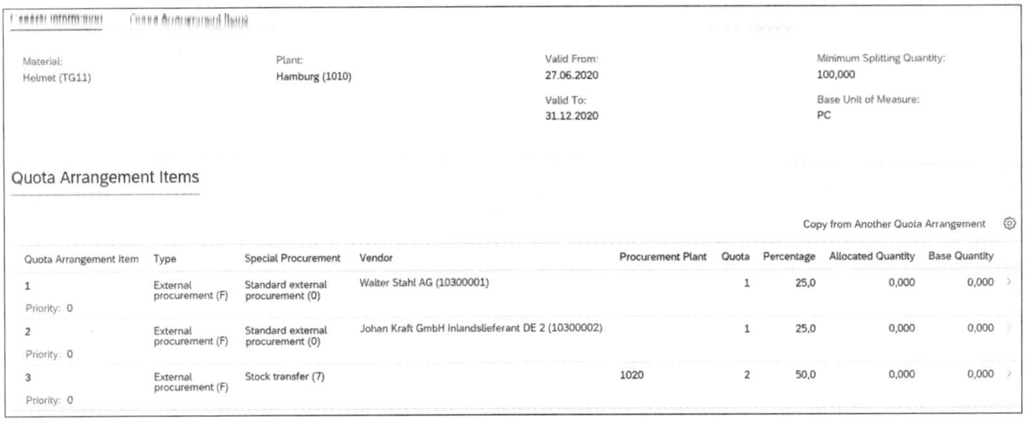

Material:
Helmet (TG11)

Plant:
Hamburg (1010)

Valid From:
27.06.2020

Valid To:
31.12.2020

Minimum Splitting Quantity:
100,000

Base Unit of Measure:
PC

Quota Arrangement Items

Copy from Another Quota Arrangement

Quota Arrangement Item	Type	Special Procurement	Vendor	Procurement Plant	Quota	Percentage	Allocated Quantity	Base Quantity
1 Priority: 0	External procurement (F)	Standard external procurement (0)	Walter Stahl AG (10300001)		1	25,0	0,000	0,000
2 Priority: 0	External procurement (F)	Standard external procurement (0)	Johan Kraft GmbH Inlandslieferant DE 2 (10300002)		1	25,0	0,000	0,000
3 Priority: 0	External procurement (F)	Stock transfer (7)		1020	2	50,0	0,000	0,000

Figure 6.31 Example of Quota Arrangement with Stock Transfer

13. Correct answer: **A**

    The base quantity is basically the manual part of the total allocated quantity. The allocated quantity of a source of supply in the quota arrangement is updated automatically if a purchase requisition is assigned to this source of supply or if a purchase order is created spontaneously without reference to a purchase requisition. The base quantity is maintained manually and is one way of bringing a new vendor, who is included in the quota arrangement in the middle of the period, up to the same level.

14. Correct answers: **A, C, E**

    For a purchase requisition to be converted automatically into a purchase order, the indicator for automatic purchase orders must be set in the material master and vendor master, and the purchase requisition must be assigned to a source of supply that need not necessarily be a contract.

15. Correct answers: **A, C**

    You can schedule the Automatic Creation of Purchase Orders from Purchase Requisitions app (Transaction ME59N) as a background job or start it online. You can use this app to convert assigned purchase requisitions into purchase orders. If a purchase requisition is assigned to a contract, a contract release order is generated from it. In contrast to a delivery schedule, a contract release is a purchase order.

16. Correct answer: **A**

    True. By means of the assignment list, you can convert several requisition items from the same vendor into one purchase order. One purchase order item is generated for each purchase requisition item.

    Note that with the Automatic Creation of Purchase Orders from Purchase Requisitions app (Transaction ME59N), if you select the **Generate Schedule Lines** checkbox, requisition items with the same material are combined to one purchase order item and one schedule line is generated for each purchase requisition.

17. Correct answer: **A**

    True. Transactions ME57 and ME58 (Assign and Process Purchase Requisitions and Create Purchase Order via Purchase Requisition Assignment List) in SAP GUI and the Manage Purchase Requisitions app (Transaction F1048) are suitable for converting bundled purchase requisitions. However, there is currently no SAP Fiori app available for mass assignment.

18. Correct answer: **B**

    If you set the blocking indicator in the vendor master, you block the vendor and are no longer allowed to order anything from him. If you wish to block this vendor for certain materials only, use the source list: set the Blocking indicator on the corresponding source list entry of the relevant materials. Caution: There must be a source list entry that does not refer to a scheduling agreement; otherwise, you only prohibit schedule lines from being generated for this scheduling agreement, but you can still order from the vendor. The source list also allows you to set a block for a certain period.

    You can use the material status to prevent materials from being procured externally but not specifically from a vendor.

    By setting the deletion indicator in the purchasing info record, you do not prevent a user from manually ordering this material from the vendor.

Takeaway

This chapter provided an overview of how you can help your customers optimize, automate, and streamline their procurement processes. Automatic source determination is an important aspect of this. At this point, it is important to understand the logic of the source determination process and to know what additional tools can be used. Automatic source determination is also an important prerequisite for a further automation step: the automatic creation of purchase orders from purchase requisitions.

Summary

In this chapter, we explored how the system can use existing sources of supply to assist the user in determining a suitable source when creating a purchase requisition. We concentrated on the creation of purchase requisitions outside the MRP run. In Chapter 8: Consumption-Based Planning, we will further explore the determination of sources of supply during the MRP run.

Here we have seen that both the source list and the quota arrangement can be used as additional instruments in source determination. We have also examined the other functions and possible uses of these objects. Furthermore, we discussed which options are available for converting requisitions into purchase orders and under what conditions this process step can be automated.

In Chapter 8, which is dedicated to consumption-based planning, we focus on an essential upstream step in the automation of the purchasing process: the MRP run enables the automatic creation of purchase requisitions.

Chapter 7
Special Functions in Purchasing

Techniques You'll Master

- Set up release strategies for purchase requisitions and purchasing documents.
- Configure the output of purchasing documents.
- Set up prerequisites for monitoring procurement processes.
- Remind outstanding deliveries.
- Remind outstanding order confirmations.
- Get an overview of new functions in SAP S/4HANA sourcing and procurement.

In the previous chapters, we gained an overview of the procurement process in SAP S/4HANA. We then covered the special features of procurement for inventory and for consumption. We proceeded to learn about the possible sources of supply, and in the previous chapter, we learned how to automate the procurement process.

This chapter covers functions of a general nature, such as the release of purchase requisitions and purchase orders, the output and transmission of purchasing documents to the supplier, and the monitoring of a procurement process. This chapter also provides a list of new features in SAP S/4HANA.

Real-World Scenario

Purchase orders, contracts, and other purchasing documents are not created for internal purposes but must be communicated to the suppliers. The medium used for the transmission of purchase orders often depends on the supplier: perhaps one supplier can only receive emails, another still works with fax, and others prefer a collaboration platform. But other criteria can also influence the method of transmission: an urgent order may need to be sent by email, while framework agreements can be sent in the mail. Sending messages is therefore a very customer-specific matter and accordingly very flexible in configuration. In addition to the classic and comprehensive method of condition technique, SAP S/4HANA offers a new form of output management that we would like to introduce here.

The situation is similar for the release of purchase requisitions and purchasing documents: every customer has their own requirements. For example, one company may want to convert purchase requisitions created by MRP automatically and without approval, while another company may want a multi-level approval process. Accordingly, the system must offer a high degree of flexibility. SAP S/4HANA offers both a classic, comprehensive procedure based on classification and a new, user-friendly procedure.

Objectives of This Portion of the Test

In this part of the certification exam, your knowledge of how to optimize the procurement process in SAP S/4HANA will be tested. For the certification exam, you must be able to do the following:

- Set up approval procedures for purchase requisitions and purchasing documents with both classic and new procedures.
- Set up output management for purchasing documents with both classic and new procedures.
- Know how to monitor a procurement process.
- Explain the new procurement features in SAP S/4HANA.

Key Concept Refresher

In this chapter, we will explain how you can set up output management and how you can use an electronic document release strategy to implement a company's signature regulations. Finally, we discuss some monitoring functions and point out SAP S/4HANA innovations in procurement.

Output Management

SAP has introduced a new output management solution that is optional for on-premise SAP S/4HANA and is officially referred to as *SAP S/4HANA output management* or *SAP S/4HANA output control*.

The entire output management scope based on condition technique already available in SAP ERP is still available in SAP S/4HANA, and currently there are no plans to deprecate this functionality. To avoid confusion, we call the "older" output management from SAP ERP times *message determination with condition technique*.

To activate and deactivate SAP S/4HANA output management, go to **IMG • Cross Application Components • Output Control • Manage Application Object Type Activation**, which is shown in Figure 7.1. The default settings are that the new output management is active in cloud installations and inactive in on-premise installations.

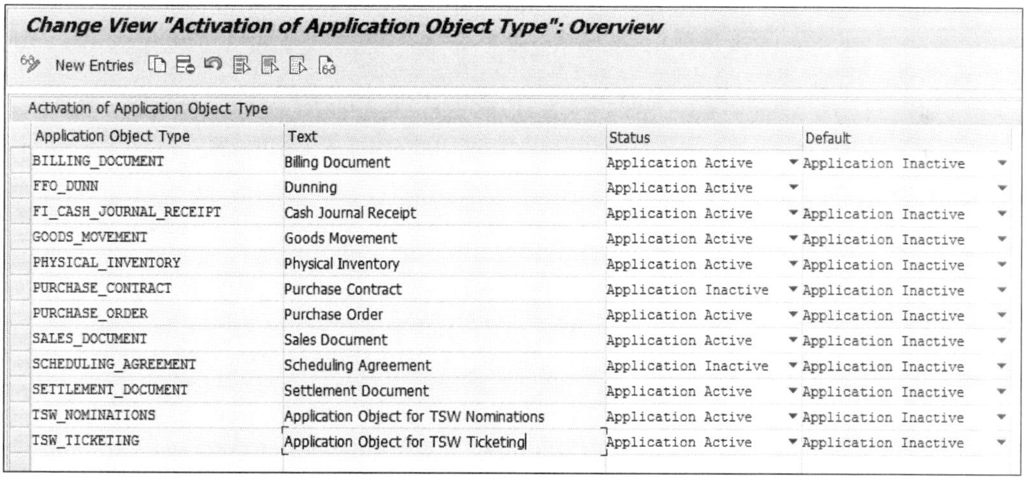

Change View "Activation of Application Object Type": Overview

New Entries

Activation of Application Object Type

Application Object Type	Text	Status	Default
BILLING_DOCUMENT	Billing Document	Application Active	Application Inactive
FFO_DUNN	Dunning	Application Active	
FI_CASH_JOURNAL_RECEIPT	Cash Journal Receipt	Application Active	Application Inactive
GOODS_MOVEMENT	Goods Movement	Application Active	Application Inactive
PHYSICAL_INVENTORY	Physical Inventory	Application Active	Application Inactive
PURCHASE_CONTRACT	Purchase Contract	Application Inactive	Application Inactive
PURCHASE_ORDER	Purchase Order	Application Active	Application Inactive
SALES_DOCUMENT	Sales Document	Application Active	Application Inactive
SCHEDULING_AGREEMENT	Scheduling Agreement	Application Inactive	Application Inactive
SETTLEMENT_DOCUMENT	Settlement Document	Application Active	Application Inactive
TSW_NOMINATIONS	Application Object for TSW Nominations	Application Active	Application Inactive
TSW_TICKETING	Application Object for TSW Ticketing	Application Active	Application Inactive

Figure 7.1 Customizing: Activation of the SAP S/4HANA Output Management

We will first explain the message determination with condition technique; then we will provide an overview of the SAP S/4HANA output control.

Message Determination with Condition Technique

If you use message determination to output purchasing documents, you use condition records to specify under what conditions, with what medium, and at what time a document can be output. You can specify these details for the message output individually for each print transaction (e.g., new creation, changes, or reminders).

During message determination, the system checks whether condition records exist for defined field combinations. If condition records exist, the system can generate and process (e.g., transmit electronically) one or more messages. Without a valid condition record, the system cannot automatically generate a message.

Tip

The setup of the message determination is quite extensive, but it is more difficult to explain than it truly is!

We recommend reproducing an example in the system by yourself. You are welcome to reproduce the example that we use in this chapter to explain the terms and logic. This practical exercise makes the theoretical explanations clearer, particularly for setting up the message determination.

We will explain message determination and the required configuration steps using an example. Imagine that you wish to set up message control depending on the purchasing organization, document type, and vendor:

- Purchasing organization 1010 is responsible for the procurement of raw materials for all plants in your company code 1010. Depending on how they are equipped, most of the suppliers receive the purchase orders with order type NB by email or by fax. Supplier 10300001, for example, receives emails, while supplier 10300002 receives faxes. However, if a supplier specifies neither an email address nor a fax, the orders for purchasing organization 1010 should be printed out.
- Purchasing organizations 1020 and 1030 are responsible for local purchasing in the respective plant 1020 and 1030—for example, purchasing of office supplies and small items, which are often requested via a blanket order. The suppliers receive the purchase orders by mail. However, if a supplier does not specify an email address, the orders for purchasing organization 1020 and 1030 should be printed out.

The system should first check whether a condition type exists for the combination purchasing organization-order type-supplier. If so, the data from this condition type is used for the order transmission. If not, the default entry for the purchasing organization should be used (printout). For this, you also need a condition record for the purchasing organization alone.

Here are the steps you need to go through:

1. Define a condition table.
2. Define the access sequence.

3. Define the output types.
4. Define the output procedure.
5. Create output records.

Now let's explore these steps in detail.

Define a Condition Table

Define your criteria for output determination using a *condition table*. Your criteria are the purchasing organization, document type, and vendor. These criteria must form a condition table. First, check whether there is a standard table that meets your criteria. If not, create a new one. In our example in Figure 7.2, the standard condition table 025 meets your requirements.

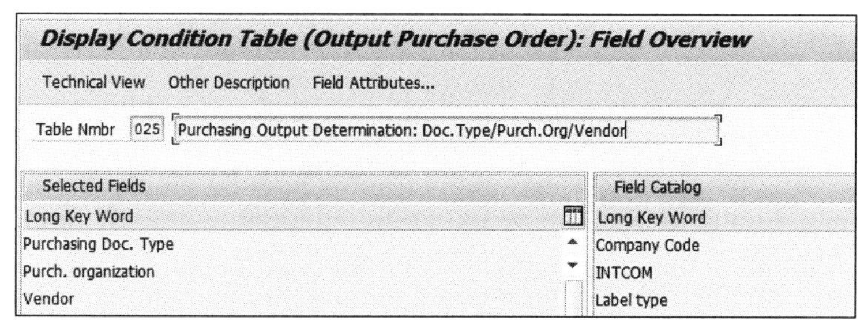

Figure 7.2 Condition Table 025

Do not forget that there should also be a condition table containing only the purchasing organization. In Figure 7.3, the standard condition table 029 meets your requirements.

Display Condition Table (Output Purchase Order): Field Overview
Technical View Other Description Field Attributes...
Table Nmbr 029 Purch.org.
Selected Fields
Long Key Word
Purch. organization
Field Catalog
Long Key Word
Company Code
INTCOM
Label type

Figure 7.3 Condition Table 029

Since message determination takes place at the header level, you can only include the fields from the document header in the condition tables. Message determination based on document item fields (e.g., the plant) is only possible indirectly in the standard system. Here *indirect* means that you can find business partners with certain partner functions at the header level, which are determined at the item level, depending on the plant or the subrange. If you have condition records that contain these partner functions, you can issue/transmit messages to the specific partners.

You can use condition tables to create message records in which you specify output details like medium, time of output, and number of outputs.

Define the Access Sequence

You can use the *access sequence* to define the order in which the system should search for a condition type. Figure 7.4 shows an access sequence ensuring that the system first searches for condition types for the combination *purchasing organization-document-type-supplier* and then simply searches via purchasing organization. In the access sequence, you place the condition tables in the desired order.

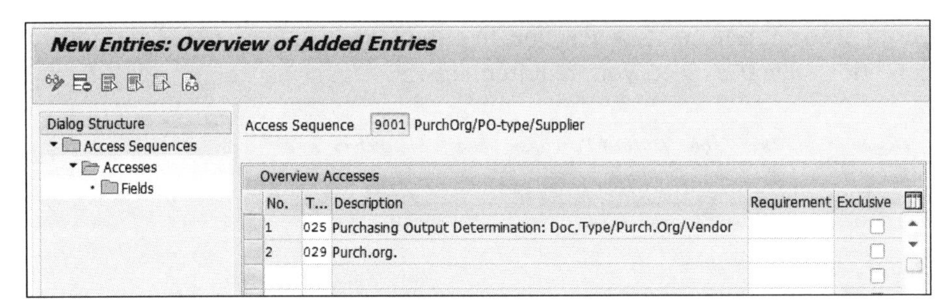

Figure 7.4 Customizing: Access Sequence

Define the Output Types

Now you must define which different types of messages are to be generated (e.g., for a purchase order) and under which conditions (e.g., purchase order output, order confirmation reminder, delivery reminder, etc.). Figure 7.5 and Figure 7.6 show where the *output types* are maintained in Customizing.

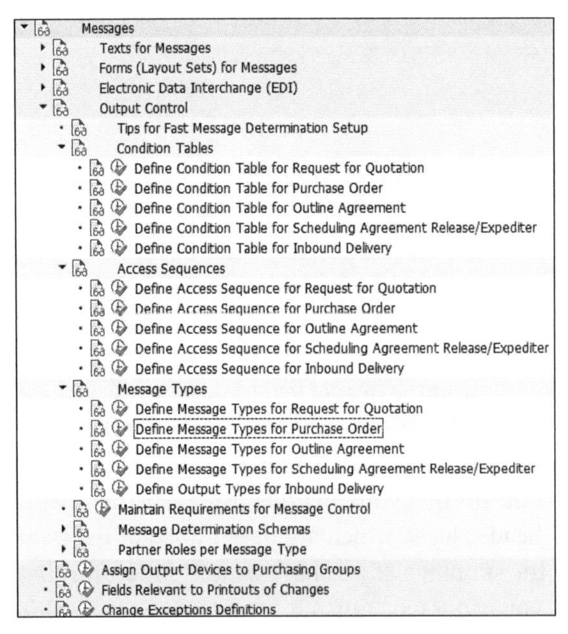

Figure 7.5 Customizing: Path to the Output Types

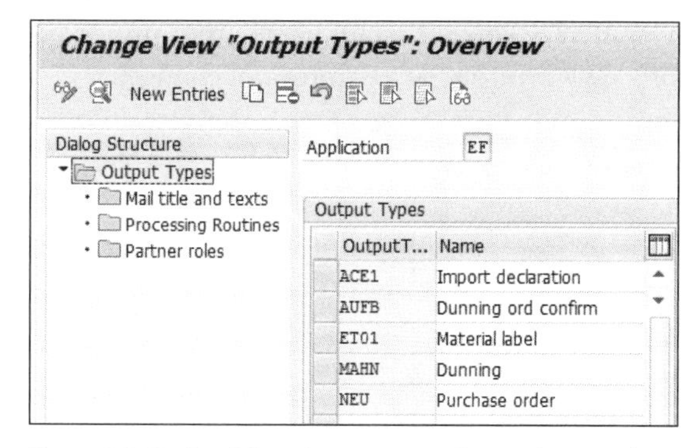

Figure 7.6 Customizing: Output Types for Purchase Orders

A message type or output type contains parameters that apply to all messages that the system assigns to it: for example, print program, form, and allowed partner functions. Examples of output types are purchase order printouts or reminders for confirmations or deliveries.

When maintaining the output types as shown in Figure 7.7, you can do the following:

- Assign access sequences.
- Define partner functions for different recipients (e.g., supplier, ordering address, and supplying plant).
- Assign processing programs and forms, depending on the output medium.
- Store default values for the messages.

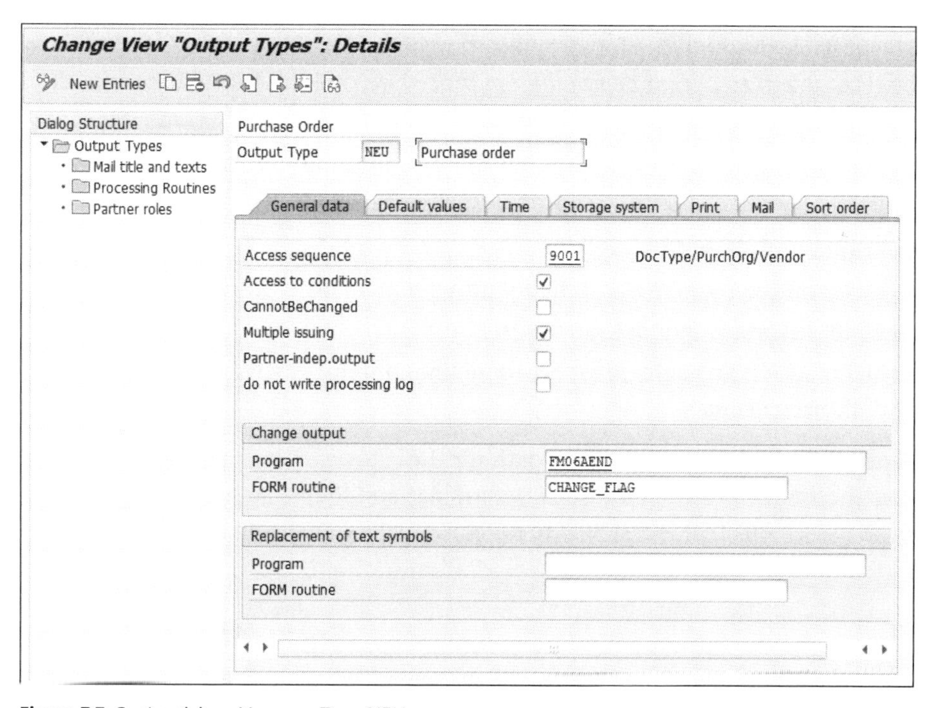

Figure 7.7 Customizing: Message Type NEU

In the fine tuned control shown in Figure 7.8, you define for which print processes the output type is intended (e.g., for a new printout or a printout of changes).

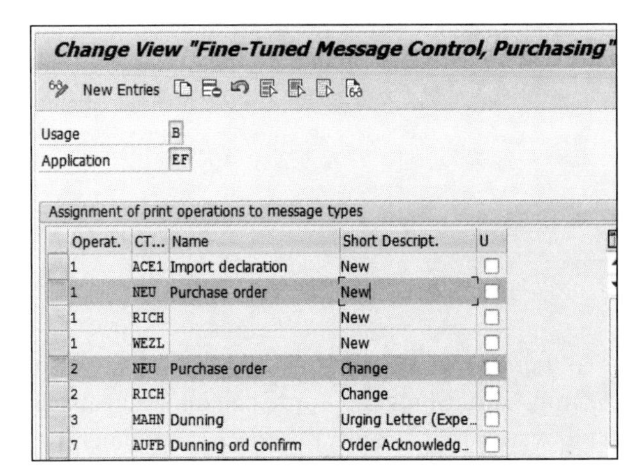

Figure 7.8 Customizing: Fined-Tuned Control for Message Types

The system only processes output types that are assigned to the output determination procedure assigned to the purchasing transaction. Standard message types are provided for each purchasing document. For example, NEU is a message type for the output of a new or changed purchase order (new printout and printout of changes).

Define the Output Procedure

The possible output types for an application object (e.g., for purchase order) must be recorded in an *output determination procedure*, and the output determination procedure is assigned to specific application objects.

In Customizing for output determination, the following so-called *applications* are available, to which you can assign an output determination procedure:

- EF for purchase orders
- EV for outline agreements (scheduling agreement and contract)
- EA for requests for quotation (RFQs)
- EL for scheduling agreement schedule lines, forecast delivery schedules, or JIT delivery schedules

Figure 7.9 shows the customizing path for the message determination schema. If you choose the **Maintain Message Determination Schema** activity, you get to the message schema that is assigned to purchase orders, as shown in Figure 7.10.

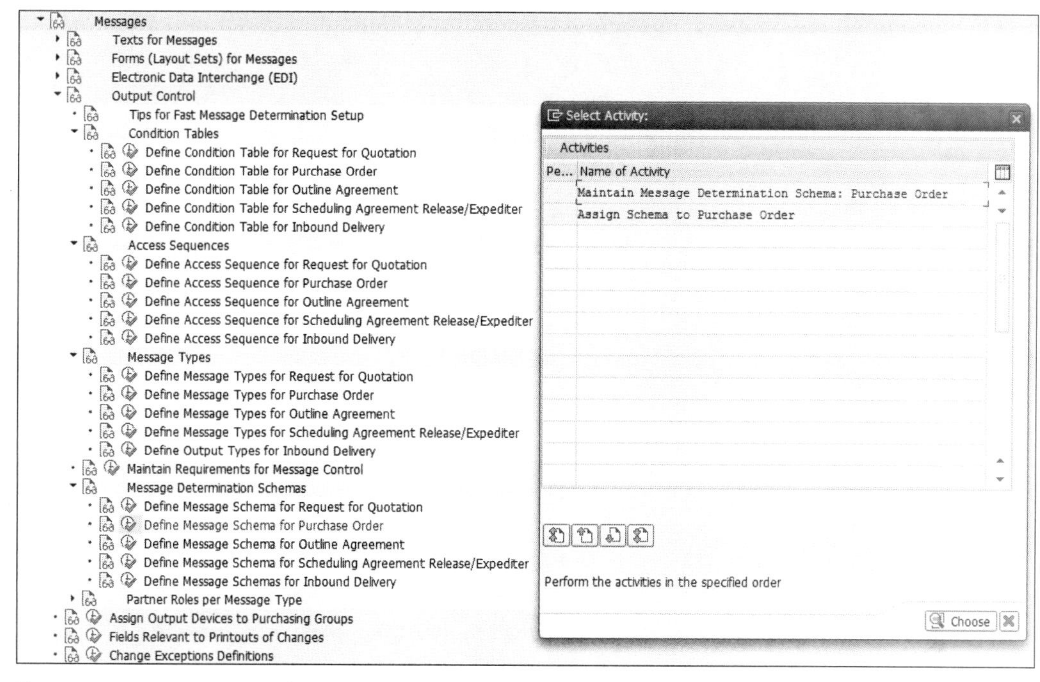

Figure 7.9 Customizing: Path for the Message Determination Schema

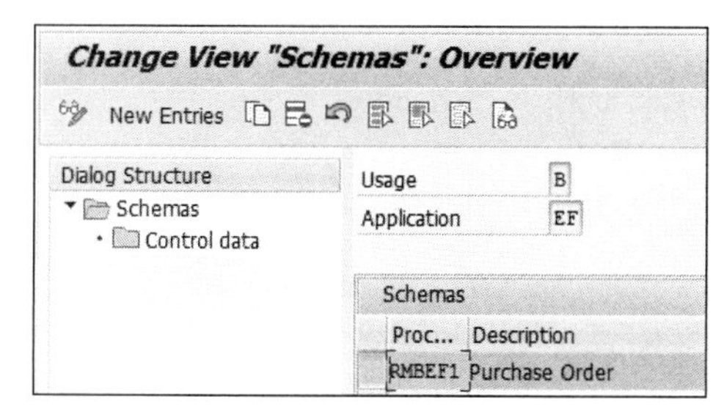

Figure 7.10 Customizing: Define Message Determination Schema

If you choose **Control Data**, note that the schema contains a list of message types, as shown in Figure 7.11. These message types should be determined under certain conditions. The prerequisites are formulated in the form of a small program, which can be found with a number in the **Requirements** column. For example, program (requirement) 101 contains a coding indicating that message type NEU is to be used if a purchase order is to be transmitted to the supplier, program 107 contains the conditions for creating an order confirmation reminder with message type AUFB, and so forth.

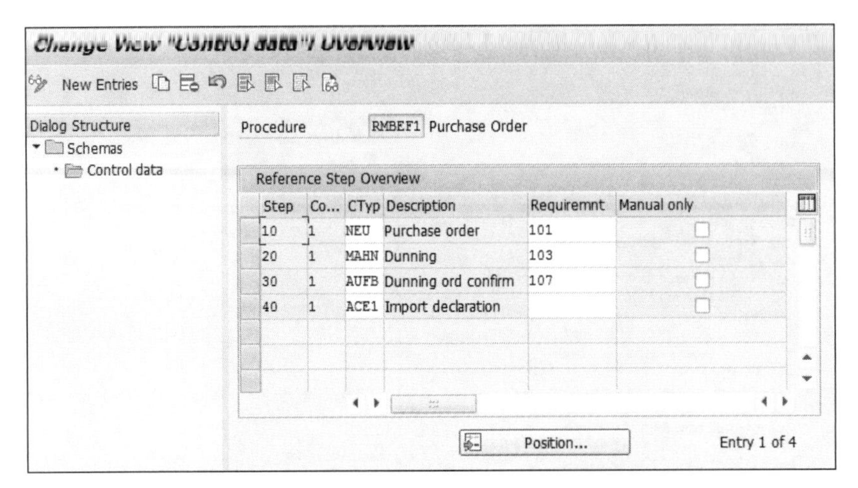

Figure 7.11 Customizing: Define Message Determination Schema

Figure 7.12 shows where the schema RMBEF1 is assigned to the purchase order.

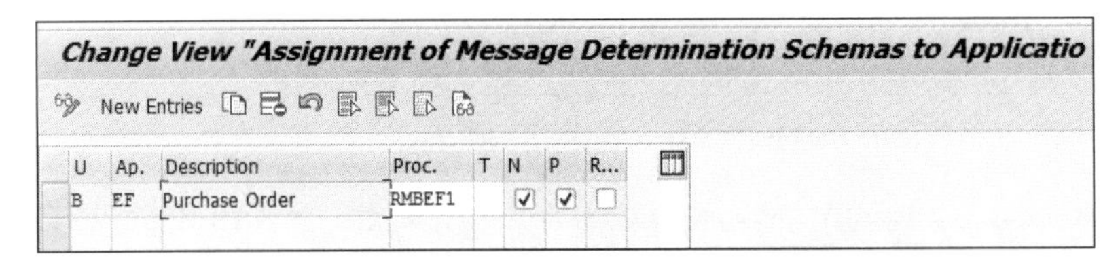

Figure 7.12 Customizing: Assignment Message Schema to Purchase Orders

Note

We use the terms *message type* and *output type* synonymously. In the same way, we say *message determination schema* as well as *message determination procedure* or *message schema*.

Note

Both the condition tables and the access sequences, the condition types and condition schema, are configured separately for purchase orders, requests for quotations, outline agreements, scheduling agreement releases, and inbound deliveries.

Create Output Records

Once the criteria for output determination (condition tables and access sequence) have been determined, and once the different types of output formats (output types) have been defined and permitted (output determination procedure), only *output records* need to be created so that the application can determine and issue the correct message according to your requirements.

Message records for purchase orders can be created with Transaction MN04, which is shown in Figure 7.13. There is one transaction/app per object (e.g., order, request, outline agreement, etc.). The system allows you to select one of the message types contained in the schema assigned to the application object (e.g., purchase order). After selecting an output type (e.g., NEU), the *key combinations* are proposed according to the access sequence assigned to the output type and its condition tables.

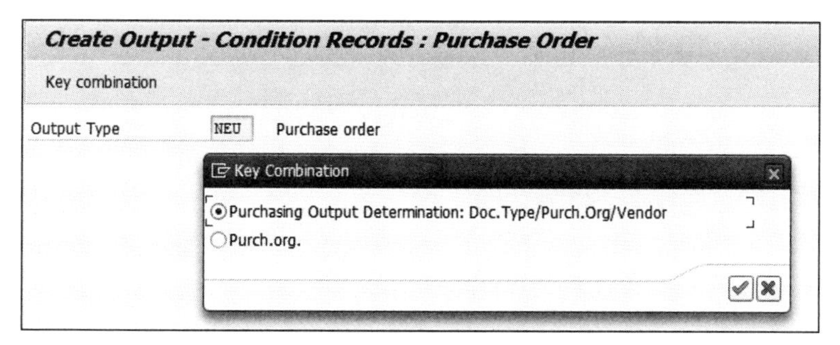

Figure 7.13 Creating a Message Record for Message Type NEU in Transaction MN04 (I)

Remember that we defined requirements at the beginning of the process. The message record that follows specifies that vendor 10300001 should receive an email (medium external send) when a purchase order is placed with them with document type NB in purchasing organization 1010. In addition, this message will be generated immediately when the purchase order is saved (the **Dispatch Time**, **Date**, and **Time** fields).

Figure 7.14 shows how the message record is created according to the provided requirements:

- The **Medium** field: 4 for external send (available media are shown in Figure 7.15)
- The **Dispatch Time** field: 4 for immediately (possible dispatch times are shown in Figure 7.16)

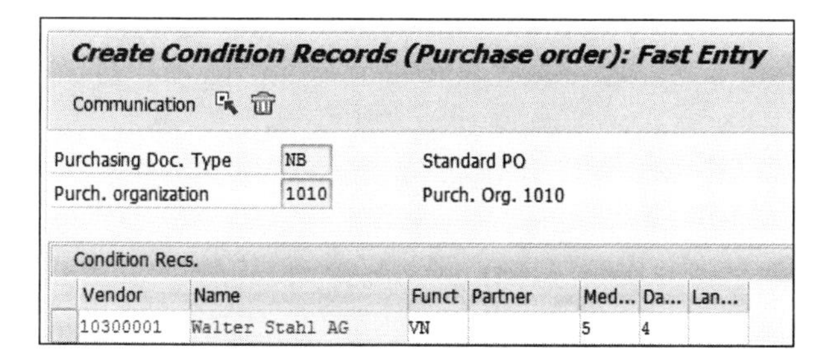

Figure 7.14 Creating a Message Record for Message Type NEU in Transaction MN04 (II)

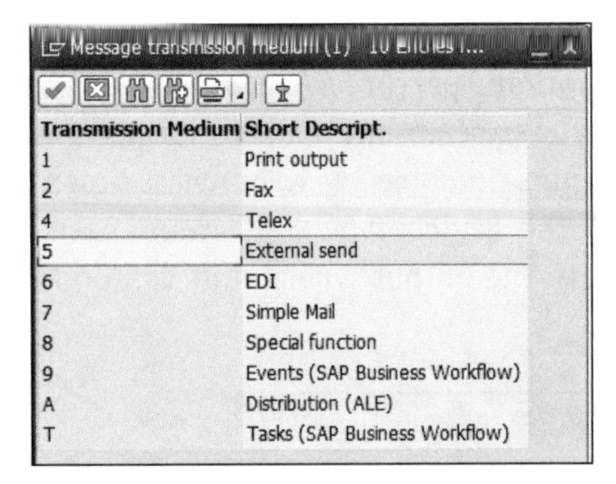

Transmission Medium	Short Descript.
1	Print output
2	Fax
4	Telex
5	External send
6	EDI
7	Simple Mail
8	Special function
9	Events (SAP Business Workflow)
A	Distribution (ALE)
T	Tasks (SAP Business Workflow)

Figure 7.15 Possible Media for the Message Record

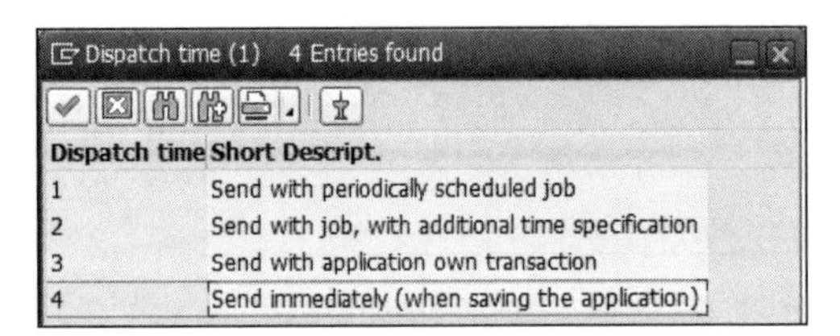

Dispatch time (1) 4 Entries found

Dispatch time	Short Descript.
1	Send with periodically scheduled job
2	Send with job, with additional time specification
3	Send with application own transaction
4	Send immediately (when saving the application)

Figure 7.16 Dispatch Times for the Message Record

Note

If you select the medium print output, you can store a printer in the message set itself. Otherwise, you can define a printer for each purchasing group in Customizing. The system can also take the default printer maintained in the user master record.

Tip

If you make changes to a purchasing document that has already been transmitted to the vendor, the system generates a change message to inform the business partner of the changes.

In the Customizing activity **Assign Message Schema,** you can choose whether the system should carry out a new message determination process. If so, you can use a different message type for changes than for the first issue, via a different medium. If not, the system uses the same message type and medium for both the first transmission and the change message.

In Customizing, you can also define for which fields changes are message relevant.

When you create a purchase order, the system proceeds as follows to create a message:

- Identify the message *schema* that is assigned to the application.
- Determine the appropriate *message type* based on the sequence in the schema and the requirements.
- Determine the *access sequence* of the selected message type.
- Determination of the *message record* for the access sequence with the field values from the purchase order.

You can use *determination analysis* like the one shown in Figure 7.17 to check the result of the message determination on the message overview screen of the advanced apps (or the corresponding SAP GUI transactions).

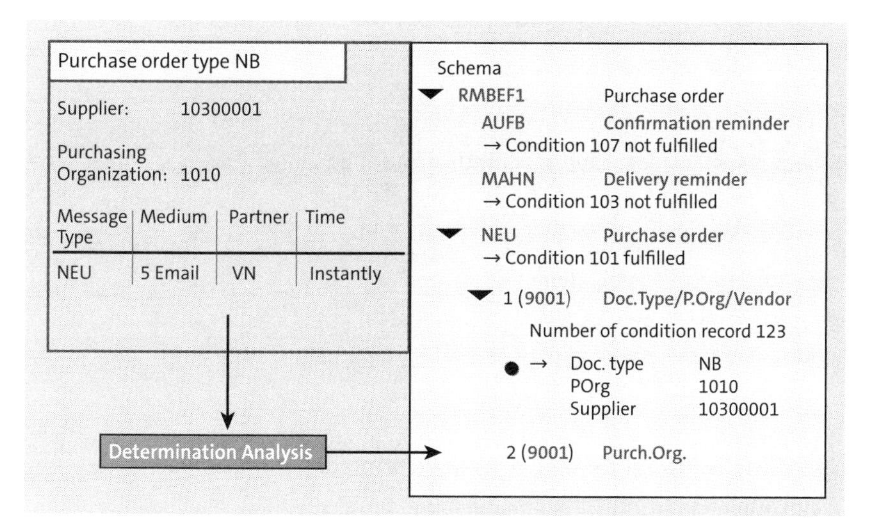

Figure 7.17 Message Determination Analysis

SAP S/4HANA Output Management

With SAP S/4HANA, there is a new output management in addition to message determination based on condition technique.

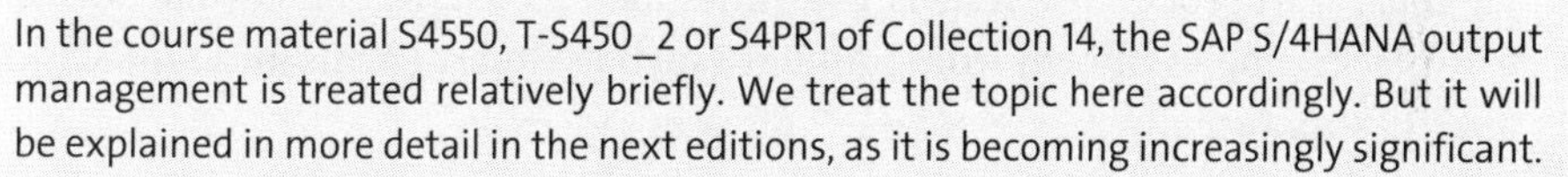

Warning

In the course material S4550, T-S450_2 or S4PR1 of Collection 14, the SAP S/4HANA output management is treated relatively briefly. We treat the topic here accordingly. But it will be explained in more detail in the next editions, as it is becoming increasingly significant.

First, SAP S/4HANA output management must be activated in Customizing. As with message determination based on the condition technique, you must define your requirements—for example, that messages for purchase orders with document type NB to business partner 10300001 in purchasing organization 1010 are to be sent by email.

The configuration is in Customizing under **Cross Application Components • Output Control** and is much more intuitive than with the condition technique. Output determination is based on Business Rule Framework Plus (BRF+) and uses decision tables.

You select the application object to be sent as shown in Figure 7.18, and go through the following options:

- **Output type**
  This specifies the document to be output (e.g., purchase order). In this step, you also define the dispatch time.

- **Receiver**
  This specifies the partner who is to receive the message.
- **Channel**
  This specifies whether the message is to be printed, mailed, or sent electronically.

Note

The new output management supports only the following channels:

- Print
- Email
- XML (for Ariba Network Integration)
- IDoc

Fax is not supported.

- **Printer settings**
  This specifies the printer for messages to be printed.
- **Email settings**
  Here you assign the email template and how the sender email address is to be determined.
- **Email recipient**
  If you don't want to use the default email address of a business partner, you can define other or additional email addresses.
- **Form template**
  This specifies the form to be used. You can also assign country-specific form languages. In procurement, it is recommended to use Adobe Forms, but the form technologies known from SAP ERP, such as Smart Forms and SAP Script, are still supported.

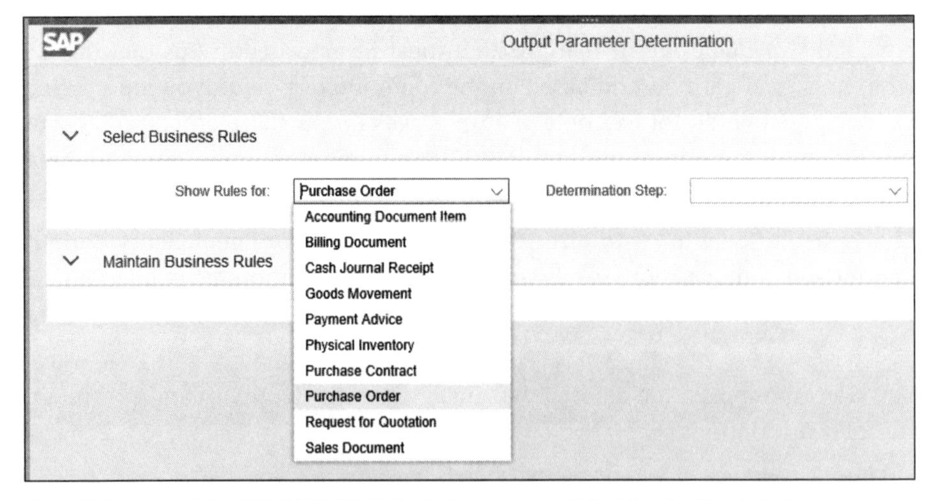

Figure 7.18 Customizing SAP S/4HANA Output Management: Select Application Object

If conditions (as seen in Figure 7.19) are missing (in our example, the purchasing organization), select **Table Settings** to add a column, as shown in Figure 7.20.

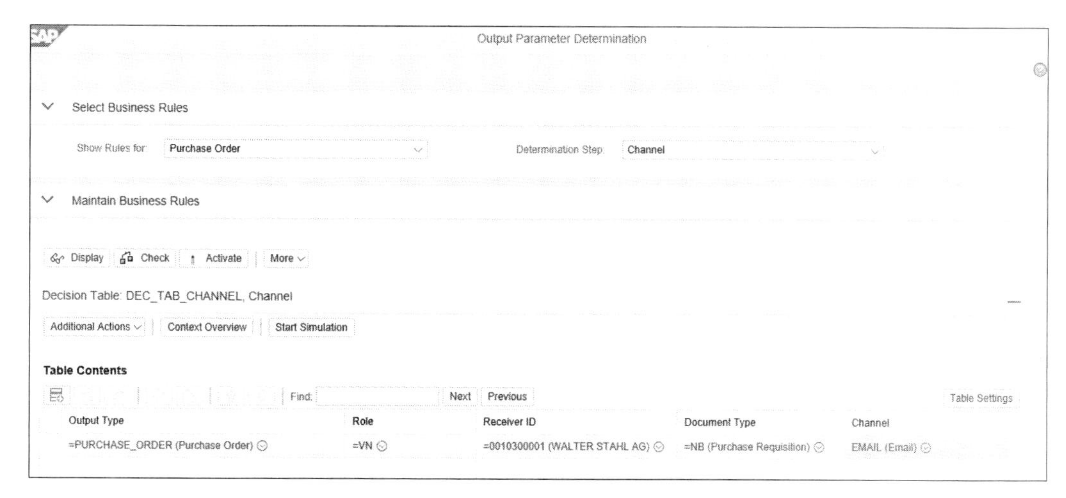

Figure 7.19 Customizing SAP S/4HANA Output Management: Conditions

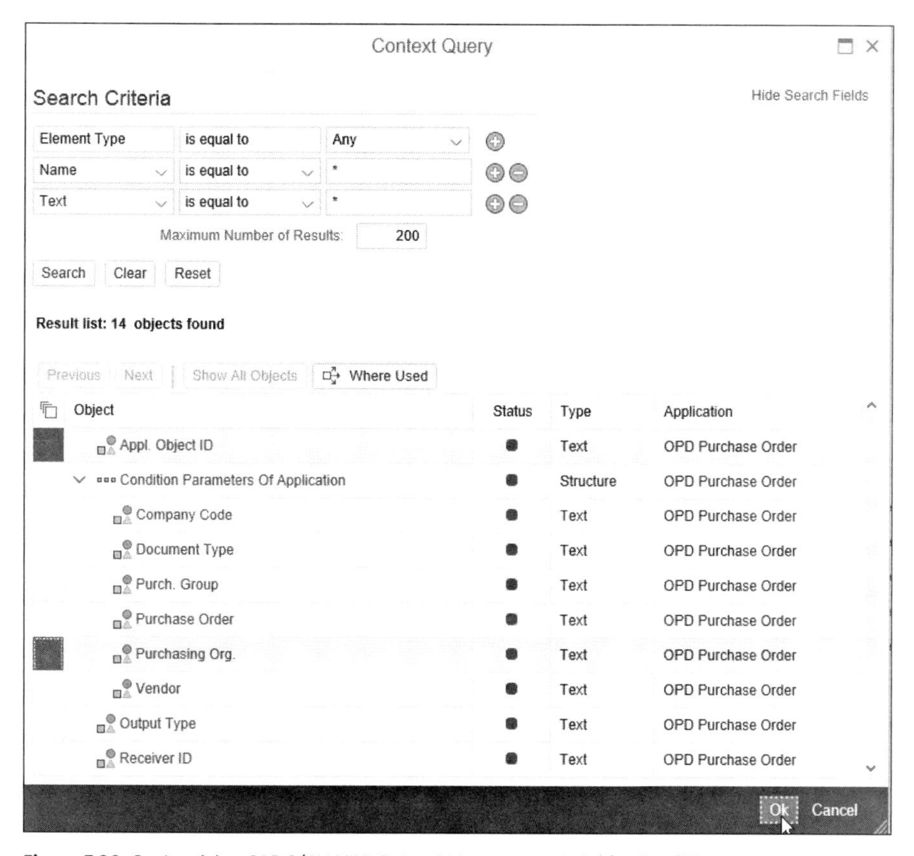

Figure 7.20 Customizing SAP S/4HANA Output Management: Add a Condition

Warning

There is no determination analysis for the SAP S/4HANA output management.

Release Procedure for Purchasing Documents

You can set up approval procedures for purchase requisitions and purchasing documents (e.g., purchase orders, contracts, scheduling agreements, and RFQs).

If a purchase requisition or external purchasing document meets certain conditions (e.g., a requisition item is assigned to a cost center), the document must be approved (by the person responsible for the cost center, for example) before further processing can take place. The release procedure automatically blocks the relevant documents for further processing:

- A blocked purchase requisition item cannot be converted into follow-on documents, such as RFQs, purchase orders, and outline agreements.
- A blocked purchase order, RFQ, or outline agreement cannot be transmitted to the supplier.

SAP S/4HANA offers two procedures for the release of purchase requisitions and purchasing documents: the classic release procedure with classification, which was already available in SAP ERP, and the flexible workflows, a new S/4HANA function that can replace or enhance the release procedure with classification. Depending on the document type, you can decide whether to use the flexible workflow or the classic release procedure.

To activate the flexible workflow for a purchase order document type, for example, go to **IMG • Materials Management • Purchasing • Purchase Order • Flexible Workflow for Purchase Order Orders • Activate Flexible Workflow for Purchase Orders**, as shown in Figure 7.21.

Document Types Purchase order Change

New Entries

Type	Doc. Type Descript.	Scenario based workflow
DB	Dummy Purchase Order	☐
ENB	Standard PO DFPS	☐
EUB	DFPS, Int. Ord. Type	☐
FO	Framework Order	☐
NB	Standard PO	☐
NB2	Enh. Rets to Vendor	☐
NB2C	Enh. Rets STO CC	☐
NBC7	CC SiT Enh. Rets STO	☐
NBR8	IC SiT Enh. Rets STO	☐
NBXE	XLO Inter Com PO	☐
NBXI	XLO Intra Company	☐
UB	Stock Transp. Order	☐
UB2	Enh. Rets STO IC	☐
ZNBF	PO f. flex. WF	☑

Figure 7.21 Customizing: Activate Flexible Workflow for Release Procedure

Classic Release Procedure

The classic release procedure with classification offers maximum flexibility. The first step is to define the blocking criteria for each document category. As is often the case, a case study can help enhance our understanding. Let's imagine the following requirements:

- In your company, purchase orders for stock materials are automatically generated out of purchase requisitions from MRP and can be sent to suppliers without approval. They have document type NB.
- Manual created purchase orders under 500 USD can be processed without approval.
- Manual purchase orders between 500 and 2,000 USD must be approved by the purchasing manager.
- Manual purchase orders that exceed 2,000 USD must be approved by the purchasing manager as well as the managing director.
- Manual purchase orders are created with the customer-specific document type ZNB.

We can deduct from the described requirements that the criteria for the release strategy are the document type and the total amount. We also see from the requirements that we need a two-level release strategy, since the purchase orders assigned to an account with a total amount greater than 2,000 EUR must be released by two persons. In the following, we explain how we can map this described strategy, so you can refresh the procedure, logic, and terms.

Define the Criteria

You must have created a *characteristic* for each criterion for your release strategy. To create, maintain, or display characteristics in Customizing, follow **IMG • Materials Management • Purchasing • Purchase Order • Release Procedure for Purchase Order Orders • Edit Characteristics.**

You are free to choose how you name the characteristic, but you must link the characteristic to a field of either database table CEBAN (purchase requisitions) or CEKKO (purchase orders). This means that the possible criteria are contained in the tables CEBAN and CEKKO. Figure 7.22 shows, among other things, the link of a characteristic to the table CEKKO.

As shown in Figure 7.23, you can assign the attribute **Multiple Values** to a characteristic so that you can enter several different values for the characteristic. You can allow interval values for characteristics of data types NUM (numeric format) or CURR (currency format). You cannot use interval values for characteristics of data type CHAR (character format).

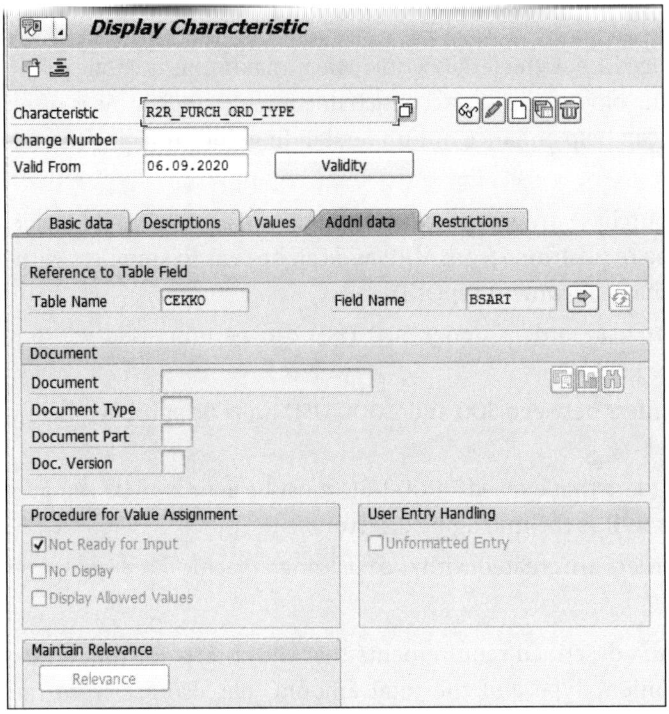

Figure 7.22 Example of a Characteristic

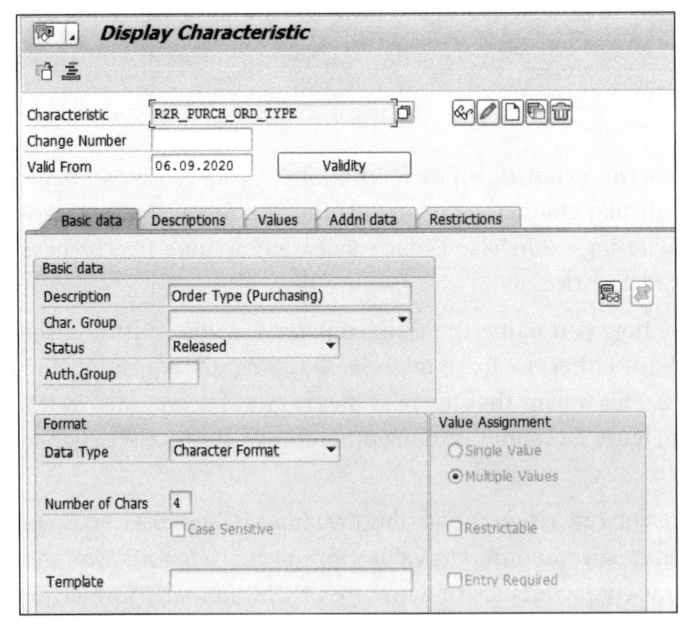

Figure 7.23 Assigning Multiple Values

You then must group together the characteristics that you want to use for your release strategy—in our example, document type, and total value—in one *class*.

You must assign the class to class type 032 (release strategies),as shown in Figure 7.24. You are free to choose the description for the class.

Note

A purchasing document (e.g., a purchase order) can only be blocked or released as an entire document. For a purchase requisition, you can opt for a complete release or a release per item. Therefore, two classes can be defined for purchase requisitions, while only one class can be defined for each purchasing document category.

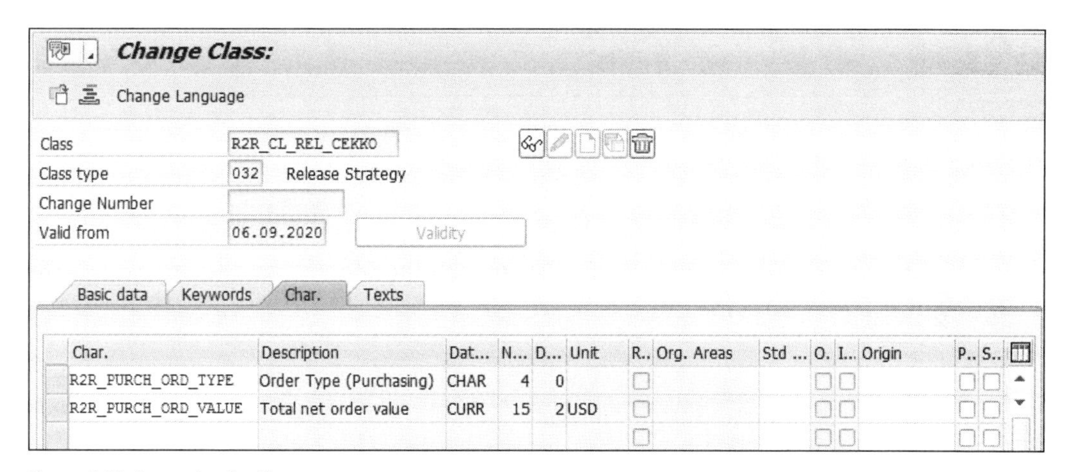

Figure 7.24 Example of a Class

Once the characteristics are defined and stored in a class, you can now configure the strategy and perform the value assignment. To do this, follow **Customizing IMG • Materials Management • Purchasing • Purchase Order • Release Procedure for Purchase Order Orders • Define Release Procedure for Purchase Order Orders.**

Assign the Class to the Release Object Using a Release Group

When you have defined the characteristics and placed them in a class, assign the class to the release object. For purchase orders, you can only assign one single class, as shown in Figure 7.25. For purchase requisitions, you could assign two classes: one for the overall release strategy and one for the item-wise release strategy. This assignment of two classes to an object is made possible by the release group.

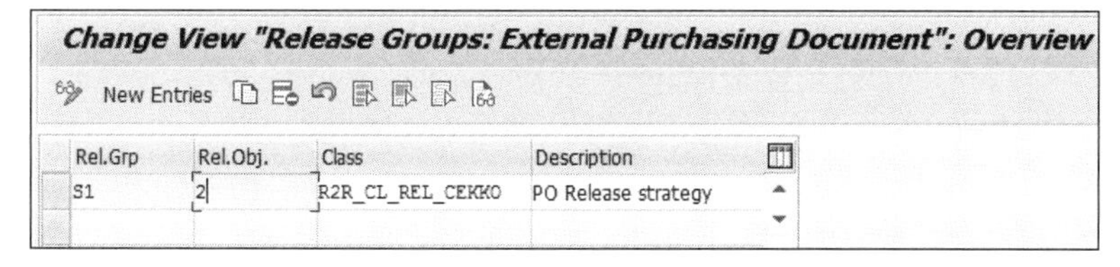

Figure 7.25 Release Group

Define Release Codes

You create all the release codes you need for your release strategies together with a suitable description (e.g., name or function of the person responsible for release) and assign them to the desired release group, as shown in Figure 7.26. If a release code should trigger a workflow, set the appropriate indicator.

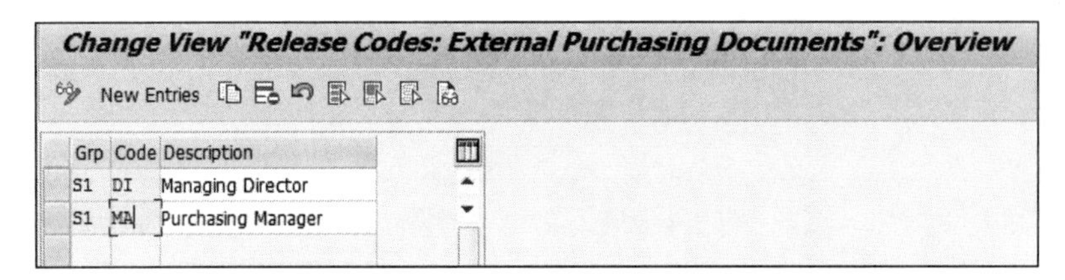
Change View "Release Codes: External Purchasing Documents": Overview

Grp	Code	Description
S1	DI	Managing Director
S1	MA	Purchasing Manager

Figure 7.26 Release Codes

The respective release codes will be assigned to the persons responsible for release by authorization: The authorization object M_EINK_FRG contains the release code and release group and specifies which purchasing documents you can release with which release codes.

Note
The number of release codes is limited to eight.

Define Release Indicators

Release indicators describe the release status of a purchasing document or purchase requisition. You require at least one indicator for the initial status and one indicator for the final status. The **Changeable** indicator in Figure 7.27 enables you to determine the consequences of changes to the document after the start of the release procedure. For value changes, you can set a percentage that will be tolerated without consequences.

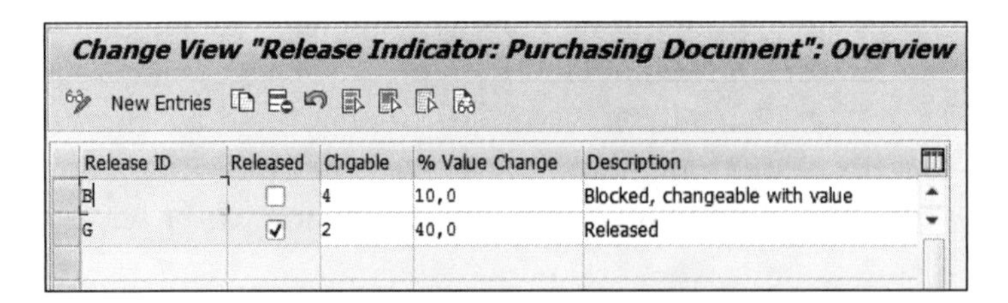
Change View "Release Indicator: Purchasing Document": Overview

Release ID	Released	Chgable	% Value Change	Description
B	☐	4	10,0	Blocked, changeable with value
G	☑	2	40,0	Released

Figure 7.27 Release Indicators

Define Release Strategies

To configure our example, we must create two strategies: one for manual purchase orders with a value between 500 and 2,000 USD and one for manual purchase orders with a value higher than 2,000 EUR.

In Figure 7.28 and Figure 7.29, you see the one-level strategy as follows:

- One releasing person: One release code
- Two release status (or release indicators): One initial, one final after release
- Characteristics values with document type ZNB and total value between 500 and 2,000 USD

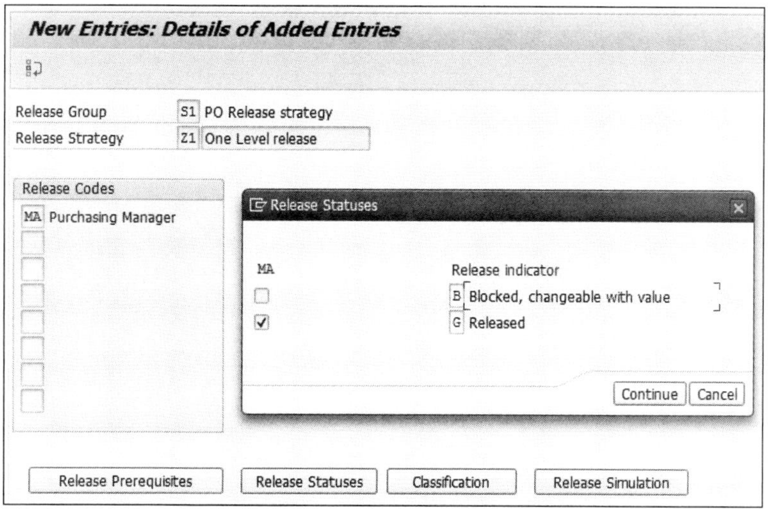

Figure 7.28 One-Level Release Strategy: Release Codes and Release Status

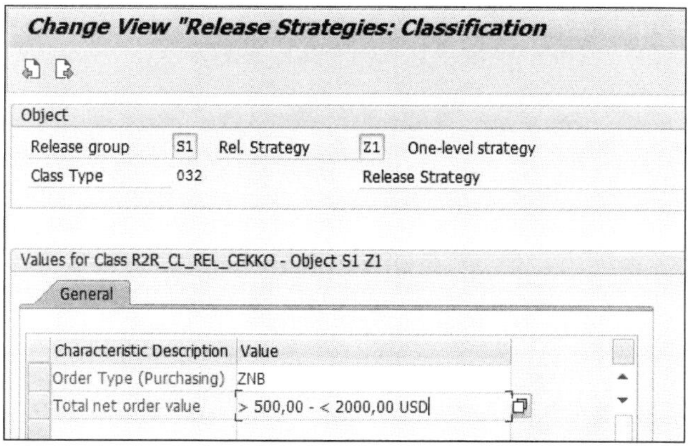

Figure 7.29 One-Level Release Strategy: Classification

The following figures illustrate the two-step strategy: If the total amount of a purchase order with document type ZNB exceeds 2,000 USD (in the **Classification** window), then the purchase order should first be released by the purchasing manager (release code MA) and second by the managing director (release code DI). The purchase order is only released after the purchasing manager and subsequently the managing director have released it (**Release Prerequisites** and **Release Statuses** windows).

Figure 7.30 shows the two release codes for the two [illegible] and the prerequisites. Figure 7.31 shows the two release statuses. Figure 7.32 shows the classification values.

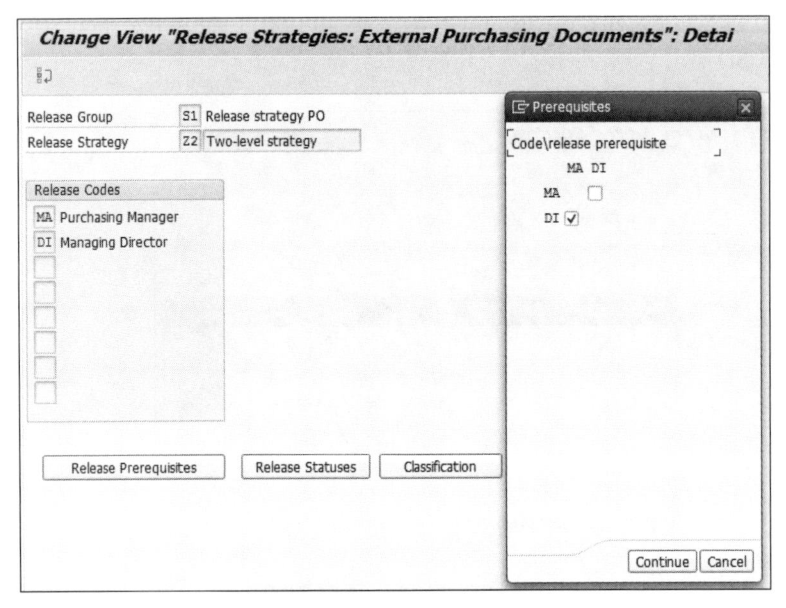

Figure 7.30 Two-Level Release Strategy: Release Codes and Prerequisites

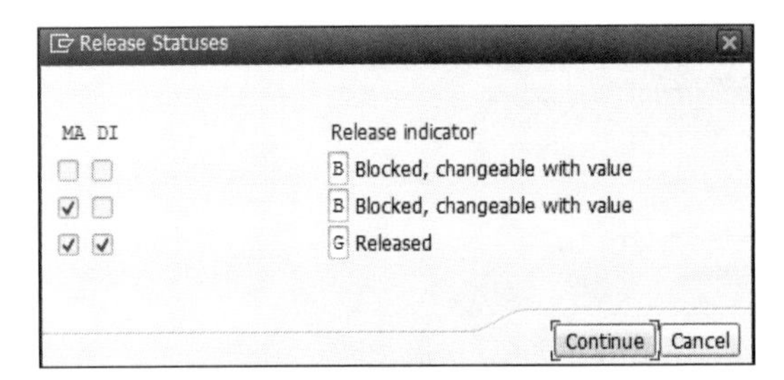

Figure 7.31 Two-Level Release Strategy: Release Codes and Release Status

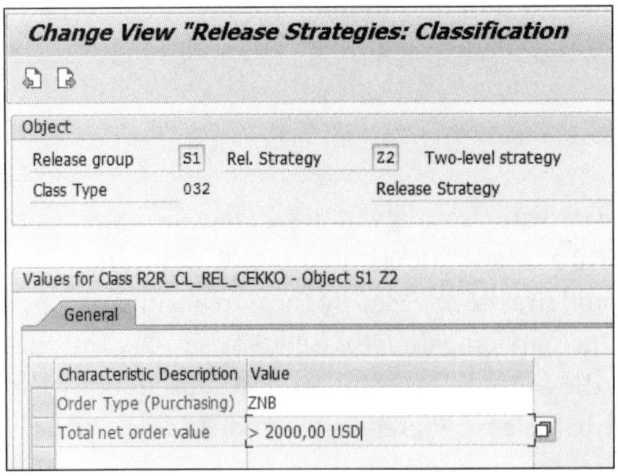

Change View "Release Strategies: Classification

Object

Release group S1 Rel. Strategy Z2 Two-level strategy

Class Type 032 Release Strategy

Values for Class R2R_CL_REL_CEKKO - Object S1 Z2

General

Characteristic Description	Value
Order Type (Purchasing)	ZNB
Total net order value	> 2000,00 USD

Figure 7.32 Two-Level Release Strategy: Classification Values

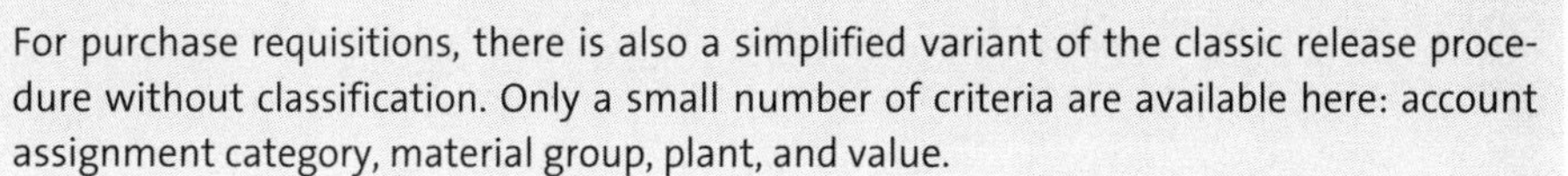

Note

For purchase requisitions, there is also a simplified variant of the classic release procedure without classification. Only a small number of criteria are available here: account assignment category, material group, plant, and value.

Flexible Workflow

The flexible workflow can be set up for one or several of the following characteristics:

- Cost center account assignment category
- Project account assignment category
- Company code
- Material group contained in at least one purchase order item
- Currency
- Document type
- Purchasing group
- Purchasing organization
- Total net amount of the document

You configure the flexible workflow for purchase orders with the Manage Workflows for Purchase Orders app on the SAP Fiori launchpad. There is a dedicated app for each object (e.g., order request, contract, etc.).

Figure 7.33 shows a workflow for purchase orders. The prerequisite for the workflow is that the purchase order has the purchasing group Z03. The purchase orders with purchasing group Z03 must be approved by user S4520-03.

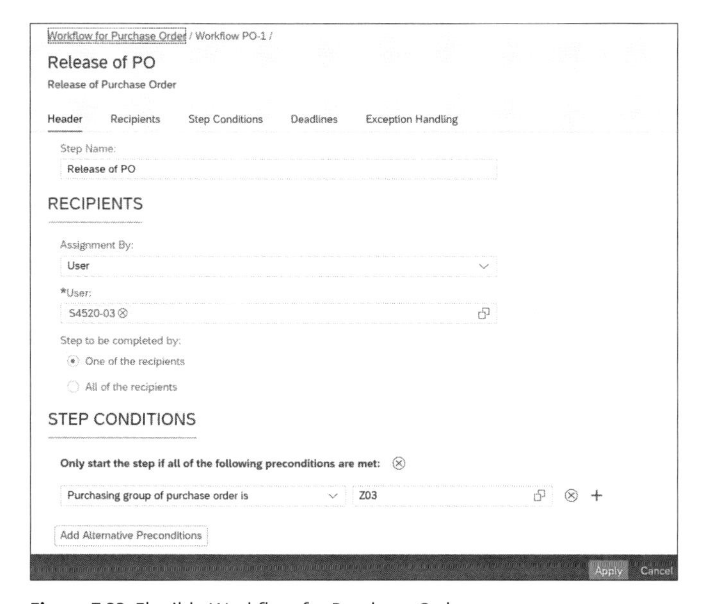

Figure 7.33 Flexible Workflow for Purchase Orders

In addition to the release and output of purchasing documents, the monitoring function also plays an important role in procurement.

Procurement Process Monitoring

In this section, we will deal with some functions for monitoring outstanding order confirmations and deliveries.

Monitoring Order Confirmations

When working with confirmations, two options are available.

The first is the simplified process. Here, you expect your vendor to send a confirmation by setting the **Acknowledgement Required** indicator at the purchase order item level, as shown in Figure 7.34 and Figure 7.35. When the acknowledgment arrives, you simply enter, for instance, a confirmation number or a date in the **Order Acknowledgement** field. You cannot enter a confirmed quantity or a confirmed date separately.

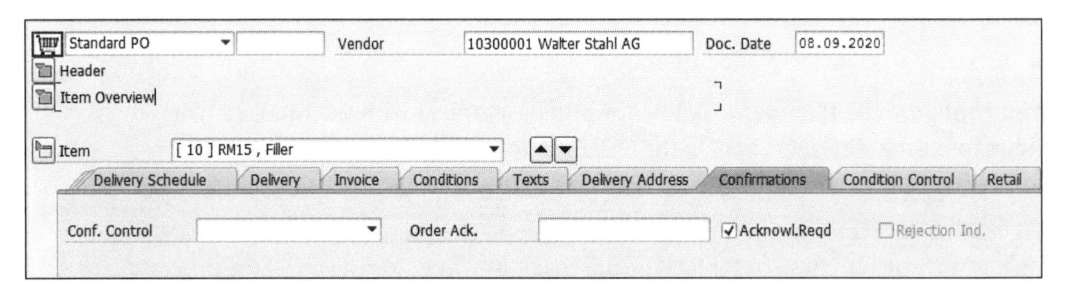

Figure 7.34 Acknowledgement Required Indicator and Order Acknowledgement Field in Transaction ME21N

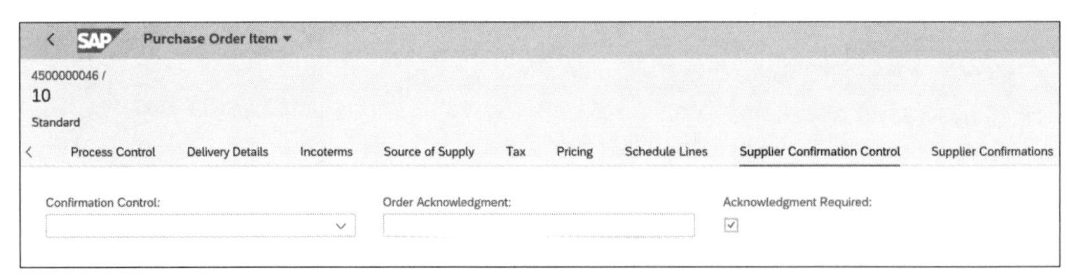

Figure 7.35 Acknowledgement Required Indicator and Order Acknowledgement Field in the Manage Purchase Order App

You can preset the **Acknowledgement Required** indicator as follows if you wish it to be generally defaulted:

- In the purchasing data of the business partner
- In the purchasing info record
- In the material master as part of the purchasing value key

- Using the EVO user parameter (see Chapter 12 for more details)
- In the personal settings for Transaction ME21N

Setting the **Acknowledgment Required** indicator allows you to monitor pending order confirmations: the system checks whether there is an entry in the **Order Acknowledgment** field during monitoring. If there is no entry, you can generate an order confirmation reminder.

The second option is an extended process, with the option to enter a sequence of several confirmation types. In this case, when creating the purchasing document, you must enter a confirmation control key in the item details. The key determines the confirmations you expect from your vendor and the sequence in which they are to be entered. The key also controls which confirmation types are relevant for materials requirements planning and goods receipt.

You can preset the confirmation control key as follows if you wish it to be generally defaulted in the purchasing document item:

- In a purchasing info record
- In the purchasing data of the business partner

If you work with a confirmation control key, you can enter confirmations as follows, depending on the confirmation categories contained in the key:

- You enter the confirmation directly in the purchasing document item.
- You create a separate document—an inbound delivery.
- You have the vendor transmit the acknowledgment electronically.

If the vendor rejects the purchase order set the rejection indicator; the deletion indicator is then set in the purchase item.

Note

If you work with inbound deliveries, you must enter the goods receipt with reference to the inbound delivery.

To generate order confirmations reminders, the following requirements must be met:

- The **Acknowledgment Required** indicator is set in the purchase order item.
- For purchase order items without a confirmation control key, you have not yet made an entry (confirmation number, for example) in the **Order Acknowledgment** field.
- For items with confirmation control keys, you have only partially or not yet confirmed the ordered quantity.
- The purchasing document has been sent out.
- You have completely set up message determination for the order acknowledgment reminder.

Use the Monitor Order Acknowledgments app (Transaction ME92F) to monitor outstanding purchase orders confirmations and send reminders. If you only want to monitor the outstanding confirmations but *not* generate reminders, you can use Transaction ME2A or the Monitor Supplier Confirmations app on SAP Fiori launchpad. With these apps, you can monitor confirmations of all confirmation categories, but you cannot create reminders for them.

Configuring Confirmation Control Key

For the configuration of the confirmation control key, follow **Customizing IMG • Materials Management • Purchasing • Confirmations • Set Up Confirmation Control**.

The confirmation control key is an alphanumeric key with up to four digits. On the first configuration screen shown in Figure 7.36, adjust the following settings:

- The **Create Inb. Delivery** indicator determines that inbound deliveries can automatically be created from purchase orders, even if shipping notifications are not required by the key.
- You use the **POD Relevant** indicator to indicate that a confirmation control key is relevant for the proof of delivery. The proof of delivery is used to inform suppliers of the receipt of goods.

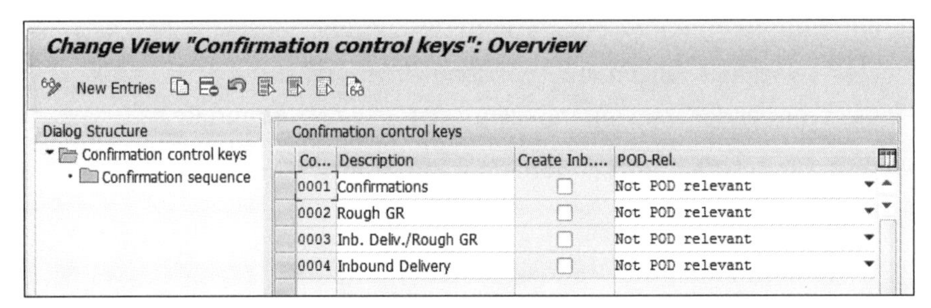

Figure 7.36 Customizing: Confirmation Control Key Setup

Be sure to define the sequence of expected confirmations. In the example of confirmation control key 0001, an order confirmation is expected first, followed by a shipping notification (see Figure 7.37). For each confirmation, define the following:

- **MRP-Relevant indicator**
  This specifies that the confirmed quantities and dates are considered in MRP.
- **GR-Relevant indicator**
  This specifies that the confirmation (e.g., inbound delivery) is reduced for MRP by a goods receipt for the corresponding order item. If several confirmations are expected, you can only mark one of them as GR-relevant.
- **GR Assignment indicator**
  In conjunction with the **GR-Relevant** indicator, this ensures that the system only proposes the confirmed quantity and that this confirmation category is a prerequisite for the goods receipt posting.

- **Monitoring Period and Reference Date columns**
  This controls up to which date you expect a confirmation. You can define this period with reference to the order date or the delivery date. If you do not enter a confirmation within this period, a reminder can be issued automatically, provided the **Sbj. to Rem.** (subject to reminder) indicator is set.

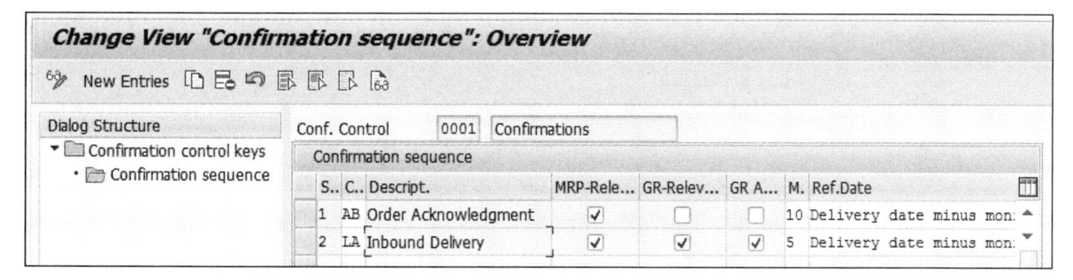

Figure 7.37 Customizing: Confirmation Sequence in Confirmation Control Key Setup

Delivery Monitoring

To send reminders for due and overdue purchase orders, scheduling agreement schedule lines, and RFQs, the following requirements must be met:

- You have entered one or more dunning levels (reminder periods) in the line item.
- The purchasing document has already been sent.
- You have completely set up message determination for delivery reminders and created message records for it.

You can enter reminder periods manually in the purchase order item. You can also define default values via the purchasing value keys, which you can enter in the Purchasing view of the material master record or for the material group in Customizing.

You create reminder messages or dunning notices for each document category with a separate transaction: Transaction ME91F for orders, Transaction ME91A for inquiries, and so forth.

Figure 7.38 shows an order item with reminder periods. Negative numbers are for reminders, and positive numbers are for dunning.

> **Note**
> ME91* and ME92* functionalities are not able to work with SAP S/4HANA output management. Only message determination with condition technique makes these transactions available.

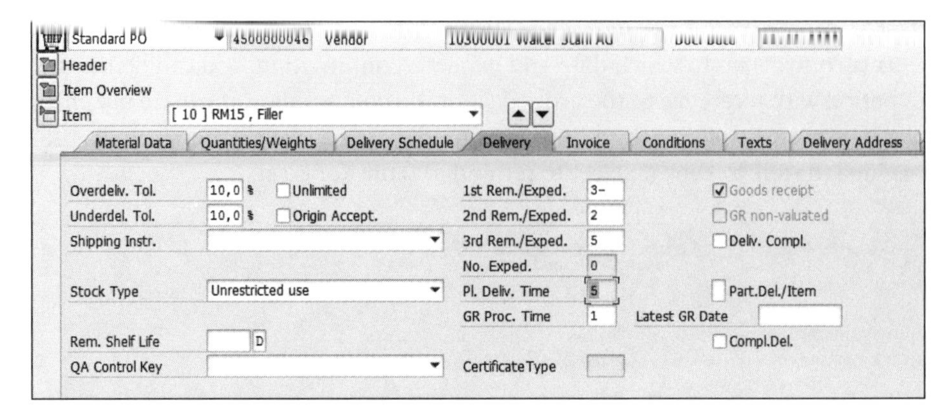

Figure 7.38 Reminder Periods for Purchase Order Item

Analytics in Procurement

Analytics supports you in monitoring your procurement processes. Embedded analysis functions such as contract analysis, spend visibility, supplier evaluation, and so forth can evaluate data from SAP S/4HANA in real time without having to extract the data into a separate data warehouse.

There are several analytics tile groups for purchasing. The tiles show you the most important information, sorted by relevance. Some are KPI tiles, which highlight key information immediately. Numbers can be color-coded to represent different levels of alert. The surface of a tile can show numbers or chart snippets. The information on a tile is updated in real time.

In addition to the various KPI tiles, the **Procurement Overview** page in Figure 7.39 shows you the most important information and tasks.

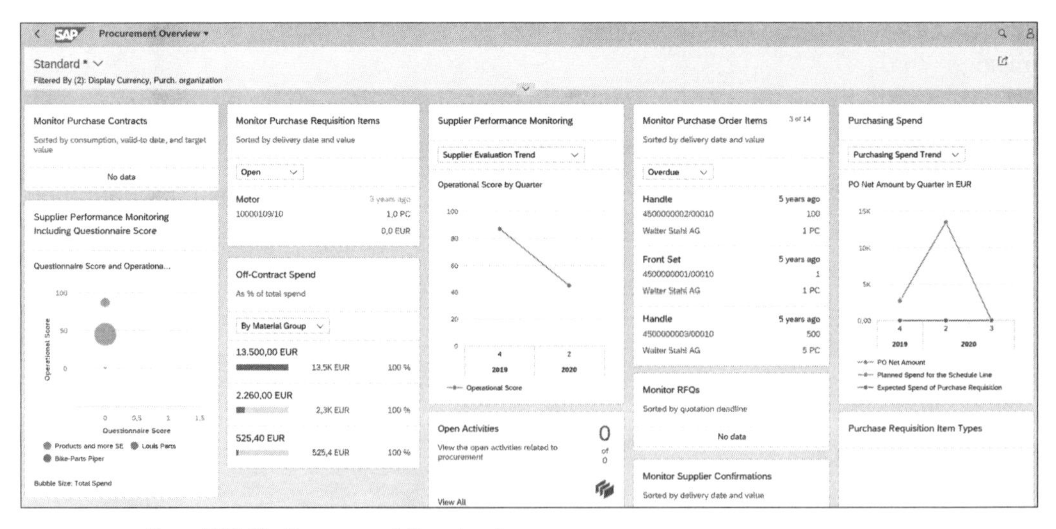

Figure 7.39 The Procurement Overview Page

The information is displayed on several actionable cards. This allows you to focus on the most important tasks, which enables faster decisions and immediate action. You can customize the overview page by rearranging the tiles, hiding, or showing them.

You can also apply a filter to the displayed information that affects all relevant tiles. For example, you can filter the content of all cards according to a specific supplier or material group.

SAP S/4HANA Innovations in Procurement

SAP S/4HANA offers a range of new features. The most important ones, such as the business partner concept, the self-service requisitioning, the new output management, the release of purchasing documents via flexible workflows, and MRP Live, the new source of supply determination logic in MRP, are discussed in this book during the process review. We will now summarize some further innovations in brief and invite you to review these points using the S4PR1 course and the Simplification List. These are not explicitly required for the SAP S/4HANA certification C_TS452_1909 but for C_TS450_1909 (Upskilling for ERP Experts).

- **New inquiry process**
  SAP S/4HANA offers a completely new inquiry process. The old process with Transactions ME4x is still available but is obsolete. Transactions ME4x are not further developed in SAP S/4HANA. They are being replaced by SAP Fiori apps, and the process logic as well as the structure of the respective documents have changed.
- **Simplified data model for inventory documents**
  Table MATDOC replaces table MKPF and table MSEG.
- **Purchasing categories**
  The purchasing category is a new business entity in SAP S/4HANA used in supplier evaluation and in purchasing real-time analytics.
- **Extended material number functionality**
  By default, both after upgrading to SAP S/4HANA and for new installations, the extended material number functionality is not activated. If you want to have a material number with 40 characters, you must activate this functionality.
- **New material type SERV**
  This is an alternative to using service master data.
- **Purchase order and sourcing collaboration**
  This is done using the Ariba Network as a supplier collaboration platform.
- **Central procurement**
  This enables the integration of an SAP S/4HANA hub system with multiple SAP ERP and SAP S/4HANA backend systems.

- **New SAP S/4HANA supplier evaluation**
  The new SAP S/4HANA supplier evaluation allowsSegmentation and classification of suppliers.

Note

The supplier evaluation is not part of the standard certification C_TS452_1909. However, if you are interested in upskilling certification (C_TS450_1909), please note that SAP S/4HANA offers two options for supplier evaluation: the classic one with the Logistics Information System and a new one with purchasing category.

You can decide which type of supplier evaluation you would like to use. If you want to use vendor evaluation by purchasing category, you must activate this function in Customizing.

Important Terminology

The following terminology was used in this chapter:

- **Condition table**
  A condition table contains the fields that you want to use as influencing factors (dependencies) for message output. You define the combination of fields for which you want to create message records.
- **Access sequence**
  The access sequence is used in the message determination with condition technique and contains the condition tables in a defined order.
- **Message type or output type**
  A message type is used within message determination with condition technique and defines the type of document to be output.
- **Message determination schema**
  Within message determination with condition technique, a schema contains all possible message types with requirements and further control parameters.
- **Message record**
  A message record is a master record and is required in message determination using the condition technique. It contains information on how and when a message is to be issued for a message type and for the field combination of a condition table belonging to the access sequence assigned to the message type.
- **Release group**
  Within the release procedure with classification, you use the release group to link the release class to the release object (e.g., purchase order).
- **Release strategy**
  The release strategy contains the criteria for blocking an item (only for purchase requisitions) or a document.

- **Release indicator**
  The release indicator is a key that indicates the current release status of the item or document.
- **Release code**
  The release code is a two-digit ID that allows an individual or a group of people to release a blocked document or to reverse a release already made. Authorizations control the use of the release code.
- **Characteristic**
  In the classic authorization procedure with classification, a characteristic corresponds to a criterion for blocking a purchasing requirement.
- **Class**
  In the classic release procedure with classification, the class contains all criteria (characteristics) that are relevant for blocking a purchasing document or purchase requisition. The class is assigned to the release group and the release object in Customizing.
- **Confirmation control key**
  The confirmation control key controls which types of confirmations are expected in which sequence for a purchase order item and whether these confirmations are relevant for goods receipt or materials planning.
- **Confirmation category**
  A confirmation category determines what type of confirmation is expected from a supplier. A possible confirmation category can be, for example, an order confirmation or a shipping notification.
- **Reminder periods**
  You can define up to three reminder periods.. Negative numbers are used for reminders; positive numbers are used for dunning. For example, an entry of 3– in the first level means that a reminder to the supplier should be generated 3 days before the planned delivery date.

Practice Questions

These questions will help you evaluate your understanding of the topics covered in this chapter. They are similar in nature to those on the certification examination. Although none of these questions will be found in the exam itself, they will allow you to review your knowledge of the subject.

Select the correct answers, and then check the completeness of your answers in the next section. Remember that on the exam, you must select all correct answers—and only correct answers—to receive credit for the question.

1. True or false: SAP S/4HANA output management is activated in on-premise SAP S/4HANA by default.

☐ **A.** True

☐ **B.** False

2. Where do you define the criteria for message determination? (This question refers to output determination with condition technique.)

☐ **A.** In message types

☐ **B.** In a message determination schema

☐ **C.** In condition tables

☐ **D.** In an access sequence

3. On which level can you assign a message determination schema in purchasing? (This question refers to output determination with condition technique.)

☐ **A.** Purchasing organization

☐ **B.** Document type

☐ **C.** Application

☐ **D.** Company code

4. Which object is assigned an access sequence? (This question refers to output determination with condition technique.)

☐ **A.** A message type

☐ **B.** A message record

☐ **C.** A message determination schema

☐ **D.** A condition table

5. What can you specify in a message record? (There are two correct answers. Note that this question refers to output determination with condition technique.)

☐ **A.** Output medium

☐ **B.** Output form

☐ **C.** Application

☐ **D.** Output time

6. Which output channels are supported in SAP S/4HANA output management? (There are three correct answers.)

☐ **A.** Printout

☐ **B.** Email

☐ **C.** Fax

☐ **D.** XML

☐ **E.** Workflow

7. Where can you preset the printer for printing a purchase order? (There are three correct answers. This question refers to output determination with condition technique.)

☐ **A.** In the message record

☐ **B.** In Customizing

☐ **C.** In the message determination schema

☐ **D.** In the business partner master record

☐ **E.** In the user master record

8. In which sequence does the system proceed when determining a message when you create a purchase order? (This question refers to output determination with condition technique.)

☐ **A.** Message type, message schema, access sequence, condition record

☐ **B.** Message schema, message type, access sequence, condition record

☐ **C.** Condition record, message schema, message type, access sequence

☐ **D.** Condition record, message type, access sequence, message schema

9. True or false: In SAP S/4HANA, you can use the classic procedure with classification for the release of purchase requisitions and flexible workflows for the release of purchase orders.

☐ **A.** True

☐ **B.** False

10. True or false: You can perform item-wise release for purchasing documents and purchase requisitions.

☐ **A.** True

☐ **B.** False

11. What can you control with the release indicator for requisitions? (There are three correct answers. This question refers to the classic release procedure.)

☐ **A.** Whether the purchase requisition may be converted into an RFQ

☐ **B.** Whether the purchase requisition may be converted into an order

☐ **C.** Whether the purchase requisition is fixed

☐ **D.** Whether the purchase requisition may be deleted

12. Which of the following authorizations must a user have in order to approve a purchase order? (There are two correct answers. This question refers to the classic release procedure.)

☐ **A.** Release group

☐ **B.** Release indicator

☐ **C.** Release code

☐ **D.** Release strategy

13. Which of the following requirements must be met to generate a delivery reminder for a purchase order? (There are three correct answers.)

☐ **A.** The purchase order item contains one or more reminder periods.

☐ **B.** The purchase order has been sent out.

☐ **C.** SAP S/4HANA output management is active for purchase orders.

☐ **D.** Message determination with condition technique has been set up for delivery reminders.

☐ **E.** The purchase order item contains a confirmation control key.

Practice Question Answers and Explanations

1. Correct answers: **B**

   False. SAP S/4HANA output management is activated in SAP S/4HANA Cloud by default, but not in on-premise SAP S/4HANA.

2. Correct answer: **C**

   The criteria for message determination are recorded in condition tables. Several condition tables can be listed one after the other in access sequences so that the system first searches for condition records sequentially according to the fields (criteria) in the first table; then, if no condition record is found, according to the fields in the second table, and so on.

3. Correct answer: **C**

   In Customizing for output determination, you can assign an output determination procedure to so-called applications. An application corresponds to one or more document categories. This means, for example, that you can only assign one output determination procedure for purchase orders of all document types. It is not possible to differentiate more precisely.

4. Correct answer: **A**

   An access sequence contains condition tables in a certain order. When you want to create a message record for an output type (NEU for example), the system proposes a key combination. This key combination corresponds to the access sequence assigned to the output type in Customizing. An access sequence is therefore assigned to an output type.

5. Correct answers: **A, D**
   The condition record contains (among other information) the medium and the time of output. The application corresponds to a document category or a group of document categories and a message schema is assigned to it. The output form is assigned to the output type in Customizing.

6. Correct answers: **A, B, D**
   The new output management supports only the following channels: print, email, XML, and IDoc (only in on-premise SAP S/4HANA). Fax is not supported.

7. Correct answers: **A, B, E**
   The printer can either be maintained in the condition record for the output (message record) or in Customizing by purchasing group, or it can be taken over from the user default.

8. Correct answer: **B**
   The correct sequence for the message determination process is message schema, message type, access sequence, condition record. Refer to the section about the determination analysis.

9. Correct answer: **A**
   True. You define per document category and document type which release procedure you want to use.

10. Correct answer: **B**
    False. Purchasing documents can only be completely blocked or released. In the case of requisitions, you can decide whether to release them in full or by item. The document type (see Figure 7.40) and release group (see Figure 7.41) determine if a purchase requisition is subject to item-wise release or overall release.

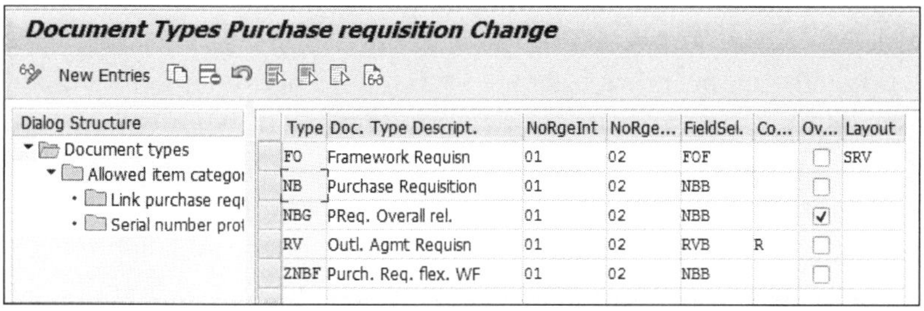

Document Types Purchase requisition Change

New Entries

Dialog Structure
- Document types
  - Allowed item categoi
    - Link purchase reqi
    - Serial number prol

Type	Doc. Type Descript.	NoRgeInt	NoRge...	FieldSel.	Co...	Ov...	Layout
FO	Framework Requisn	01	02	FOF		☐	SRV
NB	Purchase Requisition	01	02	NBB		☐	
NBG	PReq. Overall rel.	01	02	NBB		☑	
RV	Outl. Agmt Requisn	01	02	RVB	R	☐	
ZNBF	Purch. Req. flex. WF	01	02	NBB		☐	

Figure 7.40 Indicator for Overall Release of Purchase Requisitions

Change View "Release Groups: Purchase Requisition": Overview

New Entries

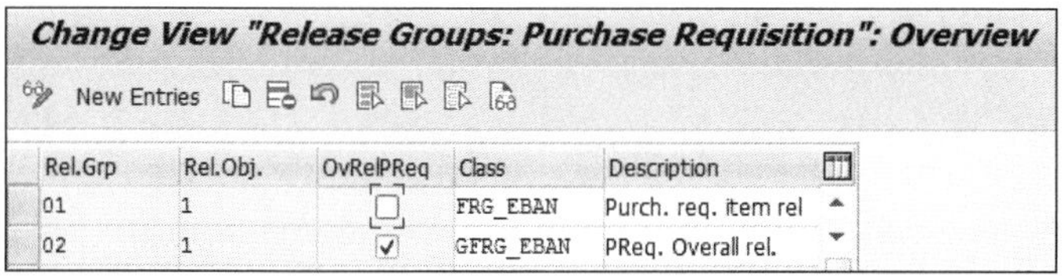

Rel.Grp	Rel.Obj.	OvRelPReq	Class	Description
01	1	☐	FRG_EBAN	Purch. req. item rel
02	1	☑	GFRG_EBAN	PReq. Overall rel.

Figure 7.41 Indicator for Overall Release of Purchase Requisitions

11. Correct answers: **A, B, C**
    You assign a release indicator to each release step of purchase requisitions. This indicator controls whether an RFQ and/or a purchase order can be created with reference to the purchase requisition. The firming indicator in the release indicator also enables you to protect purchase requisitions created in a planning run from being changed by a new planning run after the release process has begun. The release indicator provides a field selection key with which you can prevent certain fields, such as the quantity, from being changed subsequently. The **Changeable** indicator determines the consequence of a change to the document after the start of the release procedure. For value changes, you can define a percentage that is tolerated.

12. Correct answers: **A, C**
    Via the authorization object M_EINK_FRG, you specify for a user which purchasing documents he or she can release with which release codes. The authorization object contains the release group and release code.

13. Correct answers: **A, B, D**
    Prerequisites for delivery reminders are as follows: The purchase order item contains one or more reminder periods, the purchase order has been transmitted, message determination with condition technique is set up for delivery reminders, and there are message records for them. For purchase orders, message determination with condition technique should be active instead of SAP S/4HANA output management. The confirmation control key has nothing to do with delivery reminders.

Takeaway

This chapter provides an overview of how to set up message determination and release strategies. You should understand how to implement requirements and recommend one (classic) or the other (new) procedure. You should also be able to make recommendations on how to work with order confirmations. You should be able to present precise instructions on how outstanding order confirmations and deliveries can be reminded. You should be able to enumerate and describe the most important innovations in SAP S/4HANA.

Summary

In this chapter, we have learned what possibilities the system offers to create document release and message finding strategies. We dealt with order confirmations and saw how outstanding confirmations can be reminded. We also learned how to monitor and remind outstanding deliveries.

Before we change the topic in the next chapter and focus on MRP, we would like to once again recommend that all candidates for upskilling certification review the specific procurement innovations that are not necessarily dealt with in detail but have been reviewed in this chapter.

Chapter 8
Consumption-Based Planning

8

Techniques You'll Master

- Understand the procedures of material requirements planning (MRP).
- Set up an organizational structure for MRP.
- Maintain material master records for reorder point planning.
- Configure control parameters for reorder point planning.
- Execute a planning run.
- Control automatic source determination.

This chapter will focus on consumption-based planning and especially on manual reorder point planning. It will explain the most important differences between consumption-based and demand-driven planning. The MRP process, the relevant organizational structure, the MRP data in the material master record, and the available control parameters will be explained. The source determination procedure will also be covered in detail.

Real-World Scenario

In the past, the responsibilities of planners and buyers were clearly separated. Today, the roles of planners and buyers have changed. Purchasers very often concentrate on strategic tasks, such as negotiating contracts with suppliers, while the operational creation of purchase requisitions and order processing has been transferred to the material planners.

As a consultant for sourcing and procurement, it is essential that you are familiar with the upstream and downstream processes of pure procurement processing. Optimized and reliable MRP is the prerequisite for a reliable supply of both external and internal customer requirements, as well as for production.

Objectives of This Portion of the Test

The purpose of this portion of the certification exam is to test your knowledge in MRP, especially consumption-based planning with SAP S/4HANA. For the certification, you need to have a good understanding of the following topics:

- Planning procedures
- Procurement proposals
- Lot-sizing procedures
- Scheduling
- Planning levels
- Source determination
- Configuration of the planning parameters
- Running MRP and analyzing the results

Key Concept Refresher

In this chapter, we consider the processes and configuration of MRP in an SAP S/4HANA system. We look at the organizational units, the master data, and the key control parameters. We describe the steps of an MRP run, examine the different options for carrying out MRP, and finally, explore the possibilities for analyzing the planning results.

Introduction to MRP

In this introductory section, we will highlight the functions of MRP. We will then look at the possible results of an MRP run. Finally, we will explain the different planning procedures that are offered in an SAP S/4HANA system.

Function of MRP

The main function of MRP is to ensure material availability to cover all requirements and avoid delays in order fulfillment. It is the task of the department for MRP and disposition to identify, analyze, and, if possible, resolve existing shortfall situations—and to convert procurement proposals into production orders or purchase orders in good time.

Automated MRP is intended to reduce the workload of the MRP controllers as much as possible so that they can focus on those materials for which the system cannot automatically create a balance between requirements and requirements coverage (e.g., due to vendor backlogs).

MRP Results

Automated MRP focuses on shortages and supports the MRP controller in recognizing these quickly and processing them efficiently. The system automatically creates *procurement proposals* for purchasing and production. *Exception messages* are generated in the case of shortages that the planning run cannot resolve.

Planned orders, purchase requisitions, and *scheduling agreement schedule lines* are the possible procurement proposals that can be created by the automatic planning run. Planned orders and purchase requisitions are internal planning elements that can be changed, rescheduled, or deleted at almost any time. In contrast, schedule lines are fixed elements with a binding character; they are transmitted to the supplier either directly or indirectly in the form of forecast or just-in-time delivery schedules.

For each plant, you can define whether a material is produced in-house or procured externally, or whether both procurement types are possible. You determine this using an indicator in the material master record called *procurement type*. The procurement type of a material is a factor that influences which procurement proposals are created by MRP. Figure 8.1 shows a material that is procured externally (the indicator is F). An internally produced material would have an E, and if you allowed both, you would set an X in the **Procurement Type** field.

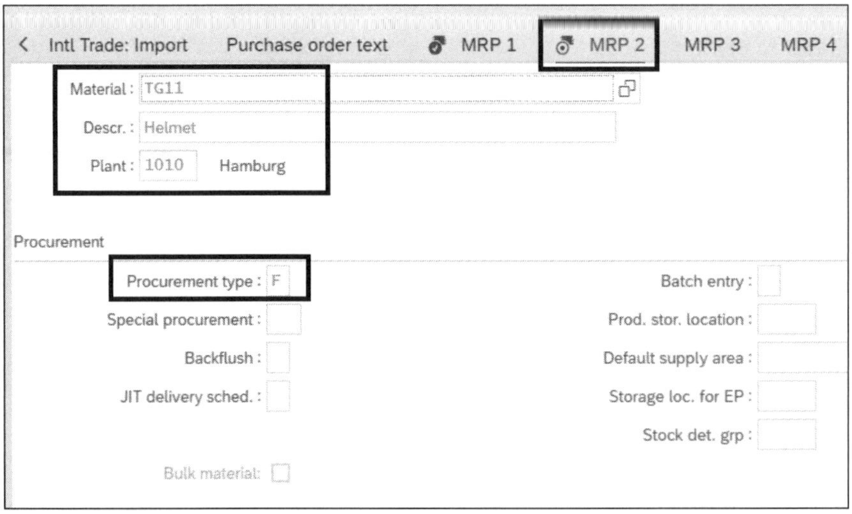

Figure 8.1 Procurement Type of a Material

The second factor influencing the creation of procurement proposals by MRP is the method you choose to run MRP. SAP S/4HANA offers two ways to start the MRP run: MRP Live (a new MRP run optimized for SAP HANA) and the so-called classic MRP, which was already available in SAP ERP. In the further course of this book, we will come back to the special features, restrictions, and significant differences of these methods where it makes sense.

Figure 8.2 summarizes the entire process and shows under which conditions procurement proposals are created:

- **In-house production**
  The system creates planned orders for materials produced in-house, regardless of whether you work with classic MRP or MRP Live. Once planning is complete, the planned orders can be converted into production orders.
- **External procurement**
  For externally procured materials, you must distinguish between MRP Live and classic MRP:
  - MRP Live always creates delivery schedule lines for externally procured material with valid delivery schedules, and it creates purchase requisitions for all other externally procured material.
  - With classic MRP, you decide when starting the MRP run whether the system creates a planned order or directly a purchase requisition for planning the externally procured quantity. If a planned order is created, the MRP controller must first convert it into a purchase requisition before the purchasing department can create a purchase order. A purchase requisition, however, is immediately available to the purchasing department. If a scheduling agreement exists for a material, it is also possible to create delivery schedule lines directly in MRP.

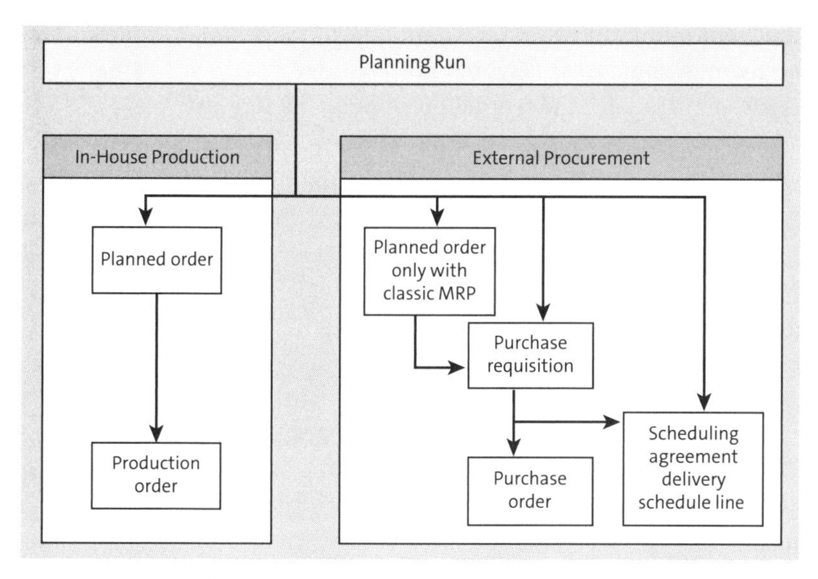

Figure 8.2 MRP Results: Procurement Proposals

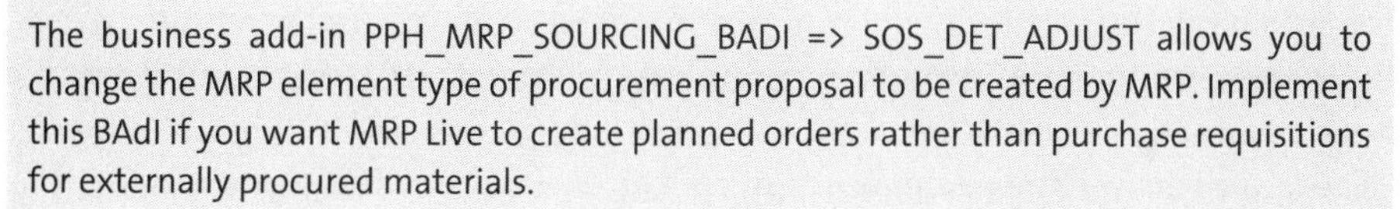

Note

The business add-in PPH_MRP_SOURCING_BADI => SOS_DET_ADJUST allows you to change the MRP element type of procurement proposal to be created by MRP. Implement this BAdI if you want MRP Live to create planned orders rather than purchase requisitions for externally procured materials.

MRP Procedures

Several planning procedures are available in SAP S/4HANA. We essentially distinguish between demand-based and consumption-based planning procedures.

In *demand-based planning*, the planning logic calculates the stock balance at any point in time, considering the available stock, the planned receipts, and issues. In the event of shortages, SAP S/4HANA creates appropriate requirements coverage in the form of procurement proposals. Planned receipts include, for example, planned orders, purchase requisitions, purchase orders, and manufacturing orders. Planned issues—which trigger requirements—include planned independent requirements, sales orders, reservations, dependent requirements, and stock transfer requirements.

Consumption-based planning procedures are based on previous material consumption. Concrete requirements such as sales orders or reservations are not usually included in the planning calculation. The following planning procedures are available in consumption-based planning:

- A very common procedure for consumption-based planning is the *reorder point procedure*. In this procedure, the system checks whether the stock available for planning falls below the reorder point from the material master. We distinguish between *manual reorder point planning*, in which the MRP controller determines the reorder point manually and *automatic reorder point planning*, in which the system calculates the reorder point using the forecast.

- Another method is *forecast-based planning*. MRP creates procurement proposals if the total forecast requirements up to a certain point in time exceed the total firmed receipt quantity up to the same point in time. The total firm receipt quantity includes stock, production orders, purchase orders, firm planned orders, and firm (or fixed) purchase requisitions. You will find a definition of fixed purchase requisitions under the "Net Requirements Calculation in Reorder Point Planning" heading.
- A variant of the forecast-based planning procedure is the *time-phased planning* procedure. In time-phased planning, future requirements are also estimated using the forecast values. In this procedure, however, planning is only carried out at fixed times within a certain time interval, for example, to consider the fixed delivery weekday of a vendor.

A low-value material with a relatively constant consumption is suitable for consumption-based methods. A high-value material with strong fluctuations in demand is more eligible for demand-driven procedures. Demand-driven planning is particularly suitable for the planning of finished products and important assemblies and components (A parts according to ABC analysis), while consumption-based planning is preferably used for areas without in-house production or for the planning of B and C parts (according to ABC analysis), as well as for operating resources.

Tip

Neither demand-driven planning nor forecasting are relevant for certification. In the following chapters, we will therefore concentrate on manual reorder point planning. Nevertheless, you must know the different planning procedures and be able to name the differences as they are explained here.

The *MRP type* is the key that you use to define the procedure for planning a material. The MRP type is part of the plant data (or the MRP area data of a material) and is entered in the material master record. This allows a material to be planned in different plants using different MRP procedures.

Many MRP types are delivered in SAP S/4HANA. In Customizing, you can configure new MRP types that better meet your requirements. Figure 8.3 shows a material that is planned with the reorder point planning procedure.

Note

The MRP type is discussed in more detail in the section "MRP Type and Associated Parameters."

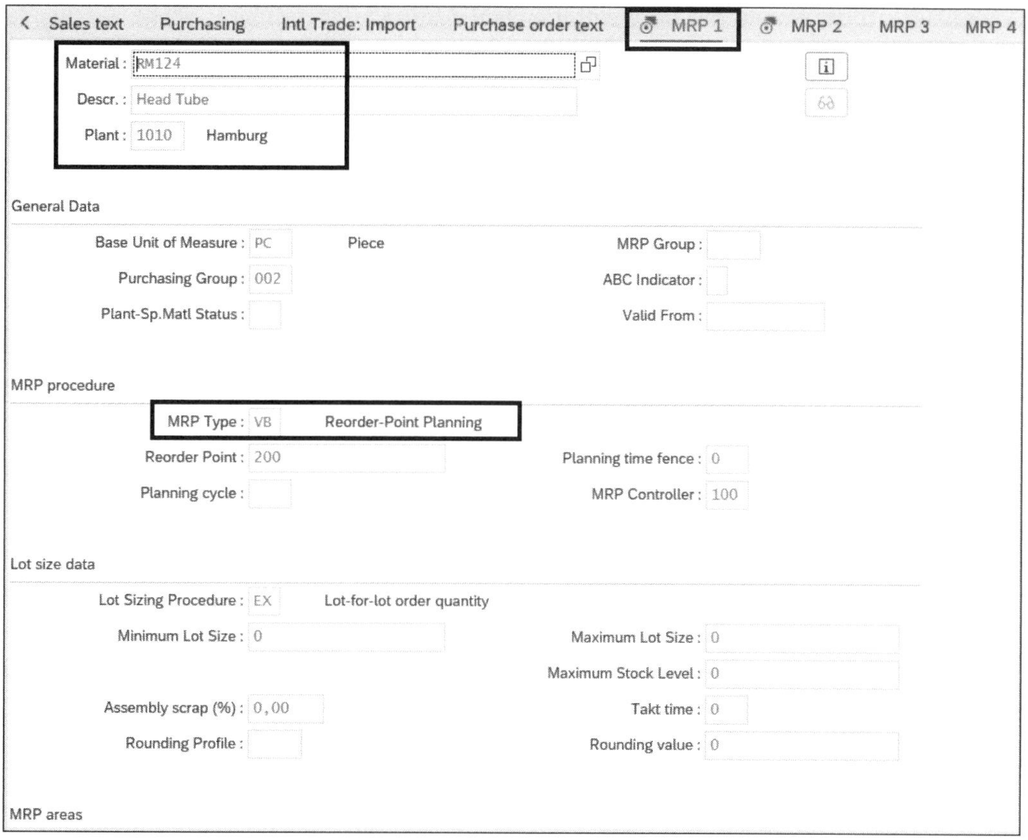

Figure 8.3 MRP Type of a Material

Planning Level in the Organizational Structure

The organizational unit relevant for materials planning in SAP S/4HANA is the *MRP area*. The MRP area corresponds to a plant or part of a plant.

We distinguish between the following MRP area types:

- Plant MRP area (type 01)
- MRP area for storage location(s) (type 02)
- MRP area for subcontractor (type 03)

When you create a plant in SAP S/4HANA, the system automatically creates a corresponding MRP area in the background with the same number. If you do not make any further settings (that is, if you do not create storage location MRP areas or MRP areas for subcontractors), the plant MRP area covers the whole plant, and MRP areas and plants are identical. If you want to carry out MRP at the plant level, you do not need any further configuration.

However, if, for example, you want to manage your standard storage location separately from your spare parts storage location within a plant, you will create MRP areas for storage locations. An MRP area for storage locations corresponds to one or more storage locations within a plant. A storage location can only be assigned to one MRP area.

Using MRP areas, you can also plan the provision of components for individual subcontractors separately. You then define an MRP area for each subcontractor. In other words, storage location MRP areas and subcontractor MRP areas enable a material to be planned differently within a plant.

The plant MRP area initially includes the whole plant with all storage locations and subcontractor stocks. If you then define MRP areas for storage locations and for subcontractors, the plant MRP area is reduced by exactly these storage locations and subcontractors, as they are now planned independently.

Figure 8.4 shows three examples of how you can define MRP areas within a plant. Here, plant 1010 corresponds to the default configuration. The entire plant stock is considered during planning, and every material maintained in this plant is planned uniformly within the plant. There are no specific settings required.

In plant 1020, the storage location 0003 is planned separately from the rest of the plant. To fulfill this requirement, you create an MRP area for storage location and assign it to the storage location 0003. The plant-MRP area is reduced, then, by exactly this storage location. Since the remaining storage locations are planned together, they can remain assigned to the plant-MRP area. No further settings are required for them.

Plant 1030 is divided into three MRP areas: two groups of storage locations, each corresponding to a storage location-MRP area, and an additional MRP area for a subcontractor.

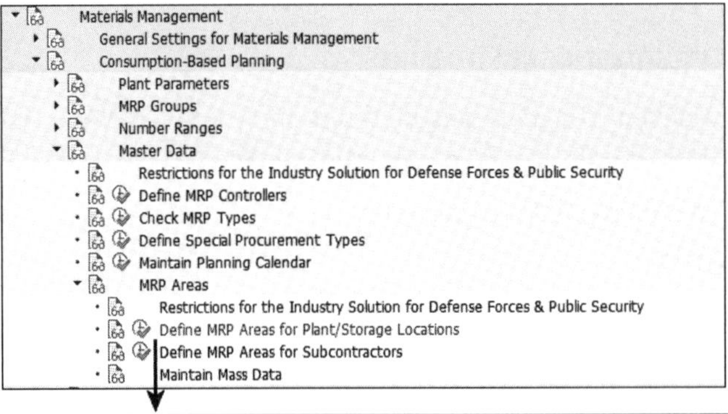

Change View "MRP Areas - Plant/Storage locations": Overview

New Entries

Dialog Structure
- MRP Areas - Plant/Storage locations
  - Plant (one entry per MRP area)
  - Assign storage locations

MRP Area	MRP Area Type	MRP Area Text	Plant	Name 1
0001	01	Werk 0001	0001	Werk 0001
0003	01	Plant 0003 (is-ht-sw)	0003	Plant 0003 (is-ht-sw)
1000	01	Berlin	1000	Berlin
1010	01	Plant 1 DE	1010	Hamburg
1020	01	Plant 2 DE	1020	Berlin
1030	01	Frankfurt	1030	Frankfurt
1040	01	Munich	1040	Munich
1710	01	Plant 1 US	1710	Plant 1 US

Figure 8.4 Planning Levels

You maintain MRP areas in Customizing. You can find the path in Figure 8.5.

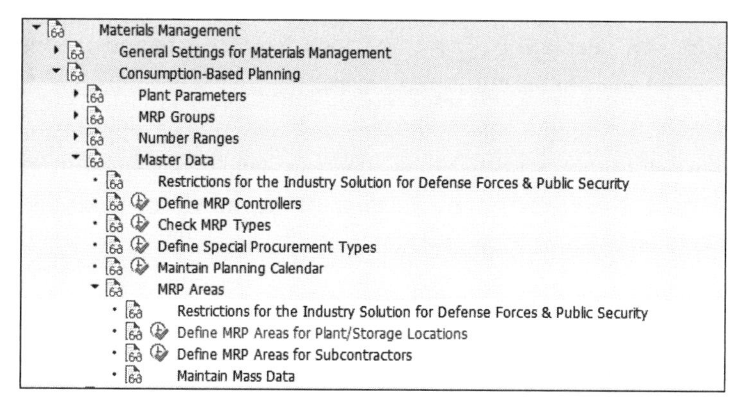

Figure 8.5 Path to Customizing of MRP Areas

Figure 8.6 shows an example of customizing MRP areas for plant/storage locations.

Change View "MRP Areas - Plant/Storage locations": Overview

New Entries

Dialog Structure
- MRP Areas - Plant/Storage locations
  - Plant (one entry per MRP area)
  - Assign storage locations

MRP Area	MRP Area Type	MRP Area Text	Plant	Name 1
0001	01	Werk 0001	0001	Werk 0001
0003	01	Plant 0003 (is-ht-sw)	0003	Plant 0003 (is-ht-sw)
1000	01	Berlin	1000	Berlin
1010	01	Plant 1 DE	1010	Hamburg
1020	01	Plant 2 DE	1020	Berlin
1030	01	Frankfurt	1030	Frankfurt
1040	01	Munich	1040	Munich
1710	01	Plant 1 US	1710	Plant 1 US

Figure 8.6 Customizing MRP Areas: Example of MRP Areas for Plant/Storage Locations

MRP areas that correspond to a plant are generated automatically and are available as standard in the SAP S/4HANA system. Their number is exactly the plant number. MRP areas that correspond to storage locations or subcontractors must be created in Customizing. Their number is assigned manually and must contain at least five alphanumeric digits.

Material Master Data for Reorder Point Planning

A material can only be planned automatically if it has been maintained in the respective plant or MRP area.

The MRP data in the material master record can be divided up as follows:

- General data that can or must be defined (MRP controller, procurement type, safety stock, etc.)
- Data that controls the MRP procedure (MRP type, reorder point by reorder point planning, etc.)

- Data that is required for scheduling (planned delivery time, goods receipt processing time, etc.)
- Data that is required for calculating the lot size (lot-size procedure and dependent data like fixed lot size or maximum stock level)

In the following section, we analyze the material master data that is relevant for manual reorder point planning and discuss its effects in detail. For certification purposes, it is important to understand the meaning of the individual fields discussed, but you only need to know their characteristics for manual reorder point planning.

Level of Material Data Relevant for MRP

The data required for MRP is basically stored in the material master record at the plant level.

If you have created storage location MRP areas and want to plan materials separately in these MRP areas, you must maintain MRP area-specific data. Otherwise, these materials can only be planned at the plant level.

If you have MRP areas for subcontractors, it is not necessary to maintain your materials for each individual subcontractor. If you do not maintain specific master data for the subcontractor MRP area, then MRP in SAP S/4HANA uses the plant data for default planning parameters.

Although it is not necessary to create an MRP area-specific material master record for each part provided by the subcontractor, it is of course possible if you want to override the default planning parameters.

MRP Type and Associated Parameters

You use the MRP type in the material master record at the plant or MRP area level to define the procedure to be used for planning the material. Depending on the MRP type, the system prompts you to enter further MRP parameters in the material master record. For example, if you select MRP type VB for manual reorder point planning, you must enter a reorder point.

Figure 8.7 shows some MRP data of a material, especially the MRP type and the reorder point on the **MRP 1** tab.

Standard SAP S/4HANA includes several MRP types that cover the most common procedures. You can create further MRP types according to your requirements in Customizing under **Materials Management • Consumption-Based Planning • Master Data • Check MRP Types**.

SAP S/4HANA provides MRP type VB for the manual reorder point procedure and type VM for the automatic reorder point procedure. The only difference between the two procedures is that in the manual reorder point procedure, the reorder level is entered manually in the material master record. In the automatic reorder point procedure, the reorder level is calculated automatically by the system using the forecast.

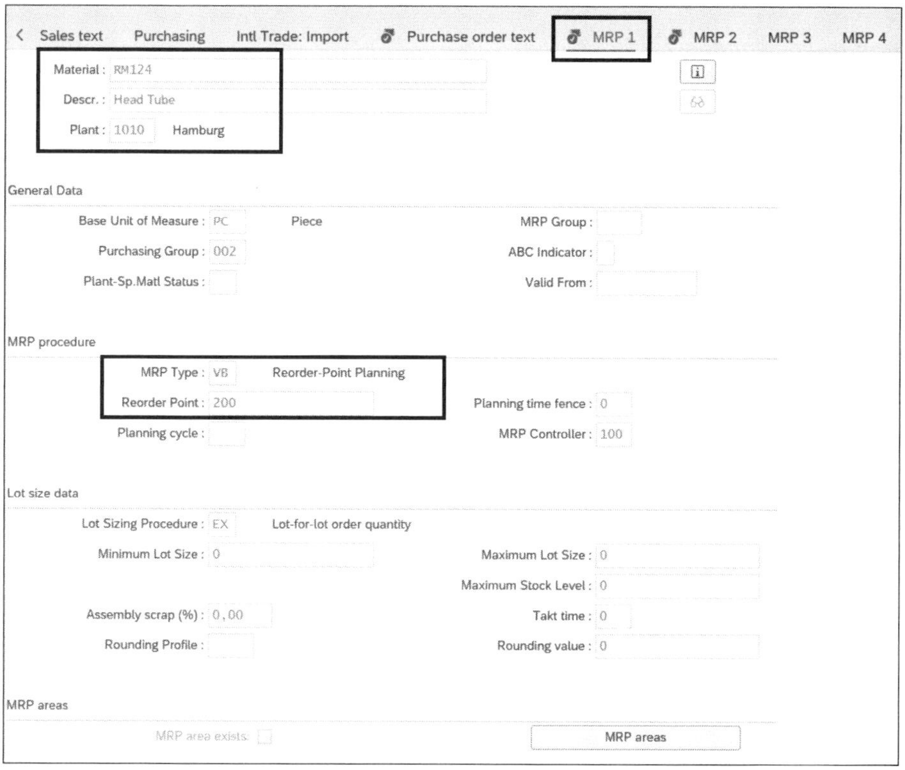

Figure 8.7 Material Master Data for MRP: MRP Type

In the following discussion, we describe the principle of reorder point planning in detail. We outline the net requirements calculation executed by the system during the planning run. In this context, we provide some definitions and explain the function and determination of reorder point and safety stock.

Net Requirements Calculation in Reorder Point Planning

Reorder point planning is based on the comparison of the *available MRP stock* with the *reorder point* of a material in a plant or MRP area.

The system calculates the available MRP stock using the following formula:

Available MRP stock = Inventory + Fixed planned orders + Fixed purchase requisitions + Purchase orders

Note

A fixed purchase requisition is either a purchase requisition that was created manually or a purchase requisition that was created via MRP but then changed manually. In both cases, such a purchase requisition cannot be changed automatically by the planning run.

The shortage quantity is the difference between the reorder point and the available MRP stock.

Reorder Point and Safety Stock

The reorder point should be defined so that it can cover the expected average material requirements during the replenishment lead time.

Optionally, you can also define a safety stock level for a material and a plant or an MRP area. Figure 8.8 shows the safety stock in a material master record on the **MRP 2** tab. This safety stock level is intended to cover both the additional material consumption that may occur during the replenishment lead time and the normal requirements in the event of delivery delays. The safety stock is therefore part of the reorder level.

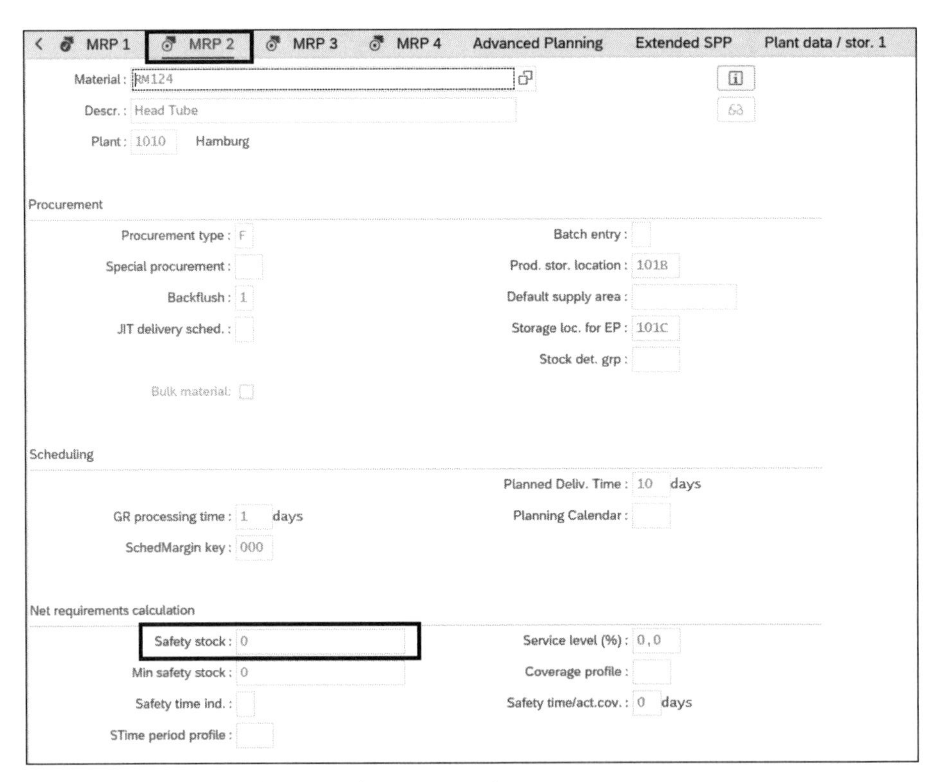

Figure 8.8 Safety Stock in the Material Master Record

The reorder level is made up of the expected material requirements during the APC and the safety stock level. Hence the formula that you should apply when determining the reorder point of a material:

Reorder point = Safety stock + Expected daily requirement × Replenishment lead time

After the net requirements calculation, the system determines the lot size using the lot-sizing procedure and associated parameters in the material master record.

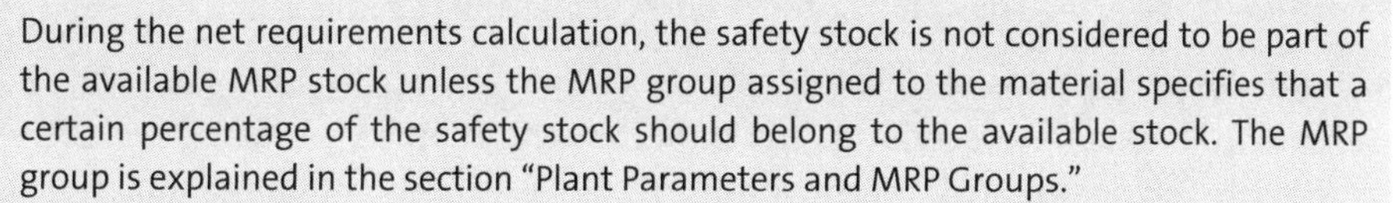

Note

During the net requirements calculation, the safety stock is not considered to be part of the available MRP stock unless the MRP group assigned to the material specifies that a certain percentage of the safety stock should belong to the available stock. The MRP group is explained in the section "Plant Parameters and MRP Groups."

Lot Size Calculation

During a planning run, the system first carries out a net requirements calculation. The system determines a shortage quantity that must be covered by receipts. A lot-size calculation is then executed, and procurement proposals are created.

You control how the system calculates the lot sizes by selecting a *lot-sizing procedure* in material master record maintenance. There are also other options in the material master record that you can use to influence the lot size, such as *rounding*, *minimum lot size*, and *maximum lot size*. Figure 8.9 shows the lot-size data in a material master record in the **MRP 1** tab.

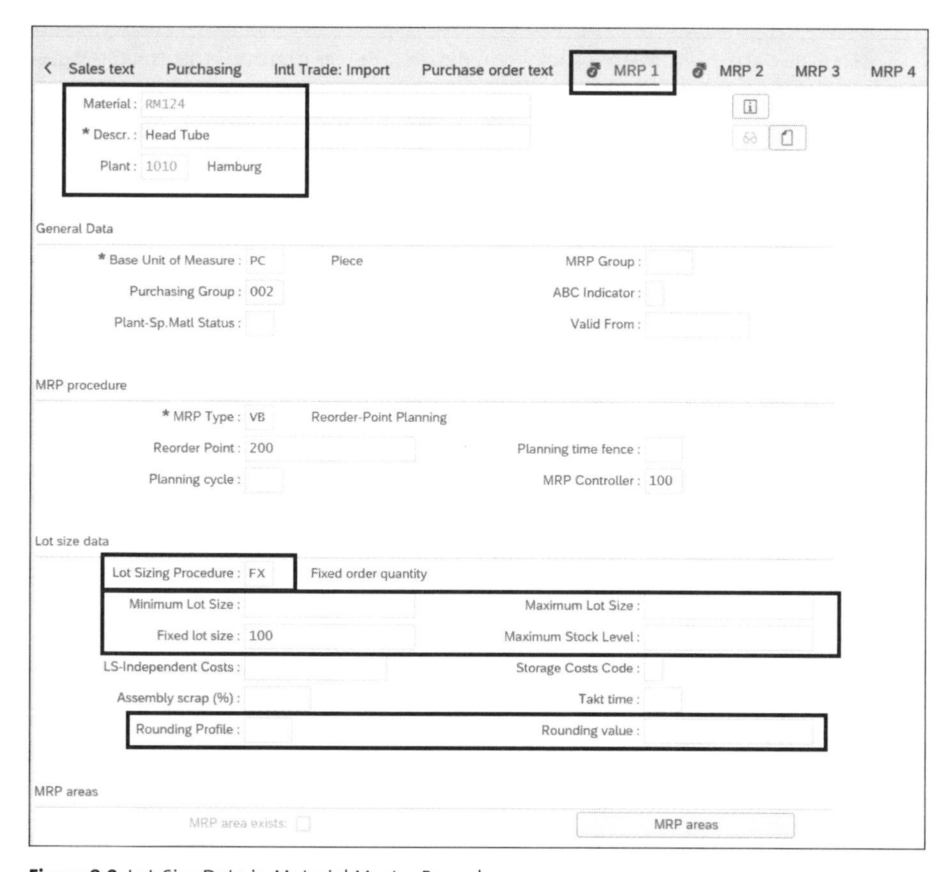

Figure 8.9 Lot-Size Data in Material Master Record

Lot Sizing Procedures

There are three categories of procedures for calculating the lot size:

- **Static lot-sizing procedures**
  In static lot-sizing procedures, the procurement quantity is calculated exclusively based on quantity specifications from the respective material master.
  - For the *exact lot size* (procedure EX), the system creates an order proposal for exactly the calculated shortage quantity. In reorder point planning, the exact lot size means that the system proposes exactly the difference to the reorder level. In reorder point planning, this procedure is therefore only used to a limited extent (e.g., for spare parts), as it can lead to frequent small order proposals.
  - With the *fixed lot size* (procedure FX), the system creates a procurement proposal for the fixed lot size in case of a shortage. If this is not sufficient to cover the requirement, the system creates several procurement proposals of the same quantity for the same date.
  - When *replenishing up to the maximum stock level* (procedure HB), the system creates a procurement proposal for the difference between the available stock and the maximum stock level defined in the material master record.
- **Periodic lot-sizing procedures**
  In period lot-sizing procedures, the system groups together several requirement quantities within a time segment to form a lot size. The period lengths can be days, weeks, months, or a period of flexible length like accounting periods, as well as freely definable periods according to the planning calendar. In consumption-based planning, periodic procedures are only suitable for forecast-based planning.
- **Optimizing lot-sizing procedures**
  In the optimizing lot-sizing procedures, requirement quantities from several periods are grouped together in one lot size, whereby an optimum cost is determined between lot size-independent costs and storage costs.

  Details about optimizing lot-sizing procedures are not further explained and are not part of the certification.

Rounding

You can use rounding to adjust the procurement quantities to the delivery, packaging, or transport units. This is recommended, for example, if materials are only delivered in packs with a fixed number of pieces or in whole pallets. There are two ways of making rounding specifications in the material master:

- **You enter a rounding value.**
  The lot size calculated by the MRP run is then rounded to a multiple of this value.

- **You enter a rounding profile.**

  Rounding profiles are defined in Customizing under **Materials Management • Consumption-Based Planning • Planning • Lot-Size Calculation • Maintain Rounding Profile**.

  A rounding profile is composed of threshold values and rounding values. The threshold value is the value from which the system rounds up. The rounding value is the value to which the system rounds up if the threshold value is exceeded. You can define several combinations of threshold and rounding values for a rounding profile.

  You can use a rounding profile to ensure, for example, that price and quantity scales are used optimally. Table 8.1 gives examples to illustrate the system behavior when you enter the profile 0001 shown in Figure 8.10 in a material master record.

Calculated Requirement Quantity	Final Procurement Quantity after Rounding
19	19
26	50
65	65
78	100

Table 8.1 Sample Rounding Values

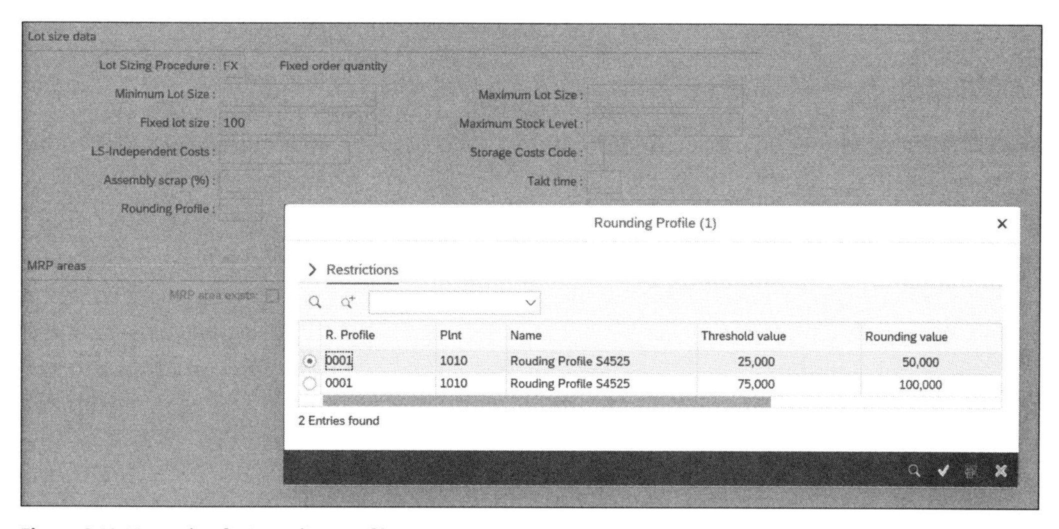

Figure 8.10 Example of a Rounding Profile

Minimum and Maximum Lot Size

You can enter a *minimum* and a *maximum lot size* as limit values in the material master record. These limit values are then considered in the lot-size calculation. The minimum lot size defines the minimum quantity possible for a procurement proposal. The maximum lot size defines the maximum possible quantity for a procurement proposal. The system then rounds up to the minimum lot size or prevents a grouping above the maximum lot size.

Scheduling

After the net requirements calculation and the lot-size calculation, the system schedules the procurement proposal during the planning run. It calculates the date on which the purchase order must be created and sent and the date on which the supplier must deliver the ordered quantity.

The material shortage date for materials planned by reorder point is the date of the planning run. During scheduling, the system determines the date on which the material will be available, starting from the date of the planning run. This procedure is called forward scheduling.

When a purchase requisition is created by MRP, the MRP date is stored in the purchase requisition as the *release date*. "Release" in this context means the release of the purchase requisition for conversion into a purchase order. If a planned order is created by MRP, the MRP date becomes the *order start date* in the planned order.

The availability date is determined with forward scheduling using the following formula:

> *Availability date = Release date + Purchasing department processing time + Planned delivery time + Goods receipt processing time*

Let's look at each of these elements. The *purchasing department processing time* is specified in working days and represents the time required by the purchasing department to convert a purchase requisition into a purchasing document. You define this processing time per plant in Customizing under **Materials Management • Consumption-Based Planning • Plant Parameters • Plant Parameters • Carry Out Overall Maintenance of Plant Parameters.**

The *planned delivery time* is specified in calendar days and represents the time required for the external procurement of a material. The planned delivery time is stored in the material master record at the plant level. You can also store the planned delivery time in contracts, scheduling agreements, or purchasing info records. MRP takes the planned delivery time from outline agreements or purchasing info records to calculate the availability date if the system can determine these sources of supply, otherwise from the material master record.

The *goods receipt processing time* is the time in workdays between the receipt of the material, and its final is put away in the warehouse. The goods receipt processing time is stored in the material master record at the plant level or in outline agreements. MRP takes the goods receipt processing time from outline agreements or purchasing info records to calculate the availability date if the system can determine these sources of supply, otherwise from the material master record.

The sum of purchasing department processing time, planned delivery time, and goods receipt processing time forms the replenishment lead time of an externally procured material.

Executing a Planning Run

In the previous chapters, we have discussed the individual steps involved in a planning run for a single material: net requirements calculation, lot-size calculation, scheduling, and determination of the type of procurement proposal. We have learned which specific fields in the material master influence and control these individual steps. Now we will discuss the options available for starting a planning run and how the scope of a planning run can be defined. First, we will discuss the meaning of the planning file.

The Planning File

The system carries out various subprocesses during a planning run. Checking the planning file entry is the first subprocess that takes place in MRP. The planning file contains all the relevant materials for a planning run. As soon as you create a material master with MRP views and a valid MRP type (except MRP type ND, which stands for "no MRP"), this material is automatically included in the planning file. The planning file controls the planning run and the scope of planning—that is, the planning file determines the materials that are to be included in the different types of planning runs.

Materials that have been subject to an MRP-relevant activity, such as the creation of a purchase order, are automatically flagged with a corresponding indicator (**NETCH** indicator) in the planning file. Here are some examples of changes that are relevant to planning:

- Posting a goods issue or a goods receipt, if this changes the requirements and stock situation of the material (for consumption-based materials, only if the reorder level is exceeded or fallen short of)
- Creation or changes of purchase requisitions, planned orders, purchase documents, sales documents, forecast requirements, dependent requirements, reservations
- MRP-relevant material changes

Figure 8.11 shows the initial screen of the planning file, and Figure 8.12 shows an example of a planning file entry. You can see that the **Net Change Planning** indicator is set (**NETCH** indicator—**Field NChge Plng**). It means that a change has occurred for this material since the last planning run and that the material will be considered in the next planning run.

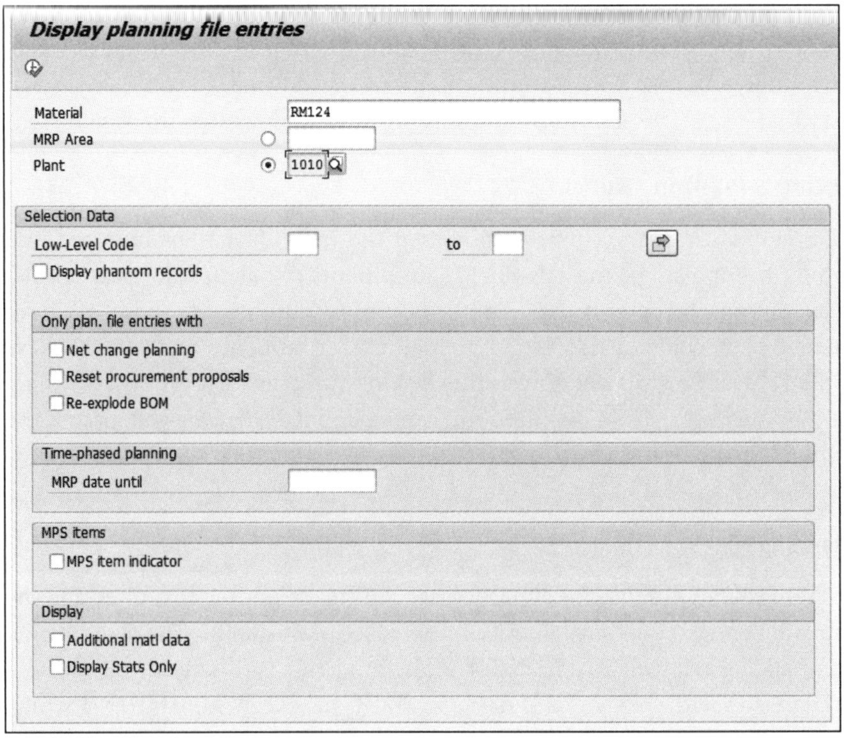

Figure 8.11 Initial Screen of the Planning File

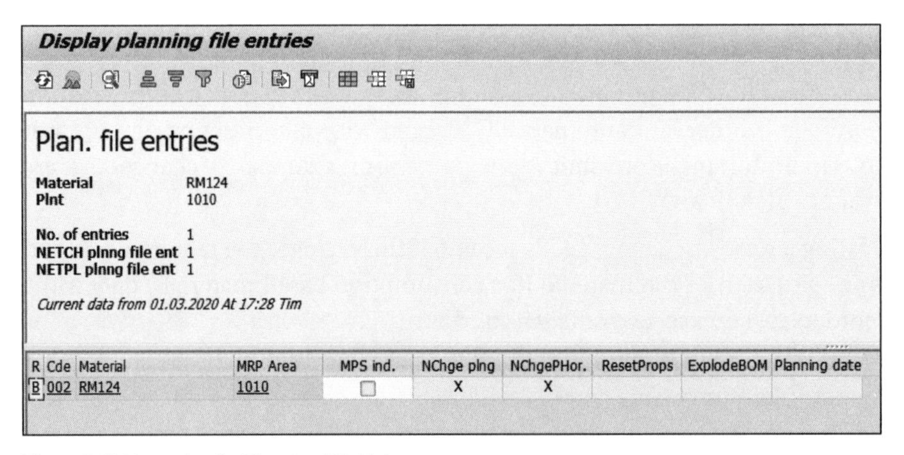

Figure 8.12 Example of a Planning File Entry

Note

SAP S/4HANA no longer supports net change planning in a limited planning horizon (processing key NETPL, Field NChgeHor). This is valid for both the classic MRP and MRP Live.

The *low-level code* (Field Cde) is also part of the planning file entry and controls the sequence in which materials are planned in an MRP run: first the materials with low-level code 0 are planned, then the materials with low-level code 1, and so on.

A material can appear in several bills of material. The low-level code is the lowest level at which a material appears in all bills of material (BOMs). If a material is not contained in a BOM, the system automatically sets the highest level (999 or blank). An entry in the planning file is permanent. Only the indicators are set or removed.

In your backend system, choose **Logistics • Materials Management • Material Requirements Planning (MRP) • MRP • Planning • Planning File Entry • Display (MD21)** to view the planning file entry.

Starting a Planning Run

Let's review how to start the planning run in both SAP S/4HANA options for MRP: classic MRP and MRP Live.

Classic MRP Execution

The classic planning run transactions (i.e., Transactions MD01, MD03, MDBT, and so forth) are still available in SAP S/4HANA.

You can carry out a planning run as a *total planning run* for one or more plants or MRP areas (one after the other, as classic MRP does not execute a cross plant planning). This procedure involves the planning of all the materials that are relevant for planning within the plants or MRP areas contained in the scope of planning.

Figure 8.13 shows the initial screen of a total planning run with a scope of planning.

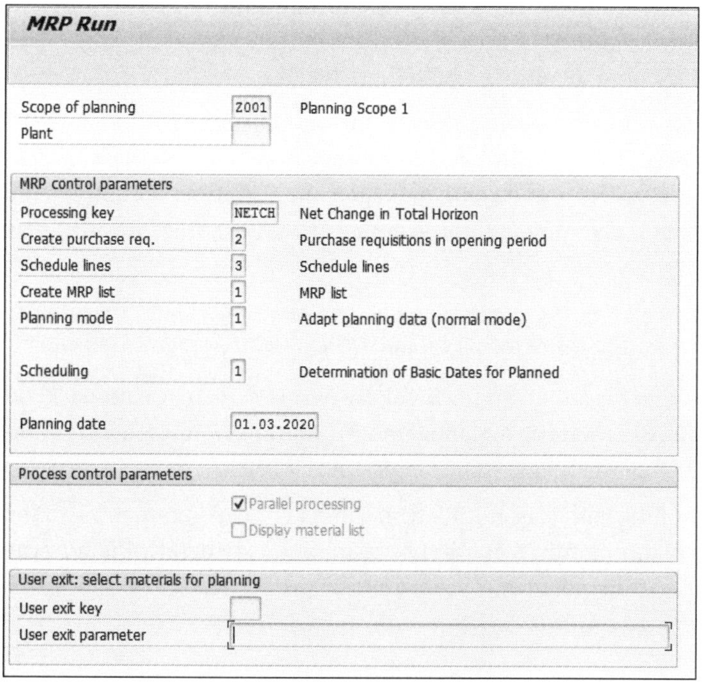

MRP Run

Scope of planning	Z001	Planning Scope 1
Plant		

MRP control parameters

Processing key	NETCH	Net Change in Total Horizon
Create purchase req.	2	Purchase requisitions in opening period
Schedule lines	3	Schedule lines
Create MRP list	1	MRP list
Planning mode	1	Adapt planning data (normal mode)
Scheduling	1	Determination of Basic Dates for Planned
Planning date	01.03.2020	

Process control parameters

- [x] Parallel processing
- [ ] Display material list

User exit: select materials for planning

User exit key	
User exit parameter	

Figure 8.13 Total Planning Run with Scope of Planning

Figure 8.14 shows the details of the scope of planning Z001.

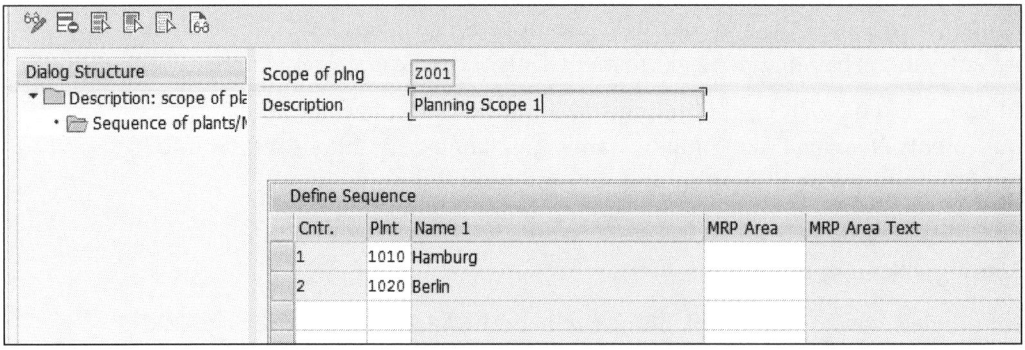

Figure 8.14 Scope of Planning: Details

You can use the scope of planning function for different purposes:

- To execute total planning for several plants and/or MRP areas, one after another
- To execute total planning for one or more MRP areas within a plant

You can define required scopes of planning in Customizing under **Materials Management • Consumption-Based Planning • Planning • Define Scope of Planning for Total Planning**.

For each planning scope, you enter a counter to specify the sequence of individual MRP levels (plants or MRP areas). This counter determines the sequence for planning. If you enter one or more MRP areas for a plant in the scope of planning, you can thereby restrict the total planning run to these levels.

You can carry out a total planning run either online or in the background processing mode. To execute a total planning, on the SAP Easy Access screen in your backend system, choose **Logistics • Materials Management • Material Requirements Planning (MRP) • MRP • Planning • Total Planning (MD01—Online or MDBT—as Background Job)**.

You can also carry out a planning run as a *single-item planning run* for an individual material. Figure 8.15 shows the initial screen of a single-item planning run.

To execute a single-item planning, on the SAP Easy Access screen in your backend system, choose **Logistics • Materials Management • Material Requirements Planning (MRP) • MRP • Planning • Single-Item, Single-Level (MD03)**.

In requirements planning with classic MRP, you can set control parameters on the initial screen of the planning run. You use these parameters to determine how the planning run is carried out and what results are to be produced.

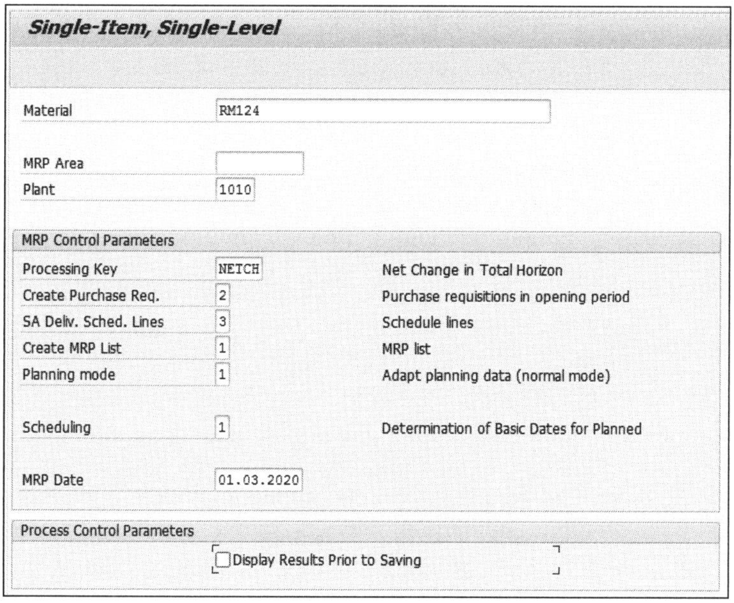

Single-Item, Single-Level

Material	RM124	
MRP Area		
Plant	1010	
MRP Control Parameters		
Processing Key	NETCH	Net Change in Total Horizon
Create Purchase Req.	2	Purchase requisitions in opening period
SA Deliv. Sched. Lines	3	Schedule lines
Create MRP List	1	MRP list
Planning mode	1	Adapt planning data (normal mode)
Scheduling	1	Determination of Basic Dates for Planned
MRP Date	01.03.2020	
Process Control Parameters		
	☐ Display Results Prior to Saving	

Figure 8.15 Single-Item Planning: Control Parameters

The control parameters include ones listed in Table 8.2.

Parameter	Possible Entries and Description
Processing key	The NEUPL processing key is used for regenerative planning. With this processing key, the system plans all materials included in the planning file, irrespective of the indicators.
	With the NETCH processing key, only those materials for which the **NETCH** indicator has been set in the planning file are planned.
Creation indicator for purchase requisitions	1. MRP always creates purchase requisitions for externally procured material.
	2. MRP created purchase requisitions for externally procured material within the opening period. Outside the opening period, planned orders are created.
	3. MRP always creates planned orders for externally procured material.
Creation indicator for scheduling agreements (SA Deliv. Sched. Lines)	1. MRP never creates scheduling agreement schedule lines for externally procured materials for which a valid scheduling agreement exists. Purchase requisitions are created instead.
	2. MRP creates scheduling agreement schedule lines for externally procured materials for which a valid scheduling agreement exists only in the opening period. Outside the opening period, purchase requisitions are created.

Table 8.2 MRP Control Parameters

Parameter	Possible Entries and Description
	3. MRP always creates scheduling agreement schedule lines for externally procured materials for which a valid scheduling agreement exists.
Create MRP list	1. MRP lists are always created.
	2. MRP lists are created depending on exception messages.
	3. MRP lists are never created.
Planning mode	1. The system reuses unfirmed procurement proposals and, if necessary, adapts the quantity and date to the new planning situation. The BOM is re-exploded only for procurement proposals that are to be adapted.
	2. The system reuses unfirmed procurement proposals and, if necessary, adapts the quantity and date to the new planning situation. The BOM is always re-exploded.
	3. The system deletes all procurement proposals that are not firmed and creates new procurement proposals as required. The BOM is always re-exploded.

Table 8.2 MRP Control Parameters (Cont.)

Note

The planning mode indicator specifies how nonfixed procurement proposals from the last planning run are handled in the next planning run. Fixed procurement proposals remain unchanged.

The planning mode is also derived from the planning file.

- Planning mode 1 corresponds to a planning file entry for a material with the **NETCH** indicator only.
- Planning mode 2 corresponds to a planning file entry with the **BOM Re-Exploding** indicator.
- Planning mode 3 corresponds to a planning file entry with the **Reset Order Proposals** indicator.

If the planning mode in the planning file differs from the planning mode in the initial screen of the planning run for a material, the system uses the mode with the highest numerical value.

MRP Live

SAP S/4HANA features MRP Live, an MRP run optimized for SAP S/4HANA. MRP Live reads material receipts and requirements, calculates shortages, and creates planned orders and purchase requisitions, all in one database procedure. This minimizes the volume of data that must be copied from the database server to the application server and back, which considerably improves performance. MRP Live also enables a more flexible planning scope: you can preselect the materials that you want to plan in an MRP Live run.

You can perform an MRP Live planning run in the system in the following ways: On the SAP Easy Access screen in your backend system, choose **Logistics • Production • MRP • Planning • MRP Live (MD01N)** as shown in Figure 8.16, or use the Schedule MRP Runs app on your SAP Fiori launchpad, as shown in Figure 8.17.

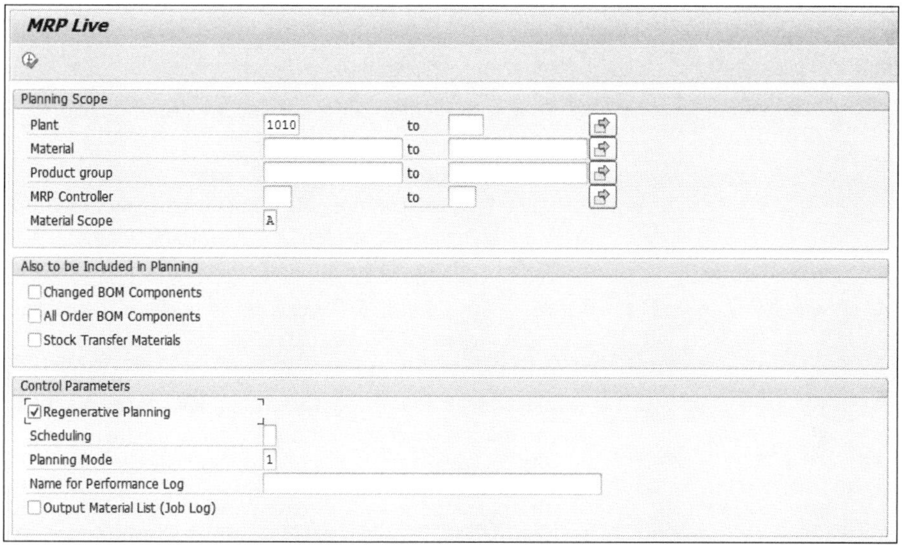

MRP Live

Planning Scope
Plant 1010 to
Material to
Product group to
MRP Controller to
Material Scope A

Also to be Included in Planning
Changed BOM Components
All Order BOM Components
Stock Transfer Materials

Control Parameters
Regenerative Planning
Scheduling
Planning Mode 1
Name for Performance Log
Output Material List (Job Log)

Figure 8.16 MRP Live (Transaction MD01N)

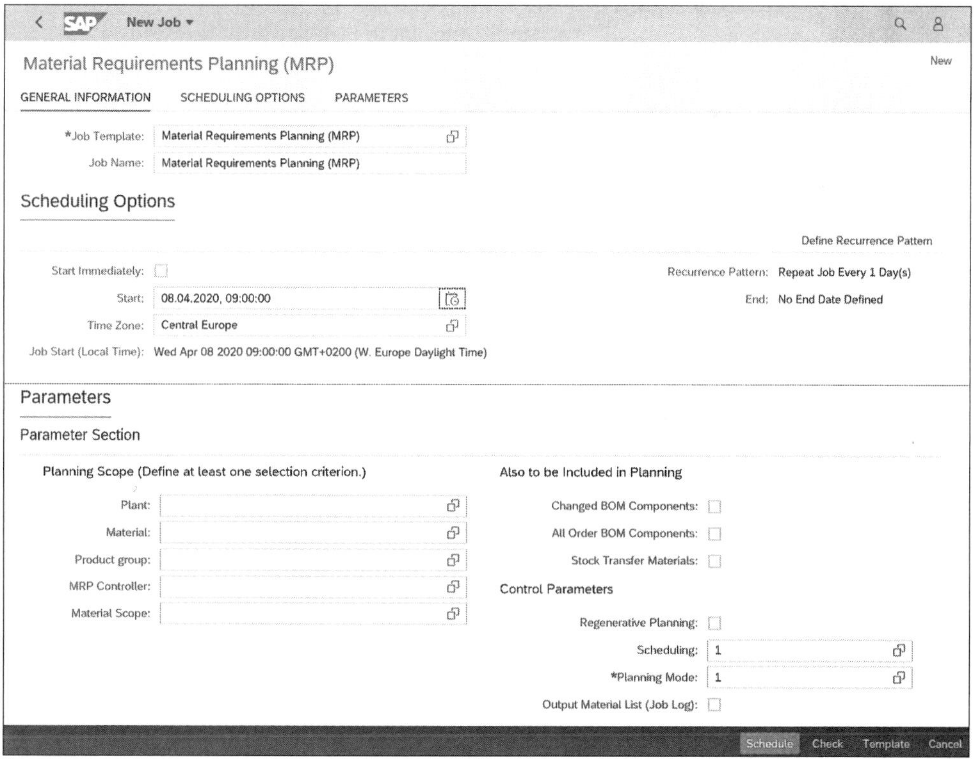

SAP New Job
Material Requirements Planning (MRP)
New
GENERAL INFORMATION SCHEDULING OPTIONS PARAMETERS
*Job Template: Material Requirements Planning (MRP)
Job Name: Material Requirements Planning (MRP)

Scheduling Options
Define Recurrence Pattern
Start Immediately:
Recurrence Pattern: Repeat Job Every 1 Day(s)
Start: 08.04.2020, 09:00:00
End: No End Date Defined
Time Zone: Central Europe
Job Start (Local Time): Wed Apr 08 2020 09:00:00 GMT+0200 (W. Europe Daylight Time)

Parameters
Parameter Section
Planning Scope (Define at least one selection criterion.)
Plant:
Material:
Product group:
MRP Controller:
Material Scope:
Also to be Included in Planning
Changed BOM Components:
All Order BOM Components:
Stock Transfer Materials:
Control Parameters
Regenerative Planning:
Scheduling: 1
*Planning Mode: 1
Output Material List (Job Log):
Schedule Check Template Cancel

Figure 8.17 The Schedule MRP Runs App

Using the MRP Live control parameters, you define how the preselected materials must be planned. For example, by selecting the regenerative planning parameter, you force the system to plan all preselected materials, even if there were no changes since the last planning run.

The main control parameters for MRP Live include the ones listed in Table 8.3.

Parameter	Possible Entries and Description
Changed BOM components	You set this indicator if you want the system to plan the BOM components in addition to the materials selected in the **Scope of Planning** section. The system only then plans the BOM components if the current planning run has changed its dependent requirements. Unchanged BOM components are ignored.
All order BOM components	You set this indicator if you want the system to plan all other BOM components in addition to the materials selected in the **Scope of Planning** section. The system plans the components in any order BOM of the planned materials, even if the component requirements have not changed.
Stock transfer materials	This indicator only affects materials that have a special procurement type "stock transfer." You set this indicator if you want the system to plan the selected materials not only in the plants that are specified in the **Scope of Planning** section but also in their supplying plants if changes relevant to planning have occurred in the supplying plants during the current MRP run.
Regenerative planning	You set this indicator if you want the system to carry out a regenerative planning run—that is, all selected materials are planned, regardless of whether they have been changed for MRP since the last planning run (independent of the **NETCH** indicator in the planning file).
Planning mode	1. Adapt planning data. The system reuses existing unconfirmed procurement proposals from the previous planning run that are to remain unchanged. Existing non-firm procurement proposals from the previous planning run that are no longer 100% valid (quantity and/or date are no longer valid) are deleted and created again.
	2. Delete and recreate planning data. The system deletes all non-fixed procurement proposals and creates new procurement proposals as required.

Table 8.3 MRP Live Control Parameters

Plant Parameters and MRP Groups

In addition to the control parameters in the planning file and on the initial screen of the MRP run, it is possible to define parameters per plant or per group of materials.

Plant parameters are control parameters for requirements planning. You can maintain plant parameters for each plant in Customizing under **Materials Management • Consumption-Based Planning • Plant Parameters • Plant Parameters • Carry Out Overall Maintenance of Plant Parameters.**

Figure 8.18 shows the initial screen for maintaining plant parameters. At the plant level, you maintain parameters such as the purchasing processing time (as shown in Figure 8.19) or the stocks that are included in the calculation of the available stock for materials planning (as shown in Figure 8.20).

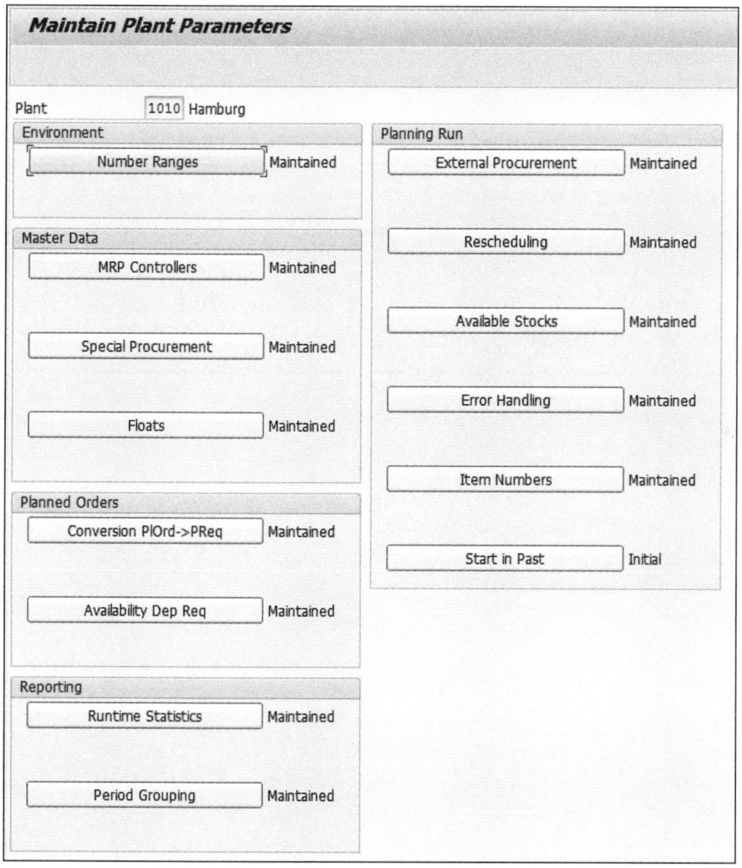

Figure 8.18 Plant Parameters

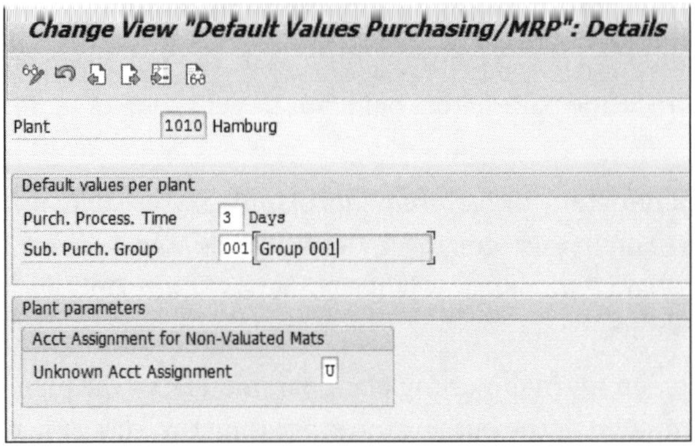

Change View "Default Values Purchasing/MRP": Details

Plant 1010 Hamburg

Default values per plant

Purch. Process. Time 3 Days

Sub. Purch. Group 001 Group 001

Plant parameters

Acct Assignment for Non-Valuated Mats

Unknown Acct Assignment U

Figure 8.19 Example of Plant Parameters: Default Values Purchasing/MRP

Change View "Available Stocks": Overview

Plnt	Name 1	Stock in transfer	Blocked Stk	Restricted-use stock
1010	Hamburg	☑	☐	☐

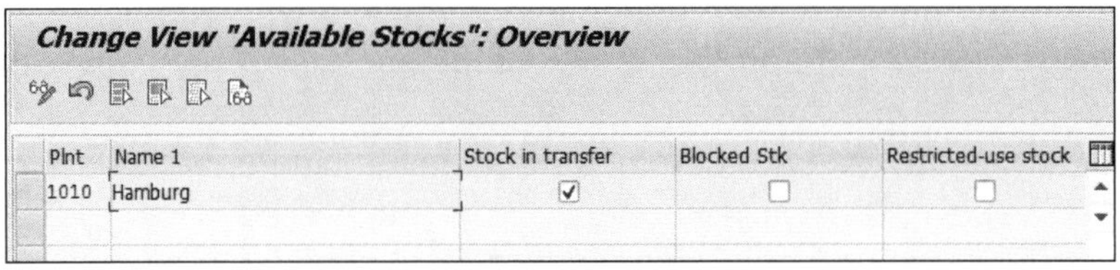

Figure 8.20 Example of Plant Parameters: Available Stocks

The *MRP group* is used to assign certain control parameters for MRP to a group of materials. You can maintain MRP groups and the corresponding parameters in Customizing under **Materials Management • Consumption-Based Planning • MRP Groups • Maintain MRP Groups • Overall Maintenance of MRP Groups**. Figure 8.21 displays the initial screen for maintaining the MRP groups.

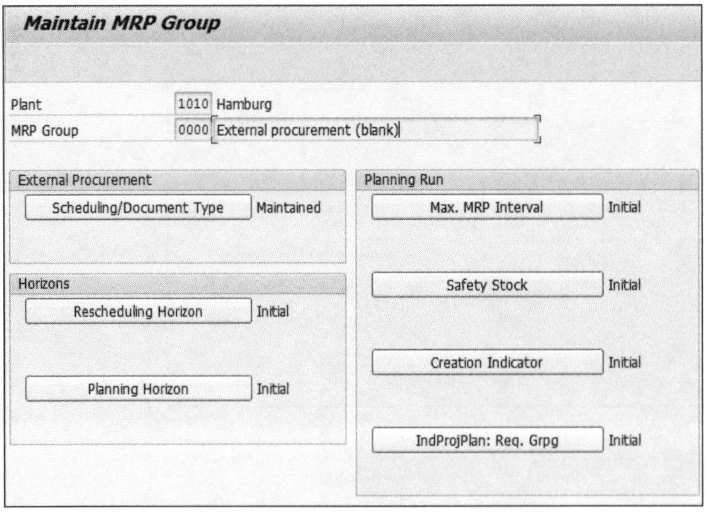

Maintain MRP Group

Plant 1010 Hamburg

MRP Group 0000 External procurement (blank)

External Procurement

Scheduling/Document Type Maintained

Horizons

Rescheduling Horizon Initial

Planning Horizon Initial

Planning Run

Max. MRP Interval Initial

Safety Stock Initial

Creation Indicator Initial

IndProjPlan: Req. Grpg Initial

Figure 8.21 Customizing MRP Groups: Initial Screen

On the MRP group level, you maintain parameters such as the creation indicator for purchase requisitions, schedule lines, or MRP list shown in Figure 8.22.

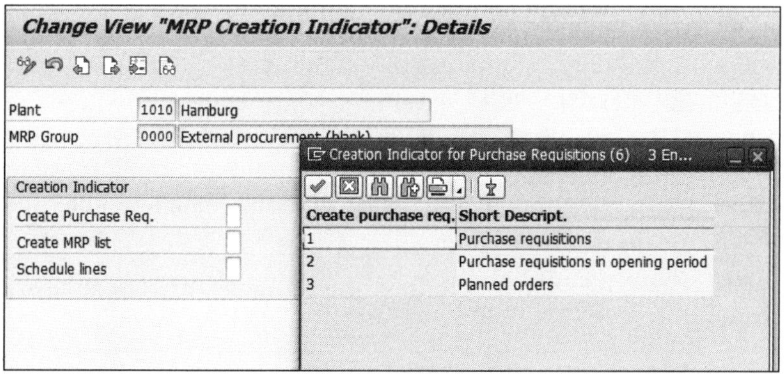

Figure 8.22 The Creation Indicator in the MRP Group

You also can define which percentage of safety stock should be considered as part of the available stock for planning (see Figure 8.23).

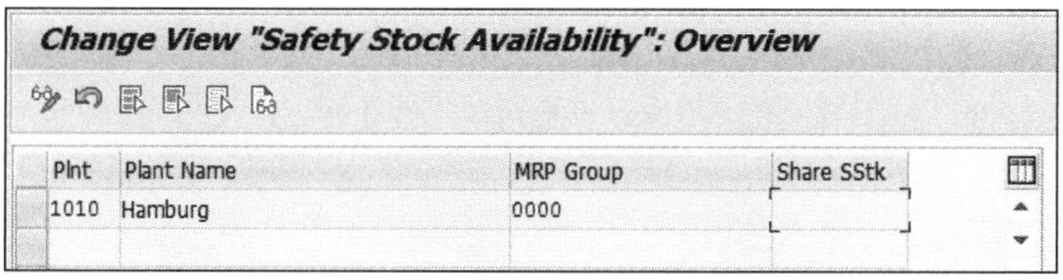

Figure 8.23 Safety Stock in the MRP Group

After you have maintained MRP groups, you can assign these to materials. Figure 8.24 shows the **MRP Group** field in a material master record on the **MRP 1** tab.

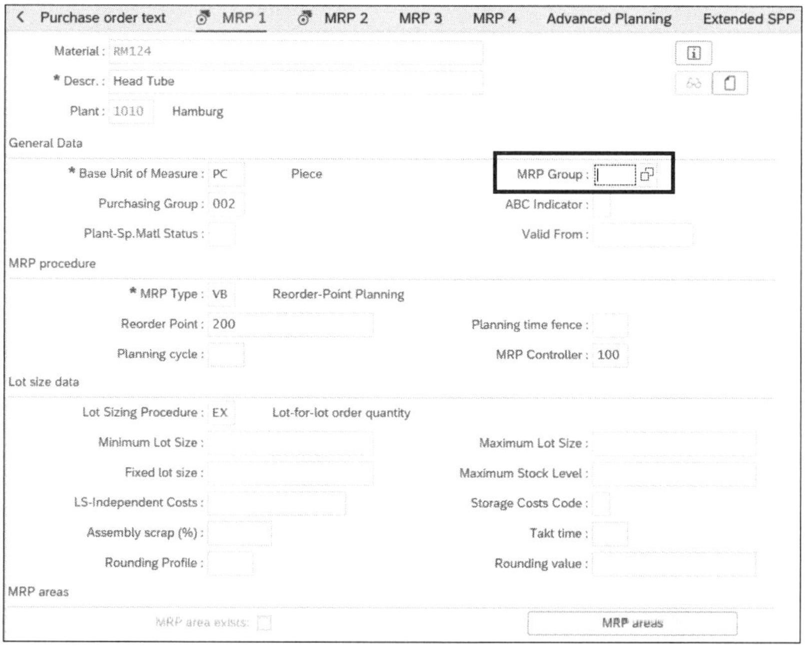

Figure 8.24 MRP Group in the Material Master Record

Some settings can be made both in Customizing for the MRP group and on the initial screen of the MRP run.

During total planning, the system checks for each material whether an MRP group is assigned to the material. If so, the settings of the MRP group have priority over the settings of the plant parameters or the initial screen. If no MRP group is assigned to the material, the system plans the material using the plant parameters and the parameters entered on the initial screen.

In single-item planning, the system always uses the parameters entered on the initial screen for planning.

Note
Maintenance of the plant parameters is a prerequisite for MRP in the respective plant.

Planning Analysis

SAP S/4HANA offers several options for monitoring the planning results. The primary tools are the MRP list and the stock/requirements list. Let's look at each.

The MRP List

The MRP list is exactly the result of the planning run for the scope of planning. The MRP list is therefore a static list with a date stamp (date and time of the MRP run). The MRP list is generated optionally in classic MRP, but also exclusively in classic MRP. On the initial screen of the planning run in classic MRP, an indicator allows you to decide whether the system should generate an MRP list.

The Stock/Requirements List

The current stock/requirements list displays the actual stocks and requirements. The stock/requirements list provides the real-time availability situation for a material. Changes that are made after the planning date are valid immediately. The list is dynamic, and is available in both classic MRP and in MRP Live.

Both individual and collective access are possible for the stock/requirements list and for the MRP list. Figure 8.25 shows the individual access to the stock/requirements list. Here you enter a single material and a plant or MRP area.

Figure 8.26 shows the stock/requirements list for the material selected.

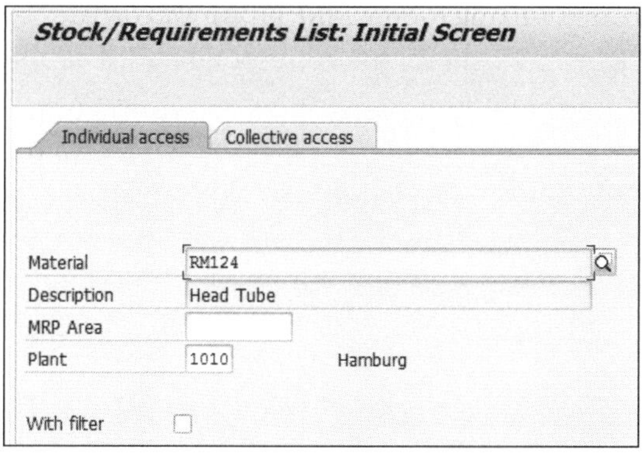

Figure 8.25 Individual Access to the Stock/Requirements List

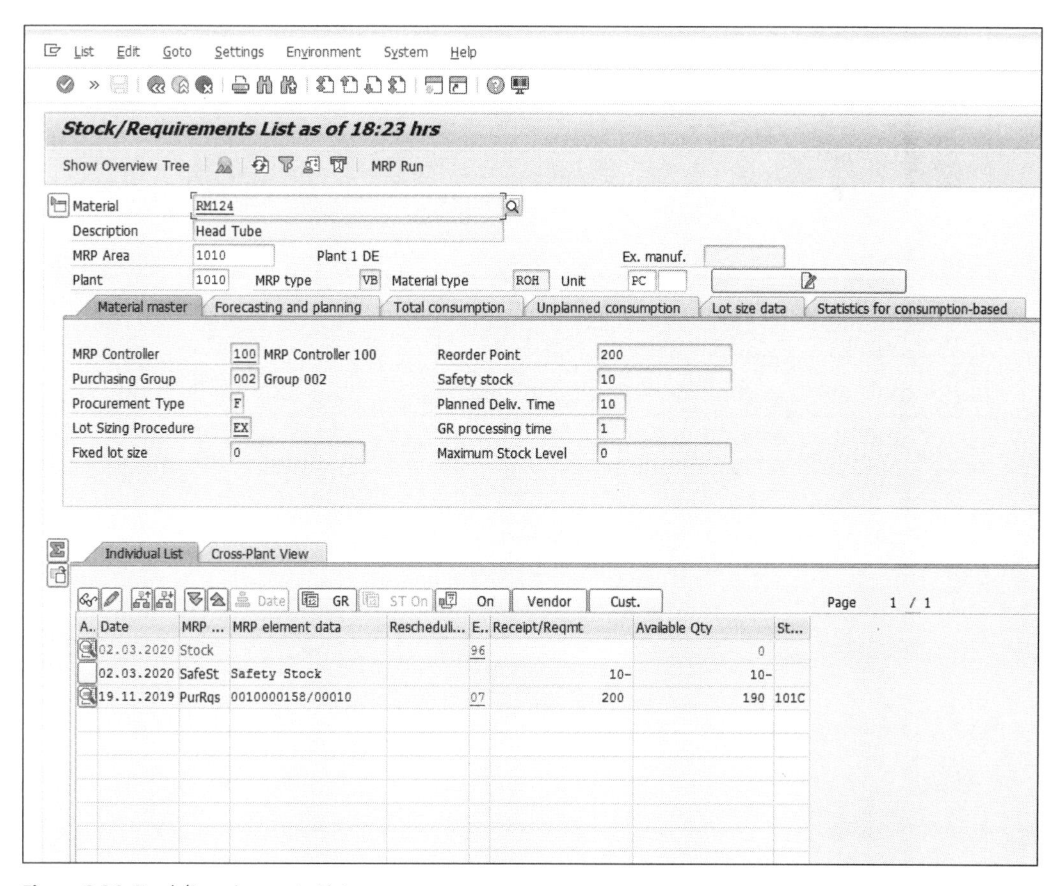

Figure 8.26 Stock/Requirements List

When you choose the *collective* access, the system selects all materials, depending on the selection made in the respective plant. You obtain a material list and can adapt it to suit your requirements (sort materials, adjust column sequence, set traffic lights). From the material list, you can branch to individual stock/requirements lists.

Figure 8.27 shows the initial screen for the collective access to the stock/requirements list with selection options. Figure 8.28 shows the corresponding list of the selected materials.

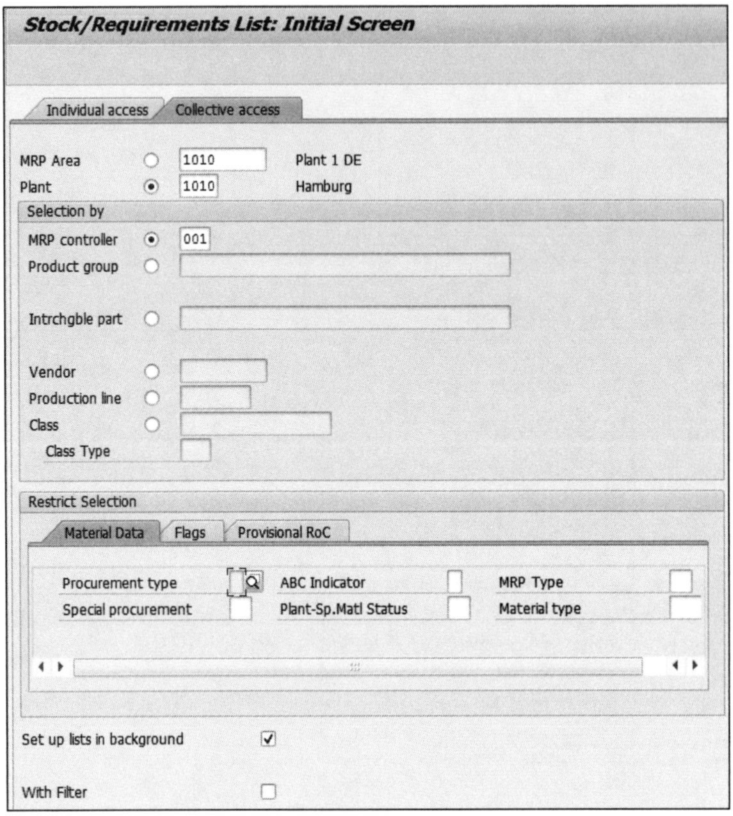

Stock/Requirements List: Initial Screen

Individual access | Collective access

MRP Area 1010 Plant 1 DE
Plant 1010 Hamburg

Selection by
MRP controller 001
Product group
Intrchgble part
Vendor
Production line
Class
Class Type

Restrict Selection
Material Data | Flags | Provisional RoC
Procurement type | ABC Indicator | MRP Type
Special procurement | Plant-Sp.Matl Status | Material type

Set up lists in background ☑
With Filter ☐

Figure 8.27 Collective Access to the Stocks/Requirements List

Stock/Requirements List: Material List

Selected Stock/Requirements Lists | Define Traffic Light | Exception Groups

Plant 1010 Hamburg
MRP Controller 001 MRP Controller 001

Light	Valid from ...	Material	MRP Area	Material description	A	StckDS	1st R...	2nd R
		M-01	1010	Ball Bearing type 0302		999,9	999,9	999,9
		T-601A300	1010	Deluxe Headlight		999,9	999,9	999,9
		T-601A301	1010	Deluxe Headlight		999,9	999,9	999,9
		T-601A302	1010	Deluxe Headlight		999,9	999,9	999,9
		T-601A303	1010	Deluxe Headlight		999,9	999,9	999,9
		T-601A304	1010	Deluxe Headlight		999,9	999,9	999,9
		T-601A305	1010	Deluxe Headlight		999,9	999,9	999,9
		T-601A306	1010	Deluxe Headlight		999,9	999,9	999,9
		T-601A307	1010	Deluxe Headlight		999,9	999,9	999,9

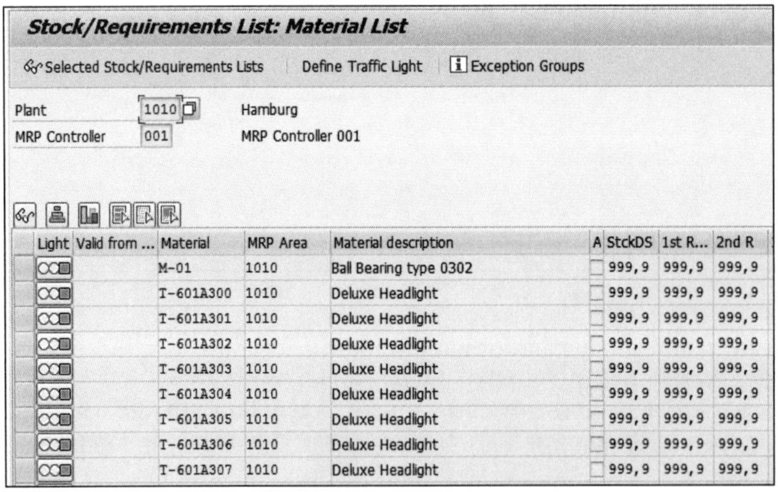

Figure 8.28 List of Selected Material in the Stocks/Requirements List via Collective Access

You have numerous options for influencing the selection and visual presentation of the data on the stock/requirements list and its collective or individual displays. This makes working with the lists more flexible and easier and improves performance.

- Adaptation of navigation with user-specific transaction calls and navigation profiles defined in Customizing
- Restriction of the selection with filters defined in Customizing
- Configuration of the header details
- Grouping exception messages and change the texts of the exception group to meet your needs better

Exception Messages

Exception messages are created during the planning run for exceptional situations that need to be checked by the MRP controller. The exception messages are divided into eight groups. The assignment of the message to the groups and the naming of the groups can be adapted for customer-specific requirements. Depending on the exception group, the materials can be assigned a traffic light.

SAP Fiori Apps for Evaluations

The following evaluations apps are available on the SAP Fiori launchpad:

- **Monitor Material Coverage Net Segments**
  With this app you can monitor all materials in your area of responsibility. You can specify a shortage definition to determine which of the materials have shortages.
- **Monitor Material Coverage Net and Individual Segments**
  With this app you can monitor all materials in your area of responsibility. This includes make-to-stock, make-to-order, engineer-to-order materials, materials for direct production and direct procurement.
- **Check Material Coverage**
  This app helps you to solve coverage issues for individual materials selected in the Monitor Material Coverage app.
- **Display MRP Master Data Issues**
  This app displays issues concerning master data that were detected during an MRP Live planning run.
- **Display MRP Key Figures**
  This app gives you detailed information about each MRP run, no matter whether it is completed, still running or has been cancelled.

Note

We recommend that you test the mentioned transactions and apps in a system in detail so that you become familiar with the possible functions.

Source Determination

During the MRP run, the system calculates the quantities to be procured (requirement quantity/lot size) and the delivery dates. In addition, the source determination process in MRP tries to assign a source of supply to purchase requisition items.

MRP first identifies material shortages. After the lot-size calculation, the planned material receipts are known with the requirements date and quantity. The source of supply determination then determines whether the required quantity is produced or purchased and, if procured externally, from which vendor under which conditions.

For externally procured materials (the procurement type of the material is F), MRP considers valid sources of supply in the following order:

1. **Scheduling agreements**
   When sourcing determines a valid scheduling agreement, MRP Live always creates schedule lines. Instead, classic MRP considers the creation indicator maintained in the initial screen of the planning run, which controls whether schedule lines should be created.
2. **Contracts**
   When sourcing determines a valid contract, MRP creates a purchase requisition assigned to this contract.
3. **Info records with the Automatic Sourcing indicator**
   If source determination does not determine a scheduling agreement or a contract but a purchasing info record relevant for automatic sourcing, a purchase requisition assigned to this source is created.

Note
You can overrule the source determination logic explained previously by using source lists or quota arrangements. Quota arrangements have higher priority than source lists.

The source list overrules the **Automatic Sourcing** indicator in the purchasing info record. If a source list entry for a purchasing info record is flagged as not relevant for MRP (the MRP indicator is blank), the corresponding purchasing info record is not considered by MRP, even if the **Automatic Sourcing** indicator is set. If the source list is flagged as relevant for MRP (the indicator is 1), the purchasing info record is considered by the MRP run, even if the **Automatic Sourcing** indicator is not set.

Quota regulations can be used if several sources of supply are to be used simultaneously. You can define a quota for each source of supply. A quota arrangement is valid for a certain period. The quota specifies how the requirement is to be distributed among the individual sources of supply. MRP always and primarily takes existing quota arrangements into account. The following quota arrangement procedures are available in MRP:

- **Assignment quota arrangement**
  Each lot is assigned to a source of supply.
- **Split quota arrangement**
  A procurement proposal is split and assigned to different sources of supply.

Note
There is a prerequisite if you want to use a splitting quota arrangement: You must use a lot-sizing procedure with a splitting quota for the material in the material master.

Important Terminology

The following terminology was used in this chapter:

- **MRP Live**
  MRP Live is the new MRP run in SAP S/4HANA.
- **MRP procedure**
  This is the principle according to which requirements are calculated: demand or consumption-based.
- **Procurement type**
  The procurement type is a key that is defined in Customizing and stored in the material master record. It controls mainly the MRP procedure.
- **Procurement proposal**
  Procurement proposals are the main results of an MRP run. Among other things, depending on the procurement type of the material, this can be purchase requisitions, delivery schedules, or planned orders.
- **MRP area**
  The MRP area is the organizational level where MRP takes place: it corresponds to a plant or part of a plant or a supplier (subcontractor).
- **Reorder point**
  The reorder point is used in the reorder point procedure. It is entered in the material master record for each plant. As soon as the stock level (available stock) falls below the reorder level, a covering of requirements is triggered.
- **Safety stock**
  The safety stock can be entered in the material master record for each plant. It should cover additional requirements or delays in delivery.
- **Lot-sizing procedure**
  This is used to calculate the lot size when a requirement is determined. Simple static procedures are fixed lot size, exact lot size, or refilling up to a maximum stock level.
- **Replenishment lead time**
  The replenishment lead time for an externally procured material is the sum of

planned delivery time, goods receipt processing time, and purchasing department processing time.

- **Plant parameters**
  These are MRP-relevant parameters that are maintained per plant.
- **MRP group**
  MRP groups are used to assign certain control parameters for MRP to a group of materials.
- **Planning file**
  The planning file contains all the relevant materials for a planning run.
- **MRP list**
  The MRP list contains the planning result for all planned materials. It is only created for classic MRP.
- **Stock/requirements list**
  This describes the current demand situation of a material.
- **Exception message**
  Exception messages are part of the results of the MRP run. They indicate an exceptional situation such as shortage.
- **Source determination**
  Supply source determination is a task of the MRP run, which tries to determine the suitable supply source of supply according to a certain logic.

 Practice Questions

These questions will help you evaluate your understanding of the topics covered in this chapter. They are similar in nature to those on the certification examination. Although none of these questions will be found in the exam itself, they will allow you to review your knowledge of the subject.

Select the correct answers, and then check the completeness of your answers in the next section. Remember that on the exam, you must select all correct answers—and only correct answers—to receive credit for the question.

1. You are implementing reorder point planning for some materials. Which of the following statements apply to the reorder point? (There are two correct answers.)

☐ **A.** The reorder point is composed of the expected material requirements during the replenishment lead time and the safety stock.

☐ **B.** The reorder point should cover additional material consumption during the replenishment lead time, as well as delivery delays.

☐ **C.** You can use the forecast program to automatically determine the reorder point in the material master.

☐ **D.** If the storage location stock falls below the reorder point, procurement is triggered during reorder point planning.

2. Which field in the material master record controls whether a material is subject to consumption-based or demand-based planning?

☐ **A.** MRP type

☐ **B.** Material status

☐ **C.** Procurement type

☐ **D.** MRP group

3. What are the possible results of an MRP run? (There are two correct answers.)

☐ **A.** Planned orders

☐ **B.** Purchase orders

☐ **C.** Delivery schedule lines

☐ **D.** Production orders

4. True or false: With MRP Live, you can choose whether schedule lines are generated instead of purchase requisitions for a material if a valid scheduling agreement exists for the material.

☐ **A.** True

☐ **B.** False

5. Which of the following statements apply to the MRP area? (There are two correct answers.)

☐ **A.** The MRP area as an organizational unit for planning is optional. If you do not activate the use of MRP areas, the plant is the planning level.

☐ **B.** The system automatically creates an MRP area for each plant in the background. If this level is enough for MRP, you do not have to configure any further MRP areas.

☐ **C.** You can define cross-plant MRP areas.

☐ **D.** You can define your subcontractors as MRP areas. However, it is not necessary to create specific material master data for these subcontractor MRP areas.

6. True or false: The available MRP stock of a material is considered for the net requirements calculation during the MRP run and includes all purchase requisitions for this material.

☐ **A.** True

☐ **B.** False

7. What is the purpose of the safety stock for a material? (There are two correct answers.)

☐ **A.** Cover extra material needs that may arise during the replenishment lead time.

☐ **B.** Cover the expected average material requirements during the replenishment lead time.

☐ **C.** Cover the normal expected requirements in the event of delivery delays.

☐ **D.** Increase calculated lot size to create a stock buffer.

8. Where do you specify the lot-sizing procedure?

☐ **A.** In Customizing for the MRP group

☐ **B.** On the initial screen of the MRP run

☐ **C.** In the material master record

☐ **D.** In Customizing of the plant parameters

9. You are testing a classic MRP run. How can you ensure that only those materials are planned that have undergone a change relevant to MRP since the last planning run?

☐ **A.** Select the processing key NETCH on the initial screen of the MRP run.

☐ **B.** Select the processing key NETPL on the initial screen of the MRP run.

☐ **C.** Select the processing key NEUPL on the initial screen of the MRP run.

☐ **D.** Select the scope of planning NETCH on the initial screen of the MRP run.

10. True or false: During the MRP run, the system first reads the planning file to determine the materials to be considered. The system carries out a net requirements calculation for each of these materials. If a material shortage is determined, the lot size is calculated. The system carries out forward scheduling to calculate the availability date of the quantity to be procured. The system issues a corresponding procurement proposal.

☐ **A.** True

☐ **B.** False

11. You want to include 50% of the safety stock of a material in the calculation of the available MRP stock. How do you proceed?

☐ **A.** Use the processing key on the initial screen of the MRP run.

☐ **B.** Use the MRP group in the material master record.

☐ **C.** Use the MRP type in the material master record.

☐ **D.** Use the planning mode on the initial screen of the MRP run.

12. True or false: The MRP list is always generated and reflects the stock/requirements situation of the planned materials directly after the planning run.

☐ **A.** True

☐ **B.** False

Practice Question Answers and Explanations

1. Correct answers: **A, C**

   The reorder point should cover the expected average material requirements during the replenishment lead time. The reorder level is made up of the expected material requirements during the replenishment lead time and the safety stock level. SAP S/4HANA provides MRP type VB for the manual reorder point procedure and MRP type VM for the automatic reorder point procedure. The only difference between the two procedures is that in the manual reorder point procedure, the reorder level is entered manually in the material master record. In the automatic reorder point procedure, the reorder level is calculated automatically by the system using the forecast.

2. Correct answer: **A**

   The MRP type controls the MRP procedure. The material status restricts the usability of a material. The procurement category indicates whether a material can be produced in-house, procured externally, or both.

3. Correct answers: **A, C**

   MRP creates procurement proposals, such as purchase requisitions or planned orders. These must be converted into a purchase order or a production order by the MRP controller. Although there is the option of automatic conversion, this is not possible through MRP. Delivery schedule lines are a special feature: although they have a binding character in contrast to purchase requisitions, they can be generated by the MRP run if a scheduling agreement exists for the material. This option enables the procurement process to be simplified and accelerated.

4. Correct answer: **B**

   False. With classic MRP, you can choose whether schedule lines are generated instead of purchase requisitions for a material with a valid scheduling agreement. A parameter is available for this on the initial screen of the MRP run. Instead, MRP Live always creates schedule lines if a valid scheduling agreement exists. The background to this feature is simplification.

5. Correct answers: **B, D**

   The MRP areas were optional in SAP ERP. You had to activate their use in Customizing. In SAP S/4HANA, we now have a uniform and simplified solution. The MRP area is the single organizational level for MRP. MRP areas can only be created within a plant. They represent a subdivision of the plant. Remember that stock of material provided to subcontractors belongs to your own plant stock.

6. Correct answer: **B**

   False. The available MRP stock of a material is considered in the net requirements calculation during the MRP run, but it only contains the fixed planned receipts—purchase orders and fixed purchase requisitions, for example. Non-fixed purchase requisitions are not included in the calculation.

7. Correct answers: **A, C**

   It is the task of the reorder point to cover the expected average material requirements during the replenishment lead time. The safety stock should cover both the additional material consumption that may occur during the replenishment lead time and the normal requirements in the event of delivery delays.

8. Correct answer: **C**

   The lot-sizing procedure is specified for each plant and material in the material master.

9. Correct answer: **A**

   By setting the NETCH processing key on the initial screen of the MRP run, you ensure that only the materials included in the planning file with the **NETCH** indicator are considered. With the indicator **NEUPL**, all materials in the planning file are considered in the planning run, regardless of their indicator. The processing key NETPL is no longer supported in SAP S/4HANA.

   The scope of planning has a completely different function: you can use it to extend the scope of the planning run to several plants in sequence or to reduce it to only some MRP areas of a plant.

10. Correct answer: **A**

    True. This sentence outlines the process of an MRP run, which you should keep in mind.

11. Correct answer: **B**

    To ensure that a certain percentage of the safety stock is included in the calculation of the available MRP stock, you must create an MRP group in Customizing, define the percentage there, and then enter this MRP group in the material master record. Otherwise, the safety stock does not belong to the available stock for MRP.

12. Correct answer: **B**

    False. The MRP list really represents the stock/requirements situation of the MRP materials directly after the planning run, but it is not generated systematically. It can only be generated optionally for classic MRP. MRP Live no longer offers this option. This is due to the change in the way MRP controllers work: in earlier times, MRP controllers used the MRP list to analyze the planning results. The performance gains now allow them to call up a real-time list.

Takeaway

This chapter describes the planning procedures offered in SAP S/4HANA and the differences between them. You should be able to explain the organizational units that are relevant for planning. You should be familiar with the control parameters for planning in both Customizing and the material master. You should also be able to enumerate and explain the various planning process steps and how sources of supply are determined. You should also know how to carry out MRP (in the classic way and with the new MRP Live) and how to analyze the planning results.

Summary

Production planners must make sure material is available when needed. MRP assists production planners with this task. The MRP run determines expected material shortages and creates planned orders, purchase requisitions, or delivery schedule lines to cover the expected material shortages.

SAP S/4HANA offers MRP Live, a new and high-performance MRP run, but classic MRP is still available. MRP Live represents a certain simplification: MRP lists are no longer created and planned orders are no longer created for externally procured materials in the standard system. However, MRP in SAP S/4HANA still offers great flexibility. Through numerous parameters in the material master, plant-specific parameters, and so forth, you can configure a very differentiated requirements calculation. As a consultant, this competence is in great demand.

Chapter 9
Inventory Management

Techniques You'll Master

- Explain the functions of inventory management.
- Understand the organizational levels that are relevant for inventory management.
- Describe the impact of a posting of a goods movement.
- Post a goods receipt, planned or unplanned.
- Describe the checks the system performs at goods receipt for a purchase order and the control options available to you.
- Post a goods issue, planned or unplanned.
- Explain the role of the movement type.
- Specify various options for a stock transfer and post a stock transfer.
- Describe a reservation and create one manually.
- Explain and set up stock determination.
- Understand how to conduct a physical inventory.

In this chapter, we will review the functions of inventory management, point out that stocks are managed on a quantity and value basis, and examine which organizational units are relevant. We will explain the posting of material movements and their effects: planned and unplanned goods receipts, goods issues, and stock transfers. We will discuss the importance of reservations and explore how they can help us plan material withdrawals. Regarding the support provided by the system, we will also mention the automatic stock determination, which can help us determine the appropriate storage location or stock.

Real-World Scenario

Inventory management is a central component of logistics, integrated in all important logistics processes and in financial accounting.

We learned in an earlier chapter that we can procure to stock. For the production process, raw materials, supplies, and components are often taken from the warehouse to be processed at the production site. In make-to-stock production, finished products are first placed in stock before they are sold. In make-to-order production, you can use the special make-to-order stock. Other special stocks allow you to execute special processes such as customer or vendor consignment, subcontracting, and so on.

After each stock change, the stock is updated in quantity and value. The quantity update has a direct link to material requirements planning (MRP), and the value update creates relevant accounting entries.

In relatively simple warehouses where materials are assigned to fixed storage bins, inventory management in SAP S/4HANA is enough for your operational activities. If you need to manage the stocks at a more detailed level than the storage location, or implement put away and picking strategies, you need to look at using Extended Warehouse Management (EWM) in SAP S/4HANA, for example, and then inventory management is the baseline.

In other words, inventory management is always there! Even if your customer uses EWM, stock postings are duplicated in inventory management and it runs in the background. It is essential that you have basic knowledge of inventory management to ensure smooth system operation.

Objectives of This Portion of the Test

This part of the certification exam tests your knowledge of the functions and configuration options of inventory management in SAP S/4HANA. To be certified, you must have a good understanding of the following topics:

- Relevant organizational units
- Posting of a goods movement
- Impact of a goods movement posting
- Creation and use of reservations
- Use and configuration of stock determination
- Procedure for carrying out a physical inventory

Key Concept Refresher

The main tasks of inventory management are as follows:

- Managing the material stock on a quantity and value basis
- Planning, posting, and monitoring goods movements
- Conducting the physical inventory

Managing the Material Stock on a Quantity and Value Basis

We distinguish between an organization's own *physical stock* and its *special stocks*.

Your own physical material stocks (i.e., normal stock) are managed on a quantity basis at the storage location level. Within a storage location, you can differentiate the material stock quantity into three different stock types: unrestricted use, quality inspection, and blocked. If a material is stocked in batches, each batch is managed individually.

Your own valuated stock is managed at the valuation area level. The valuation area is the organizational unit in which inventory value is managed and corresponds either to the plant or the company code. SAP recommends stock valuation at the plant level.

In addition to your own valuated physical stock, you can also manage so-called special stocks. Special stock is characterized by the fact that it is either valuated but not physically available in your warehouse (in the case of stock of material provided to subcontractors, for example) or physically available but not part of your property (in the case of vendor consignment stock, for example). It is also possible to manage materials in stock on a quantity basis only. For example, if you order brochures for a project (e.g., a trade fair event), you can keep the brochures in stock on a quantity basis while the project bears the costs.

For each valuation area (usually corresponding to a plant), the material type controls whether a material is kept in stock and, if so, only on a quantity basis or also on a value basis. The following series of figures shows two examples of material types: ROH (raw materials) and UNBW (non-valuated materials).

Materials of material type ROH are managed in stock on both a quantity and value basis in most valuation areas, whereas materials of material type UNBW are only managed on a quantity basis. Note that in plant 1040, the materials of both material types are not managed in stock at all. Figure 9.1 shows the two material types ROH and UNBW selected in the list of possible material types.

You can access the material types in Customizing by choosing **Logistics General • Material Master • Basic Settings • Material Types • Define Attributes of Material Types**.

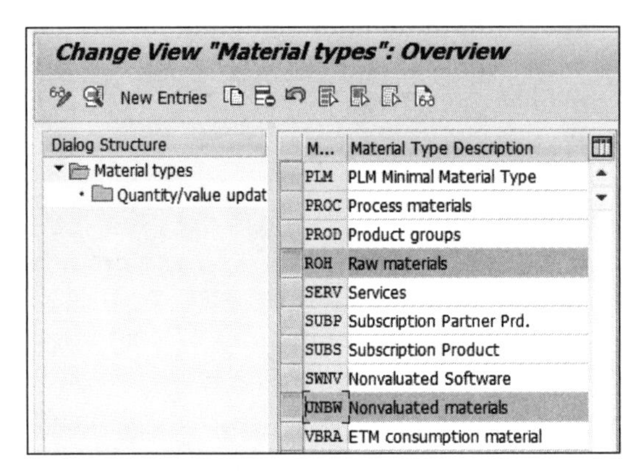

Figure 9.1 Customizing Material Types

Figure 9.2 shows that materials of material type ROH are managed in stock on both a quantity and value basis in valuation areas 1010, 1020, and 1030.

Change View "Quantity/value updating": Overview

Dialog Structure
- Material types
  - Quantity/value updating

Quantity/value updating

Val. area	Ma...	Qty updating	Value updating	Pipe.mand.
1000	ROH	☐	☐	☐
1010	ROH	☑	☑	☐
1020	ROH	☑	☑	☐
1030	ROH	☑	☑	☐
1040	ROH	☐	☐	☐

Figure 9.2 Material Type ROH (Raw Materials)

In Figure 9.3, note that materials of material type UNBW are only managed on a quantity basis in valuation areas 1010, 1020, and 1030, for example. In valuation area 1040, for example, the materials of both material types are not managed in stock at all.

Change View "Quantity/value updating": Overview

Dialog Structure
- Material types
  - Quantity/value updat

Quantity/value updating

Val. area	Ma...	Qty updating	Value updating	Pipe.mand.	Pi
1000	UNBW	☐	☐	☐	
1010	UNBW	☑	☐	☐	
1020	UNBW	☑	☐	☐	
1030	UNBW	☑	☐	☐	
1040	UNBW	☐	☐	☐	
1400	UNBW	☑	☐	☐	

Figure 9.3 Material Type UNBW (Non-Valuated Materials)

Planning, Posting, and Monitoring Goods Movements

Goods movements include both external movements (such as goods receipts from external procurement or goods issues for sales orders) and internal movements (such as goods receipts from production, material withdrawals for internal purposes, stock transfers, and transfer postings).

Note

You post goods movements directly in inventory management if you are either not using warehouse management at all (EWM in SAP S/4HANA, for example) or if you are currently processing goods movements in a storage location that is not EWM-managed. Even if you are using EWM, it is not always necessary to have all storage locations managed by EWM; in some cases (e.g., relatively small storage locations), the structure of inventory management is enough.

When you post a goods movement, the system creates documents that constitute the basis for quantity and value updates and that can be used for evaluations:

- **Material document**
  The material document provides information on the goods movement and is a source of information for all subsequent processes. The material document consists of a header and at least one item. The header contains general data, such as the date and the delivery note number. The items describe the individual movements (material number, quantity, movement type, plant, storage location, etc.).
- **Accounting document**
  If the movement is relevant for financial accounting—that is, if the movement leads to an update of the general ledger accounts—the system creates an accounting document in parallel with the material document. Using automatic account determination, the system updates the general ledger accounts that are affected by a goods movement.

Goods movements can be planned. The aim of planning in advance is to facilitate and accelerate the goods receipt process and to organize work in the goods receipt area efficiently, thereby avoiding bottlenecks, for example. MRP also uses planned goods movements to monitor and optimize stocks.

The documents for planning a goods receipt are used as reference documents when you enter the goods movement. You plan goods receipts using purchase orders and production orders. You can plan goods issues using outbound deliveries or reservations, for example.

Goods Movements

In parallel to specific SAP Fiori apps for specific goods movements, Transaction MIGO (the Post Goods Movements app) is available for all types of goods movement postings. Figure 9.4 shows the available apps for posting goods movements:

❶ Transaction MIGO for posting all types of goods movements

❷ Pre-configured Transaction MIGO for specific postings (e.g., goods receipt, stock transfer, goods invoice)

❸ SAP Fiori apps for dedicated goods movements

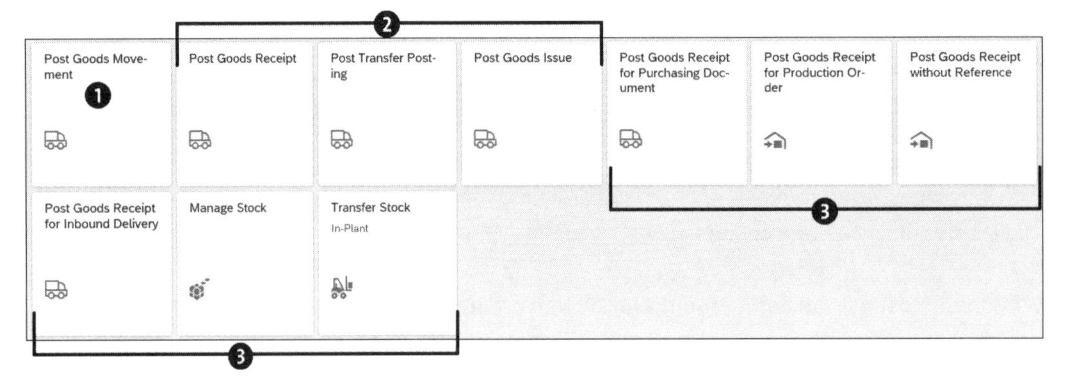

Figure 9.4 SAP Fiori Apps for Posting Goods Movements

Figure 9.5 shows the important fields and functions of Transaction MIGO in the SAP GUI environment for initiating a goods movement posting:

❶ **Transaction**
In this list field, you select the business transaction that you want to process, such as goods receipt, goods issue, and so on.

❷ **Reference document**
In this list box, you specify the kind of reference document you wish to refer to, such as order, reservation, and so on. The possible entries available in the **Reference Document** list field depend on the selected transaction.

❸ **Input field**
In this list field, you enter the number of the reference document.

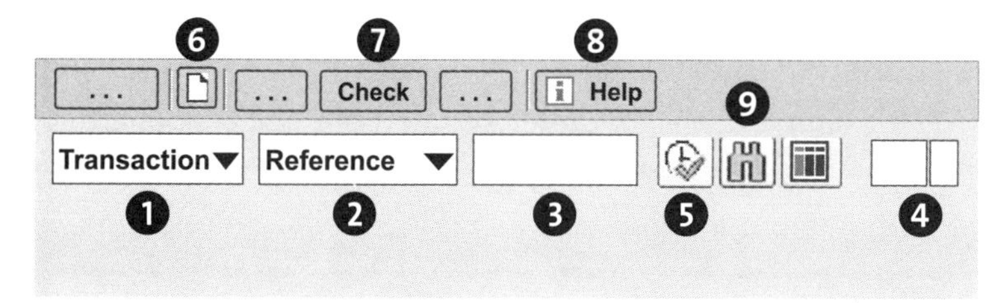

Figure 9.5 Transaction MIGO in SAP GUI

❹ **Standard values for movement type and special stock indicator**
You can use these fields to enter default values for the movement type and the special stock indicator. The system then proposes these values for all items.

❺ **Execute function**
After you have entered the business transaction (e.g., goods receipt), the category of the reference document (e.g., purchase order), and its number, you can verify the default value for movement type and then choose the **Execute** button. The system reads the data from the reference document and displays it for processing.

❻ **Restart function**
You do not need to leave Transaction MIGO to cancel the current processing and start over. Just choose the **Restart** button.

❼ **Check function**
When you process goods movements and enter data, the system does not issue any warning messages at first, so as not to disturb processing. If you want to know before the actual posting whether the system will issue messages, first choose **Check**. The system then displays all messages in a dialog box.

❽ **Help function**
If you choose **Help**, a separate screen area opens with information and user tips on Transaction MIGO. You can display the help documentation while working in the transaction.

❾ **Find function**
Depending on the reference selected, you can search for reservations, purchasing documents, or material documents using the **Find** button.

The functions and screen layout of the Post Goods Movement app on the SAP Fiori launchpad correspond to those of Transaction MIGO in SAP GUI, as shown in Figure 9.6:

❶ Transaction

❷ Reference document

❸ Number of the reference document

❹ Default values for movement type and special stock

❺ Function: *Execute*

❻ Function: *Find*

The basic navigation in the app also corresponds to the transaction navigation. However, there are small differences in the arrangement of the buttons and in the menu navigation. For example, the **Post**, **Cancel**, **Restart**, **Hold**, and **Check** buttons are located on the lower-right-hand side of Figure 9.6.

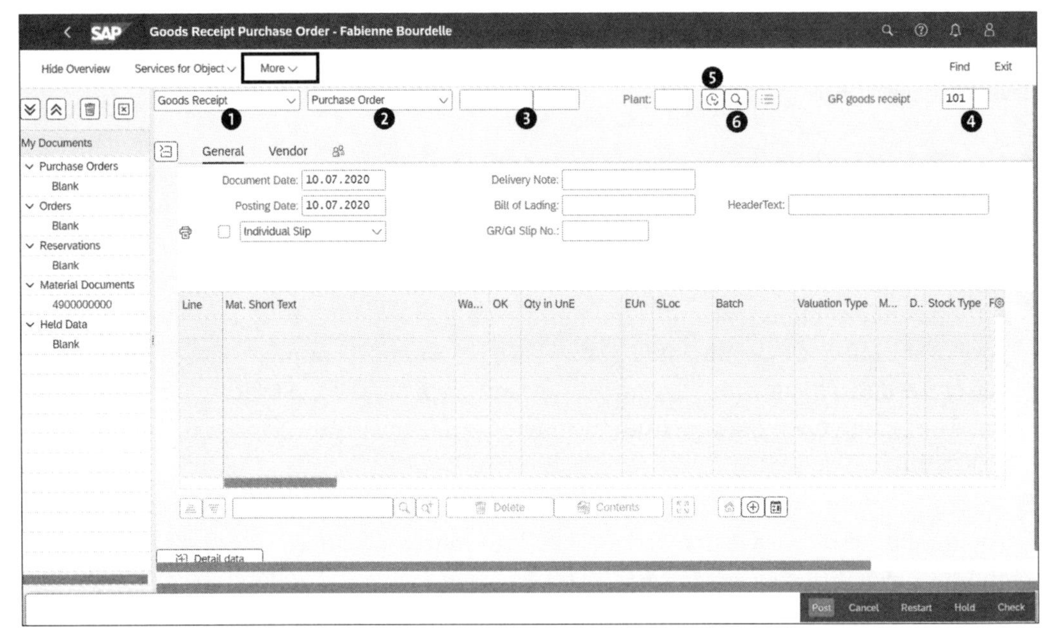

Figure 9.6 The Post Goods Movements App in the SAP Fiori Launchpad (Transaction MIGO)

Note

If you must interrupt the processing of a goods movement before posting, you can hold the data entered so far. If you want to continue entering the goods movement, you do not have to start from the beginning. Instead, you call up the held data and continue processing.

You use Transaction MIGO (or the Post Goods Movements app) to determine which type of goods movement you post by entering a business transaction, a type of reference document, the number of the reference document (if applicable), and the movement type.

This brings us to the movement type, a three-digit key that contributes to control of the type of goods movement you are posting. Let's look at the role and control capabilities of the movement type using examples. We'll start with goods receipt postings.

Goods Receipt

In Chapter 3, we discussed goods receipt with reference to purchase orders. You may refer to this chapter for an in-depth discussion of this topic; we will concentrate here on unplanned goods receipts.

An unplanned goods receipt is a goods receipt that does not refer to a preceding document (purchase order, production order, etc.). A typical example of an unplanned goods receipt is an initial entry of stock balance.

You can enter an unplanned goods receipt with Transaction MIGO (or the Post Goods Movement app) by choosing the **Goods Receipt** business transaction and the reference **Other**, as shown in Figure 9.7. You then specify what type of unplanned goods receipt you want to post by selecting the appropriate movement type.

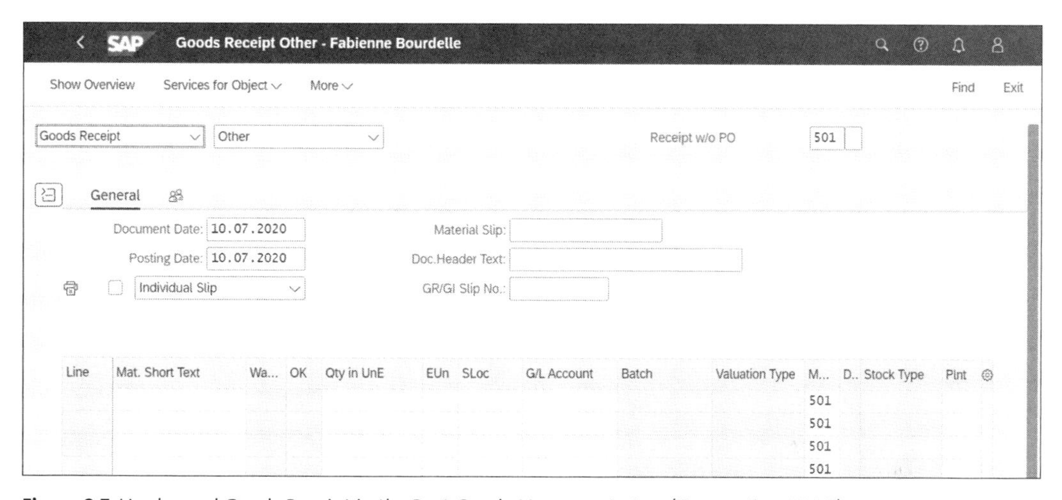

Figure 9.7 Unplanned Goods Receipt in the Post Goods Movements App (Transaction MIGO)

Let's take a closer look at the movement type.

Movement Types

For goods movements, especially those that are unplanned, the movement type has a major function and determines the kind of goods movement to be posted, as the name suggests. It also does the following:

- It controls the quantity and value update.
- It influences the account determination for the offsetting posting to the stock posting.
- It influences the field selection during goods movement entry.
- It influences the output determination.

You can configure the movement types in Customizing by choosing **Materials Management • Inventory Management and Physical Inventory • Movement Types • Copy, Change Movement Types.**

SAP S/4HANA offers a range of movement types. Sometimes a customer-specific movement type is required, due to field selection requirements, for example. Say you want to declare an additional field as a mandatory field for a certain goods movement posting. In this case, we recommend that you copy an existing standard movement type and change the newly copied movement type to suit your needs. The standard movement types then remain unmodified.

Note

The movement type entered in the upper line of Transaction MIGO is a default value for the items. This means that the movement type is an item field, which implies that you can enter items with different movement types in one goods movement posting. The movement type you enter in the items must be consistent with the business transaction and reference already selected. The system checks and, if necessary, issues an error message!

In Chapter 12, we will again talk about the movement types and discuss the configuration options in detail. Here we just introduce this important key.

The most common goods receipts without reference are:

- **Initial entry of stock balances (standard movement type 561, shown in Figure 9.8)**
  You can post an initial entry of stock in all three stock types (unrestricted use, quality inspection, blocked stock) and in special stock such as consignment stock, project stock, or sales order stock. In addition to the movement type, you simply enter the desired stock type and/or the special stock indicator.

  You can valuate the opening balance posting in one of two ways:
  - By using the valuation price in the material master record
  - By manually entering a value (external amount in the **Local Currency** field in the item details on the **Quantity** tab)

  You can enter a manual value because of the field selection control of movement type 561. If you want to prevent the entry of an external value, create a new movement type in Customizing by copying movement type 561 and set **the External Amount** field to **Not Ready for Input**.
- **Goods receipt without a purchase order (standard movement type 501)**
  If you receive an external receipt of goods without having previously created a corresponding purchase order in the SAP system, you must enter it as an *other goods receipt*.
- **Goods receipt without a production order (standard movement type 521)**
  The same scenario also applies to an internal receipt from production for which no production order has been previously created.

> **Note**
>
> The reason why preceding documents for external or internal procurement are missing should be because the corresponding application area has not been implemented. Otherwise, this type of posting is not recommended!

- **Free-of-charge delivery (standard movement type 511)**
  If you receive a free of charge delivery from a supplier without having previously created a purchase order, you post the delivery as an other goods receipt. The **Vendor** and **Text** fields are mandatory fields for the standard movement type 511.

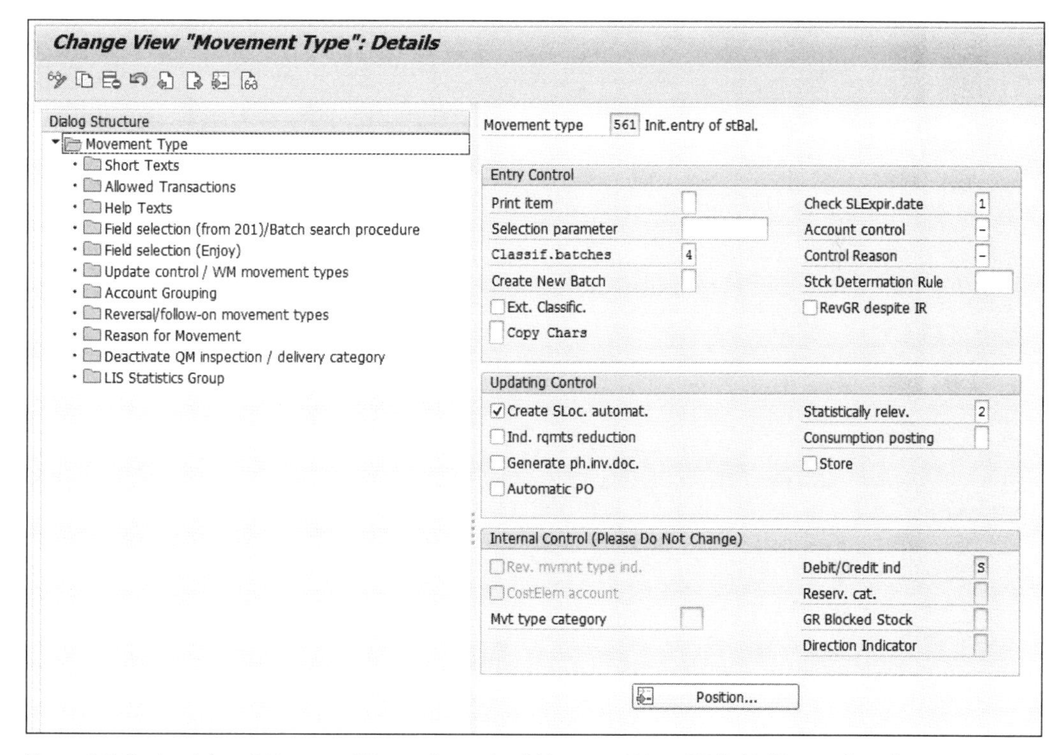

Figure 9.8 Customizing of Movement Types: Example of Movement Type 561 (Initial Entry of Stock Balance)

Unplanned Goods Receipt

Let's summarize the procedure for entering an unplanned goods receipt with Transaction MIGO (or the Post Goods Movement app):

1. Choose **Goods Receipt** as the business transaction and **Other** as the reference.
2. Check the default value for the movement type and change it as needed.
3. Enter the material number, quantity, storage location, stock type, plant, and—if it differs from the default value—the movement type for the items to be entered.

Figure 9.9 shows the Post Goods Movement app for a goods receipt without reference. Note the different movement types, storage locations, and stock types for each item.

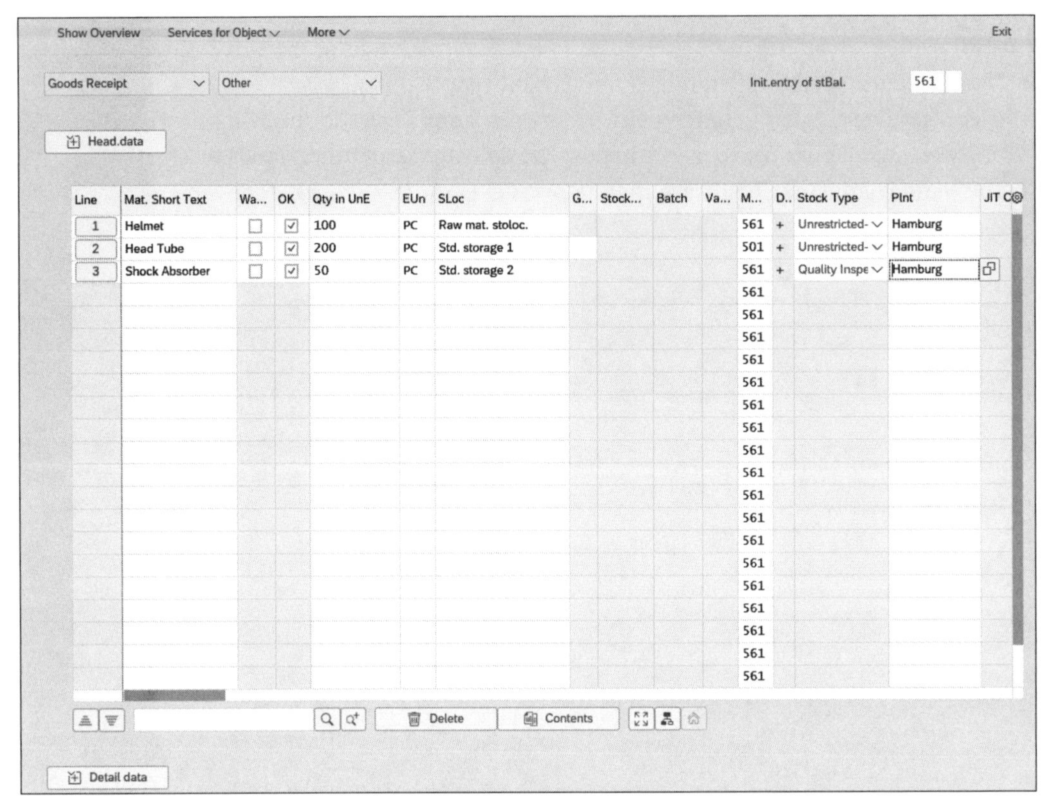

Figure 9.9 The Post Goods Movement App for a Goods Receipt without Reference (Transaction MIGO)

In addition to Transaction MIGO (or the Post Goods Movements app), the following apps are also available for posting an unplanned goods receipt (goods receipt without reference):

- Manage Stock (only for initial stock entry)
- Post Goods Receipt without Reference

We have already discussed goods receipts with reference to purchase orders in Chapter 3. Now we continue with the goods issue postings.

Goods Issue

Depending on the material, there can be various reasons for stock removal, including delivery of a finished product or trading goods to a customer, withdrawal of components or raw materials for production, withdrawal of a material for a cost center, scrapping, and so on. (Here we are not discussing the delivery of sales orders or backflushing.)

As a first step, we look at unplanned goods issue postings; then we deal with reservations, which allow you to plan goods issues and enter them with a reference.

You can post all unplanned withdrawals, using Transaction MIGO or the Post Goods Movement app, as follows:

1. Choose **Goods Issue** as the business transaction and **Other** as the reference.
2. Check the default value for the movement type and change it, if necessary.
3. Enter the material number, quantity, storage location, stock type, plant and—if it differs from the default value—the movement type for the items to be entered.
4. Depending on the selected movement type, you must enter an account assignment. For example, if as in Figure 9.10 you select the standard movement type 201, which stands for a withdrawal to a cost center, you must enter the receiving cost center. The system displays the input fields, depending on the field selection control of the selected movement type.

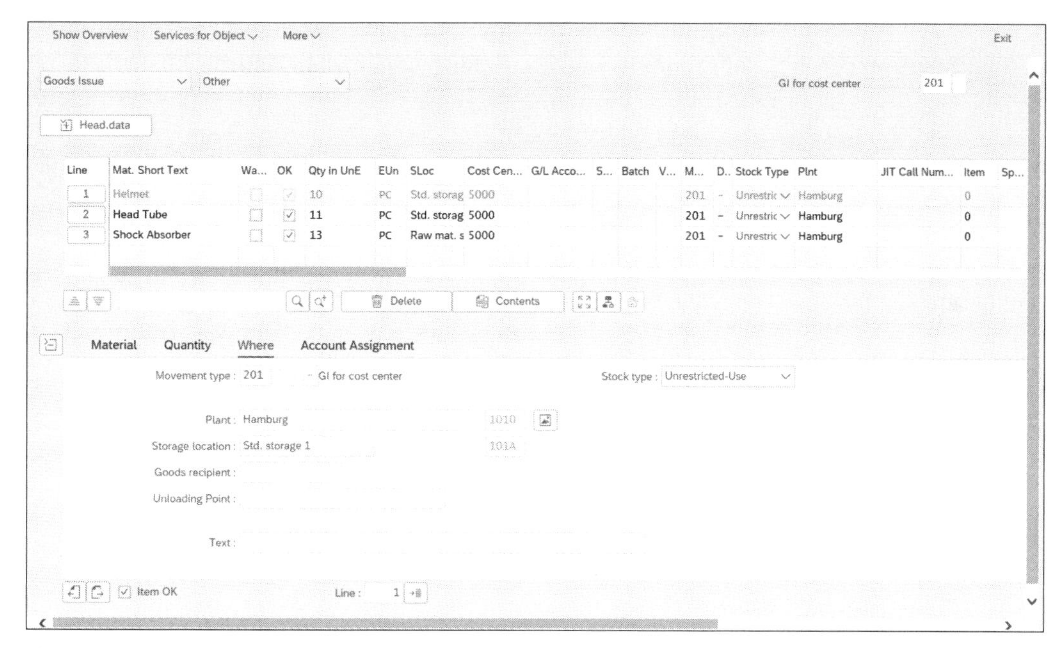

Figure 9.10 The Post Goods Movement App for a Goods Issue without Reference (Transaction MIGO)

In Figure 9.10, the **G/L Account** field is ready for input, which is due to the setting of the movement type. The user can therefore enter an account manually. This is the offsetting account to the stock account. However, if the user does not make an entry, or in the case of a movement type for which the **G/L Account** field is not ready for input, the system automatically determines the corresponding account in the background, as it does for the stock account anyway. There are various factors for automatic account determination, such as the valuation area, the material, and the business transaction, including the movement type. A detailed description of the automatic account determination and the setting options for Customizing are provided in Chapter 12.

The main effects of a goods issue posting are as follows:

- The stock quantity of the material is reduced.
- A material document and an accounting document are created.
  The system posts the value of the withdrawn material from the stock account to an offsetting account dependent on the business transaction—for example, a consumption account. The value of a goods issue posting is always as follows: *Withdrawal quantity × Current valuation price of the material in the plant.*
- The consumption statistics of the material are updated if the withdrawal is a consumption posting. Whether a goods issue posting is a consumption posting depends on the Customizing settings for the movement type.
  Figure 9.11 shows the movement type 201 (goods issue for cost center). The **Consumption Posting** indicator states that a withdrawal to a cost center is defined as a consumption posting. On the other hand, no consumption is updated for withdrawals for scrapping or for sampling, as shown in Figure 9.12.

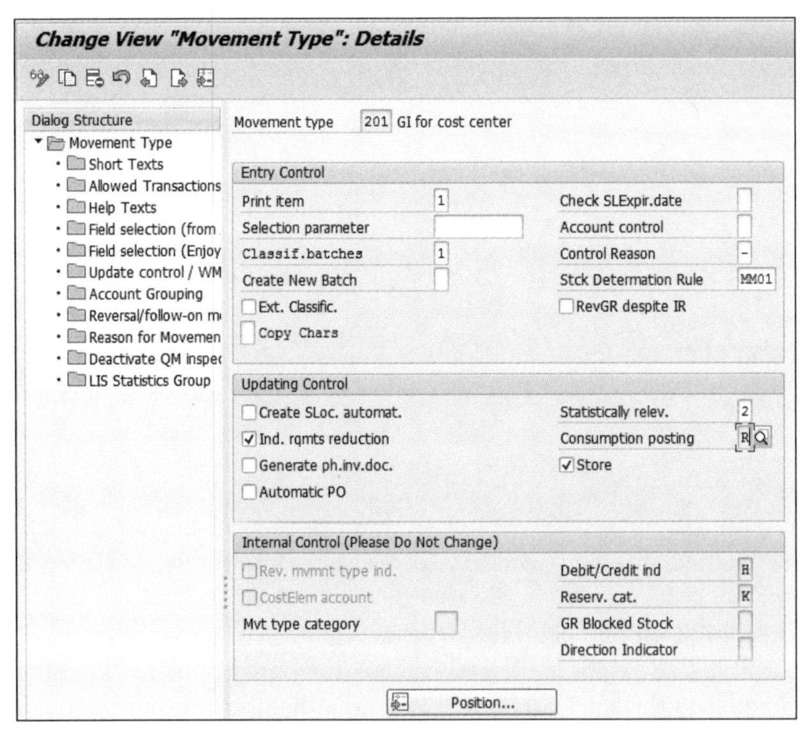

Figure 9.11 Customizing: Movement Type 201 (Goods Issue for Cost Center)

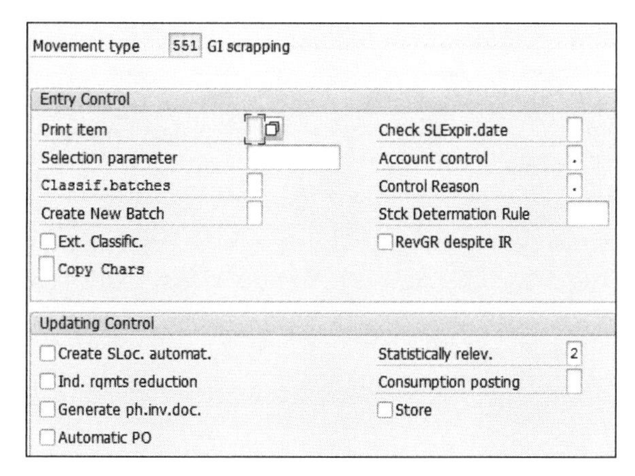

Figure 9.12 Customizing: Movement Type 551 (Scrapping)

A distinction is made between planned and unplanned withdrawals. If the withdrawal is planned, then only the total consumption is updated. In the case of an unplanned withdrawal, the unplanned consumption is updated in addition to the total consumption. Figure 9.13 shows the different values for the **Consumption Posting** indicators.

- If applicable, the system charges the specified account assignment, such as the cost center.

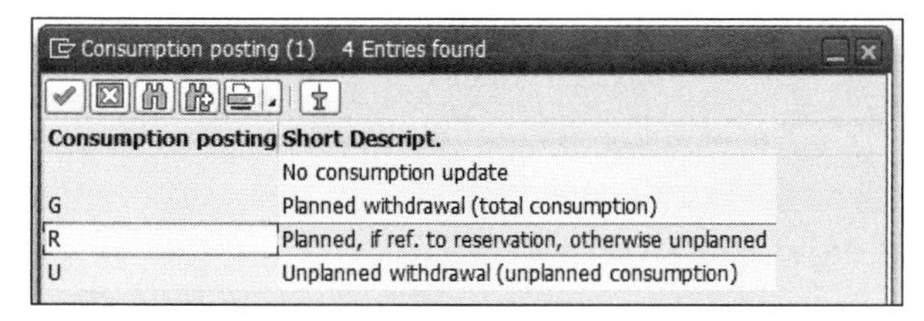
Consumption posting (1) 4 Entries found

Consumption posting	Short Descript.
	No consumption update
G	Planned withdrawal (total consumption)
R	Planned, if ref. to reservation, otherwise unplanned
U	Unplanned withdrawal (unplanned consumption)

Figure 9.13 Consumption Posting Indicator: Possible Values

Note

You can display the consumption statistics of a material in the material master record, as shown in Figure 9.14. To do this, choose **Additional Data** at the plant level and then **Consumption**.

If you have used a reservation to ensure that a material is staged on time, you should also enter the goods issue with reference to this reservation. By referencing the reservation, you simplify the entry of the goods movement, since the system can copy the data for the individual items from the reservation.

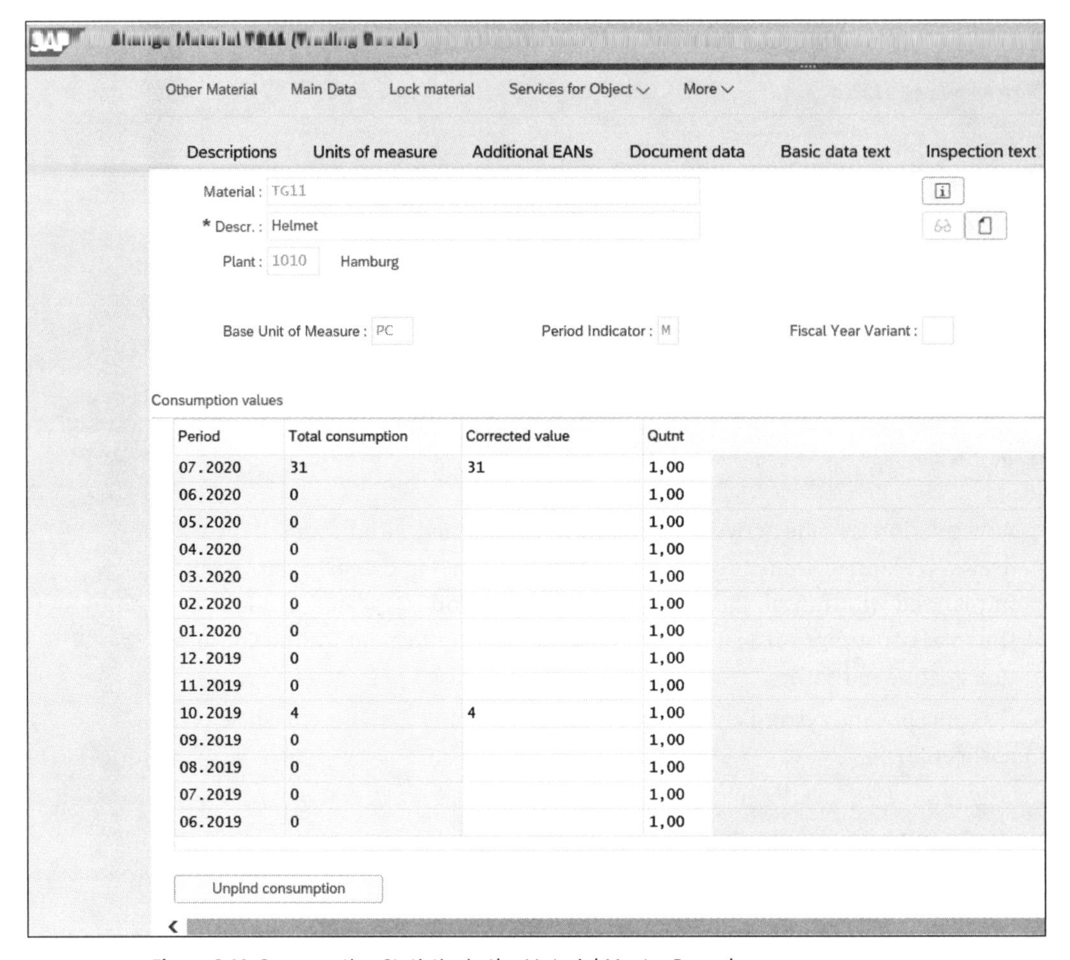

Figure 9.14 Consumption Statistics in the Material Master Record

Reservations

A reservation is a requirement to the warehouse to make a certain quantity of material available for a goods movement (usually a withdrawal) at a certain date and for a certain purpose. Reservations are considered in material planning so that a required material can be procured at the right time. A reservation also simplifies and accelerates the entry process and helps prepare work in the goods receipt department. Typically, you use reservations to plan goods issues and transfer postings, but you can also use them to plan goods receipts.

Note

You can only plan transfer postings using reservations for one-step transfer postings.

You should only plan goods receipts with reservations in exceptional cases or if you do not use the procurement and/or the production functions. If you work with purchasing and production, goods receipts are planned with purchase orders or production orders. Reservations of goods receipts are therefore not necessary.

Reservations can be created manually or automatically. There are two types of automatic reservations:

- **Dependent reservations**
  These reservations refer to orders, networks, and work breakdown structure (WBS) elements. When you create an order, network, or project, the system automatically reserves the warehouse components.
  You can display dependent reservations, but you cannot maintain them directly. For example, you cannot change a dependent reservation for an order directly but must change the components in the order. The system then updates the reservation automatically.
- **Stock transfer reservations**
  These reservations can be used to plan stock transfers between MRP areas within a plant, as an alternative to stock transport requisitions. If, for example, reorder point planning is carried out in an MRP area (which corresponds to a storage location, for example) and the available stock falls below the reorder level, the system creates a stock transfer reservation in the supplying MRP area for the replenishment quantity.

You create a reservation for a purpose. The purpose is represented by a movement type and an account assignment object such as a cost center or a production order. A reservation can only be created for one single purpose -that is, a reservation contains one single movement type.

A reservation is made up of a head and at least one position.

The movement type is part of the header. You find the following data at the item level:

- **Material**
  This is the number of the required material.
- **Quantity**
  This is the quantity of material required.
- **Date**
  This is the date on which the material quantities are required.
- **Plant and optionally storage location**
  These are the issuing organizational units. The specification of the plant is mandatory.
- **General ledger account**
  The G/L account is additional to the account assignment object.
- **Goods movement allowed indicator**
  This indicator must be set in order to post a goods receipt with reference to the reservation item. You can also use this indicator to prevent a goods receipt from being posted with reference to the reservation item. This is useful if the requirement date for the material is far in the future. You can use a report (Transaction

MBVB or the Manage Reservations app) to set this indicator automatically for a reservation whose base date lies in the immediate future.

- **Recipient, unloading point, and item text**
  These are optional free text fields.
- **Item completion indicator**
  If you withdraw the entire reserved quantity (with a goods issue posting with reference to the reservation), the system automatically sets the **Item Completed** indicator. You can also set the indicator manually if you only withdraw a partial quantity and no longer need the rest.

In the following discussion, we will examine how to create a reservation manually using Transaction MB21 or the corresponding Create Reservation app on the SAP Fiori launchpad.

You start the entry with an initial screen where you enter a base date and a movement type; in Figure 9.15, this base date is 12.07.2020. You can also enter a plant. In contrast to the baseline date and movement type, which are header fields (that is, unique per reservation), the plant you enter on the initial screen is only a default value for the items. Often, a user is only responsible for one plant, and thanks to this default value, they do not have to repeat the entry for each item. The plant, however, is an item field and can differ for each item.

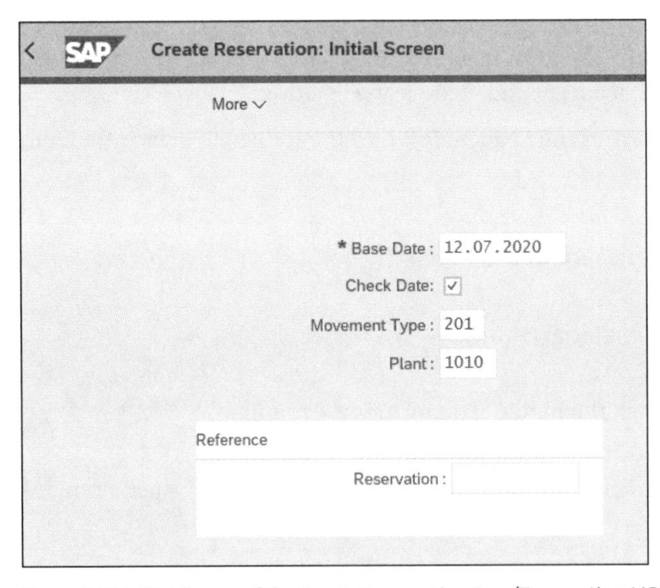

Figure 9.15 Initial Screen of the Create Reservation App (Transaction MB21)

A straightforward means of doing this is to enter the item data on the collective processing screen as shown in Figure 9.16 after you have completed the initial screen, and then post the reservation.

In Figure 9.17, you can see the item details with Transaction MB22 (the Change Reservation app).

Create Reservation: Collective Processing
New Items
Choose
Delete Item(s)
More
Exit
Movement Type : 201 GI for cost center
Goods Recipient :
Cost Center : 5000 MM-00
Items
Item Material Quantity UnE Plant Storage Loc. Batch Movement
1 TG11 2 PC 1010 101A
2 RM12 2 PC 1010
3 RM124 3 PC 1010
4 1010
5 1010
6 1010
7 1010
8 1010
9 1010

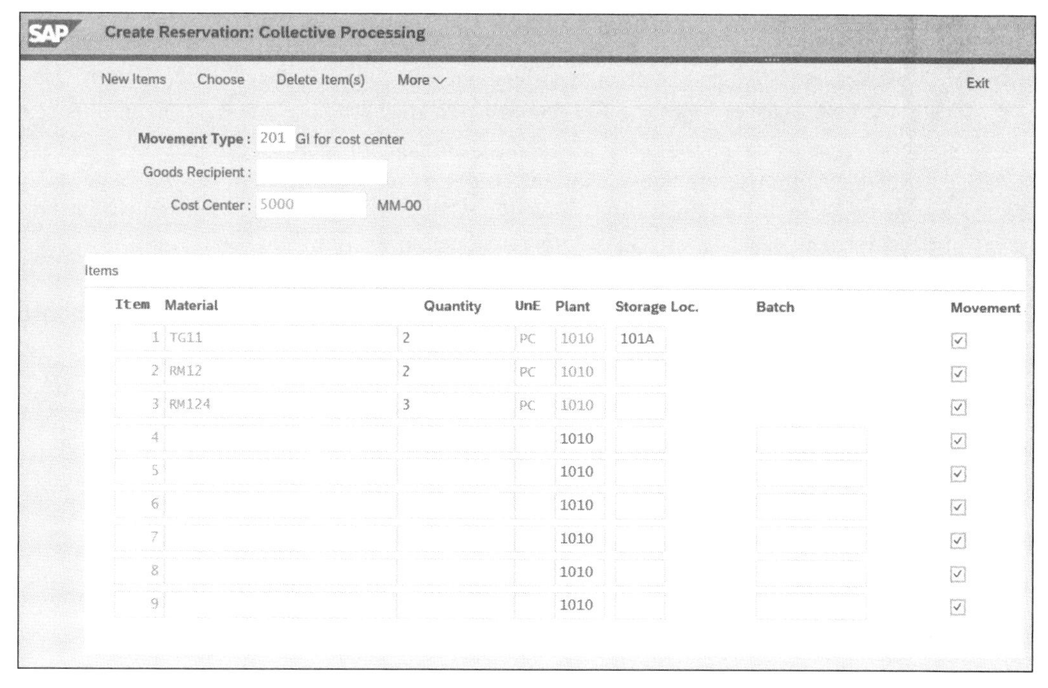

Figure 9.16 Collective Processing in the Create Reservation App (Transaction MB21)

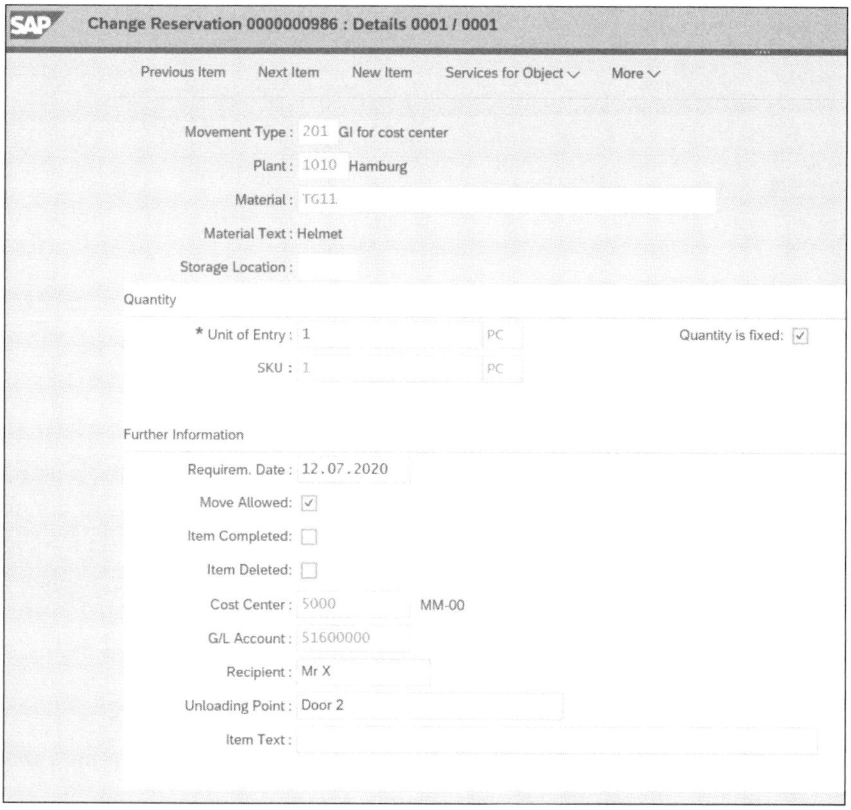

Change Reservation 0000000986 : Details 0001 / 0001
Previous Item
Next Item
New Item
Services for Object
More
Movement Type : 201 GI for cost center
Plant : 1010 Hamburg
Material : TG11
Material Text : Helmet
Storage Location :
Quantity
* Unit of Entry : 1 PC
Quantity is fixed:
SKU : 1 PC
Further Information
Requirem. Date : 12.07.2020
Move Allowed:
Item Completed:
Item Deleted:
Cost Center : 5000 MM-00
G/L Account : 51600000
Recipient : Mr X
Unloading Point : Door 2
Item Text :

Figure 9.17 Item Details in the Change Reservation App (Transaction MB22)

Note

The movement type and account assignment data (general ledger account and, in this example, a cost center) cannot be changed after you have saved the reservation.

Moreover, although the cost center appears on the item details, it cannot be defined per item when you create the reservation. It is unique for the entire reservation. The general ledger account, in turn, is determined using automatic account determination or can be entered manually for each item during creation, since it is material-dependent.

If you want to post a goods movement with reference to a reservation, choose Transaction MIGO (the Post Goods Movement app), then the appropriate business transaction (for example, **Goods Issue**), and the reference **Reservation**. If you then enter the reservation number and choose **Execute**, the system proposes the data from the reservation, as shown in Figure 9.18.

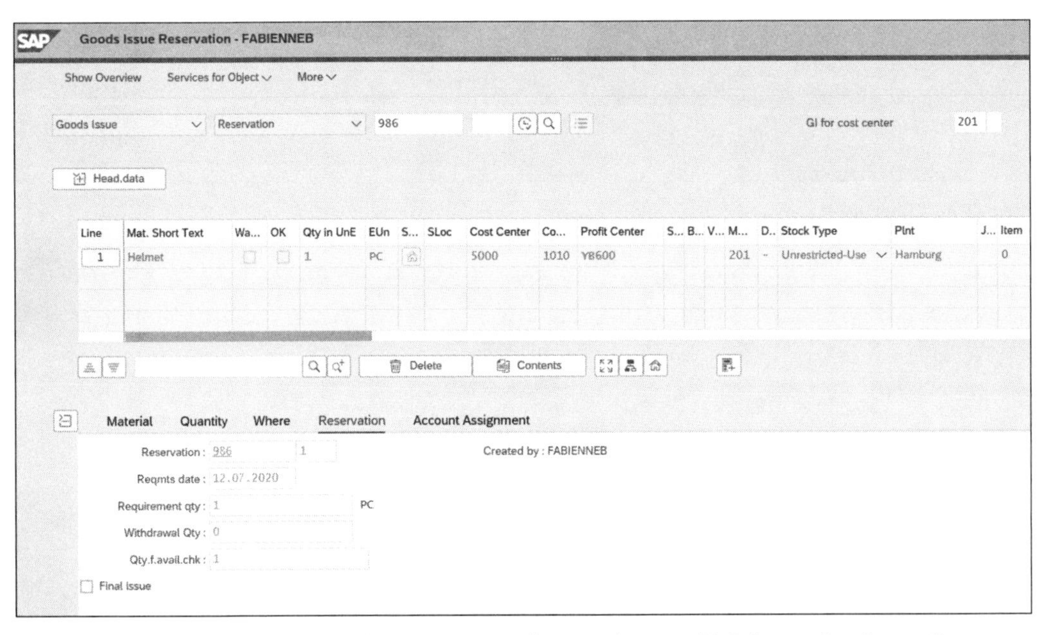

Figure 9.18 The Post Goods Movement App for a Goods Issue with Reference to a Reservation (Transaction MIGO)

So far, we have dealt with goods receipts and goods issues, and we have learned that reservations can be used to plan goods movements. The last type of goods movement that we will now consider is stock transfers posting. These can also be planned using reservations. We will review the different ways of planning and posting stock transfers.

Stock Transfers

In a company, goods movements are not limited to the form of goods receipts and issues. Depending, for example, on the organization of the company (e.g., decentralized storage), internal stock transfers might also be necessary.

Stock transfers can occur at three different levels in inventory management:

- Stock transfer from company code to company code
- Stock transfer from plant to plant within a company code
- Stock transfer from storage location to storage location within a plant

We do not discuss transfers from company code to company code in this book, since this is not relevant to the sourcing and procurement certification. Let's start with stock transfers from plant to plant within a company code.

There are four different procedures for carrying out a stock transfer between two plants within a company code:

- Stock transfer via stock transfer posting using a one-step procedure
- Stock transfer via stock transfer posting using a two-step procedure
- Stock transfer using a stock transport order without delivery
- Stock transfer using a stock transport order with delivery

This chapter discusses stock transfer by stock transfer posting using the one-step and two-step procedures. We already discussed the stock transport order in Chapter 2.

A stock transfer posting consists of a goods issue from the issuing plant and a goods receipt in the receiving plant. You can perform the goods issue and the goods receipt with a single posting (one-step procedure) or with two postings (two-step procedure). The advantage of the one-step procedure is that you enter only one transaction in the system. This simplified procedure is suitable if the two plants are close to each other. Using the one-step procedure, you can transfer stock quantities from any stock type in the issuing plant to any stock type in the receiving plant.

Note

Posting a stock transfer from any stock type in the issuing plant to any stock type in the receiving plant is only possible on the SAP Fiori launchpad with the Transfer Stock—Cross-Plant app, not with Transaction MIGO (the Post Goods Movement app).

The two-step procedure allows you to monitor stocks that are on the way. In the two-step procedure, the quantity is assigned to stock in transfer after you post the goods issue from the issuing plant. The stock in transfer belongs to the valuated stock of the receiving plant. Using the two-step procedure, you can only transfer material from unrestricted-use stock in the issuing plant to unrestricted-use stock in the receiving plant.

If material valuation is carried out at the plant level (recommended), a cross-plant stock transfer posting (one- or two-step) results in a value update to the stock accounts. Valuation takes place at the time of the goods issue from the issuing plant. An accounting document is generated parallel to the material document for the goods issue. The stock transfer posting is valuated at the valuation price of the material in the issuing plant.

The advantage of a stock transfer posting over a stock transport order is the quick and straightforward execution of a stock transfer between plants. With the two-step procedure, you even have the option of monitoring the stock on the road, as you can do with a stock transport order.

Note
With the two-step stock transfer posting procedure, the stock on the road is called *transfer stock* while it is called *stock in transit* for a stock transport order.

However, this method presents some limitations regarding the procedure with a stock transport order:

- You cannot enter delivery costs or a forwarder.
- This is merely a posting; planning and purchasing are not involved.
- You have a limited possibility to monitor the process.

To post a stock transfer between two plants in one company code, you can use Transaction MIGO or the Transfer Stock—Cross-Plant app, which is more user-friendly in this case. Figure 9.19 and Figure 9.20 show how to post a stock transfer between two plants in one step with the Transfer Stock—Cross-Plant app:

❶ Select a material.

❷ Select the plant and stock type from which you want to withdraw quantities.

❸ Select the plant and stock type to which you want to transfer the quantity.

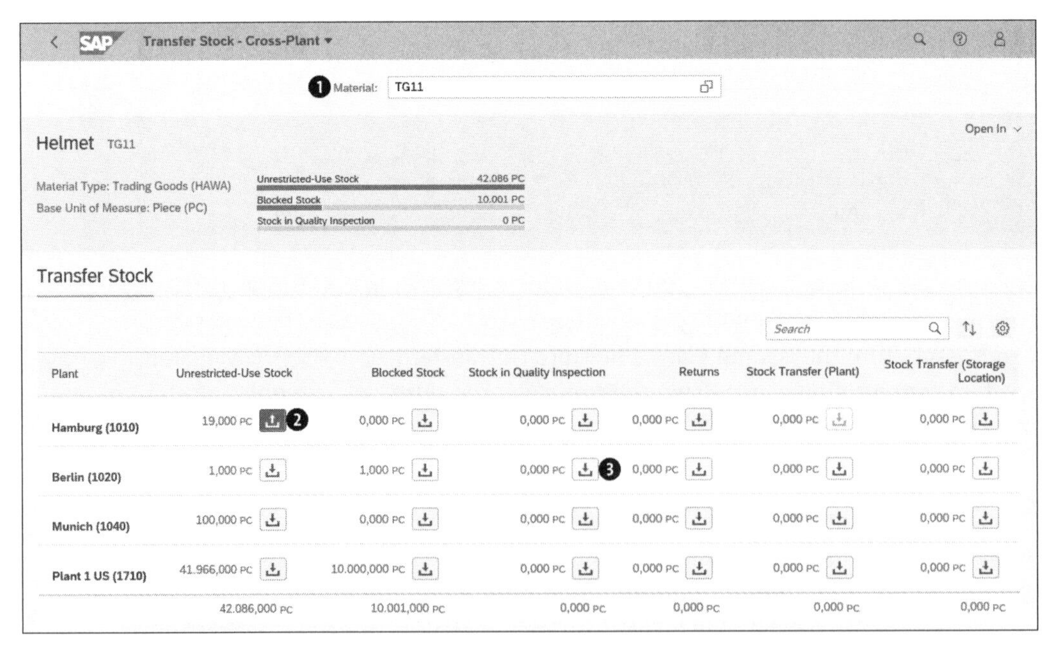

Figure 9.19 Posting a Stock Transfer in One Step with the Transfer Stock—Cross-Plant App (I)

The system then opens the dialog window shown in Figure 9.20. Here, take the following steps:

❶ Select the issuing and receiving storage location in the respective plant.

❷ Enter the quantity you want to transfer.

❸ Check and, if necessary, change the document date and the posting date.

❹ Post the transfer.

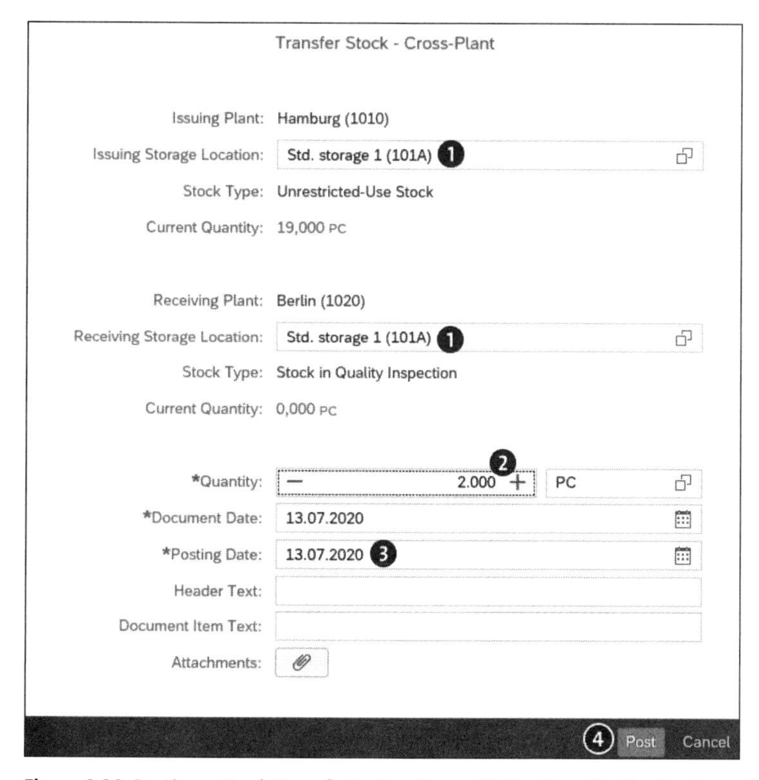

Figure 9.20 Posting a Stock Transfer in One Step with the Transfer Stock—Cross-Plant App (II)

Note

To post a stock transfer in two steps with the Transfer Stock—Cross-Plant app, choose **Stock Transfer (Plant)** as the target destination at goods issue posting.

Like cross-plant stock transfers as outlined in Table 9.1, you can also post *plant-internal stock transfers between storage locations* using Transaction MIGO, as outlined in Table 9.2. Otherwise, you can utilize the user-friendly Transfer Stock—In Plant app shown in Figure 9.21, which operates similarly to the Transfer Stock—Cross-Plant app.

As with a stock transfer between two plants, only unrestricted-use stock quantities can be transferred between storage locations using the two-step procedure. However, in the one-step procedure, any type of stock can be transferred into any type of stock, even using Transaction MIGO.

In a plant internal stock transfer using the two-step procedure, the quantity removed from storage is posted directly to the receiving storage location—not to unrestricted-use stock but to the stock in transfer of the storage location. An internal plant stock transfer is not valuated.

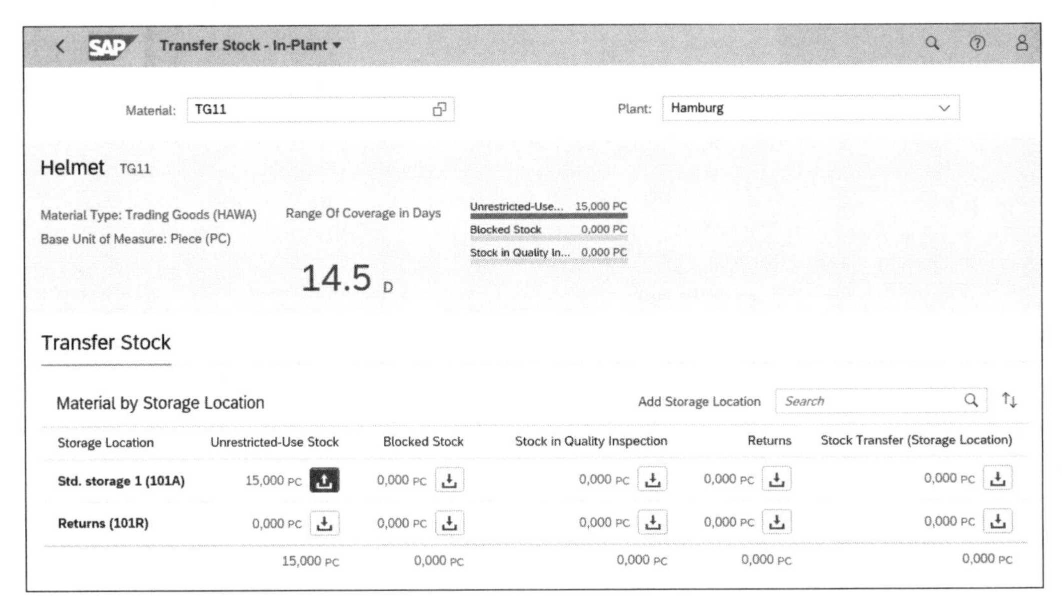

Figure 9.21 The Transfer Stock—In-Plant App

		Business Transaction	Reference	Movement Type
One-step		**Transfer Posting**	**Other**	311, 323, 325, etc. (from all stock types to all stock types)
Two-step	Goods issue	**Remove from Storage**	**Other**	313 (from unrestricted use)
	Goods receipt	**Place in Storage**	**Material Document**	315 (to unrestricted use)

Table 9.1 Posting a Stock Transfer from Plant to Plant with Transaction MIGO (the Post Goods Movement App)

		Business Transaction	Reference	Movement Type
One-step		**Transfer Posting**	**Other**	301 (only from unrestricted use to unrestricted use)
Two-step	Goods issue	**Remove from Storage**	**Other**	303 (from unrestricted use)
	Goods receipt	**Place in Storage**	**Material Document**	305 (to unrestricted use)

Table 9.2 Posting a Stock Transfer from Storage Location to Storage Location within a Plant with Transaction MIGO (the Post Goods Movement App)

In addition to physical stock transfers between plants or between storage locations within a plant, there are also transfer postings that do not involve a physical movement:

- **Transfer postings between stock types**
  For this, you can use Transaction MIGO (the Post Goods Movement app) or the Transfer Stock—In Plant app.
- **Transfer postings between materials**
  You can transfer a quantity of stock from one material to another using Transaction MIGO (Post Goods Movement app), the **Transfer Posting** business transaction, the **Other** reference, and movement type 309. A quantity transfer posting from material to material is only possible in one step and for unrestricted-use stock. The materials must have the same base unit of measure.

Stock Determination

We have learned that material stock quantities are managed at the storage location level. Consequently, you must always enter a storage location when you post a goods movement.

If you store your materials in several storage locations, you can define a strategy in Customizing according to which the system automatically determines a storage location for a goods issue or stock transfer. This also applies if you manage materials both in your own stock and on a consignment basis. You can create a strategy so that the system proposes your own stock first.

Inventory determination strategies can be implemented not only in inventory management but also in other applications like sales, production, and so forth. Each application uses its own logic. In inventory management, stock determination strategies can be defined and configured depending on the materials and movement types.

Let's analyze the logic and the stages of setting up a stock determination strategy in inventory management. Choose in Customizing **Materials Management • Inventory Management and Physical Inventory • Stock Determination**, as show in Figure 9.22.

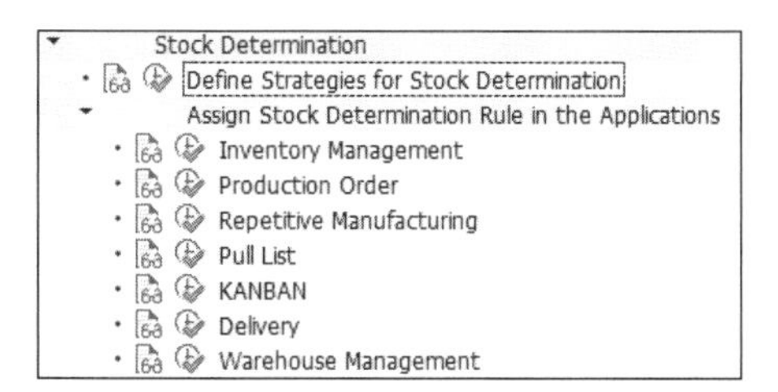

Figure 9.22 Stock Determination Strategy in Customizing

You assign *stock determination rules* to movement types in Customizing. In this way, you can assign the same rule to several movement types and treat them in the same way from the point of view of stock determination. You are free to name the stock determination rules. In Figure 9.23, we see that a rule MM01 has been assigned to movement type 201 (withdrawal to cost centers).

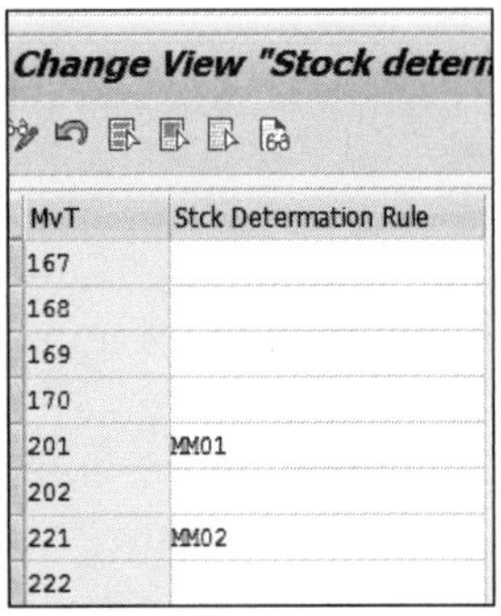
Change View "Stock detern

MvT	Stck Determation Rule
167	
168	
169	
170	
201	MM01
202	
221	MM02
222	

Figure 9.23 Customizing: Stock Determination Rules Assigned to Movement Types

You assign a *stock determination group* to the materials for which you want to use stock determination. You do this in the material master record, Plant Data/Storage 2 view, **General Plant Parameters** tab shown in Figure 9.24. Plant-dependent stock determination groups must first have been created in Customizing. All materials that are to be handled according to the same logic from the stock determination point of view are assigned the same group.

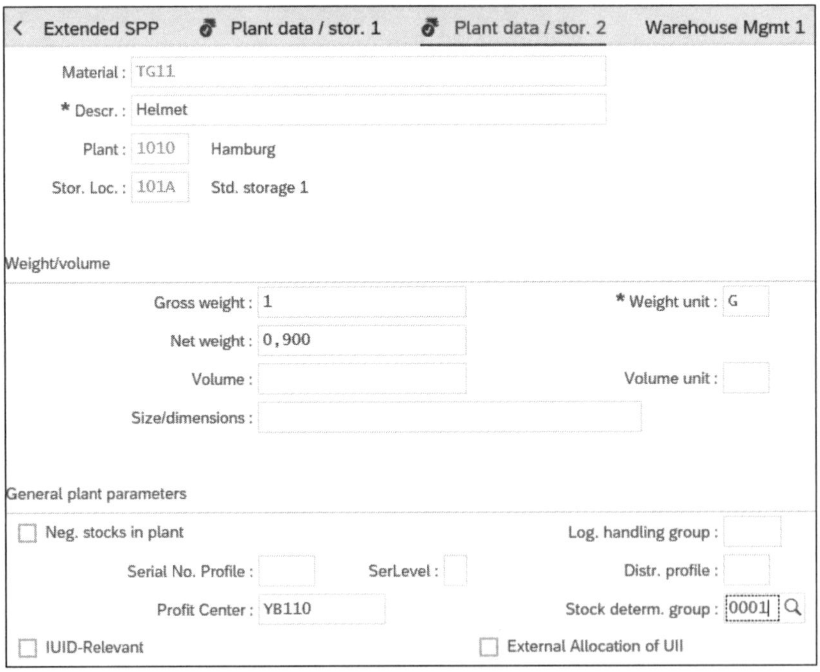

Figure 9.24 Material Master Record: Stock Determination Group

Now you can define a strategy for stock determination in Customizing for each combination of plant, stock determination group (representing materials), and stock determination rule (representing movement types). Figure 9.25 shows the combination of plant-group-rule for which you can define a strategy.

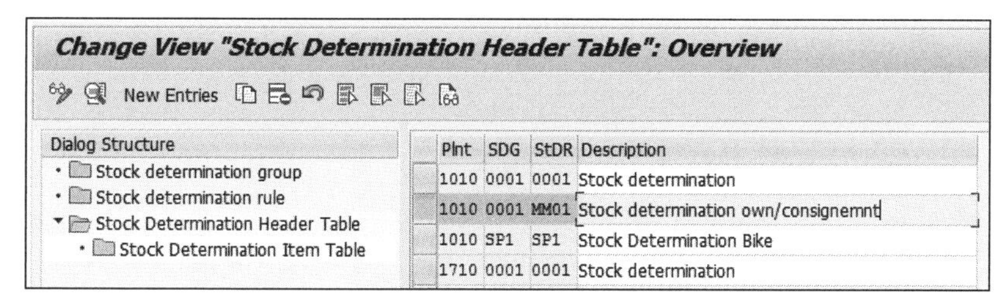

Figure 9.25 Example of a Combination of Plant–Group–Rule

It is best to maintain the item table first. In the item table, you enter the stocks (unrestricted-use stock or consignment stock) and the storage locations to be considered in stock determination. You can use the priority indicator to sort the stocks in ascending or descending order and thus put them in the desired sequence.

Figure 9.26 shows a simple example of a stock determination strategy for plant 1010, stock determination rule MM01, and stock determination group 0001. It applies to storage location 101C: first (**Priority Indicator** 1) stock is to be withdrawn from the unrestricted company's own stock (special stock indicator F in column **S**), followed by supplier consignment stock (special stock indicator K in column **S**).

Change View "Stock Determination Item Table": Overview

New Entries

Dialog Structure
- Stock determination group
- Stock determination rule
- Stock Determination Header Table
  - Stock Determination Item Table

Plant 1010 Hamburg
Stock determ. group 0001 Stock determination group 00
Stck Determation Rule 0001 Stock determination rule 000
Description Stock determination

S	SLoc	Priority indicator	Valuation Type
F	101C	1	
K	101C	2	

Figure 9.26 Example of a Stock Determination Strategy in Customizing

After you have maintained the item table, you can make adjustments in the header table.

Among other fine-tuning options, you can adjust the sorting criteria under **Sorting**, including ranking from the item table, price, or quantity. You decide in which order the criteria are to be considered by the system. For example, you can specify that the system first looks at the stock quantity and withdraws from the storage location/stock with the smallest quantity, and if the quantities are the same everywhere, only then does it consider the sequence of the item table. In Figure 9.27, the sequence from the item table (meaning own stock for consignment) has priority.

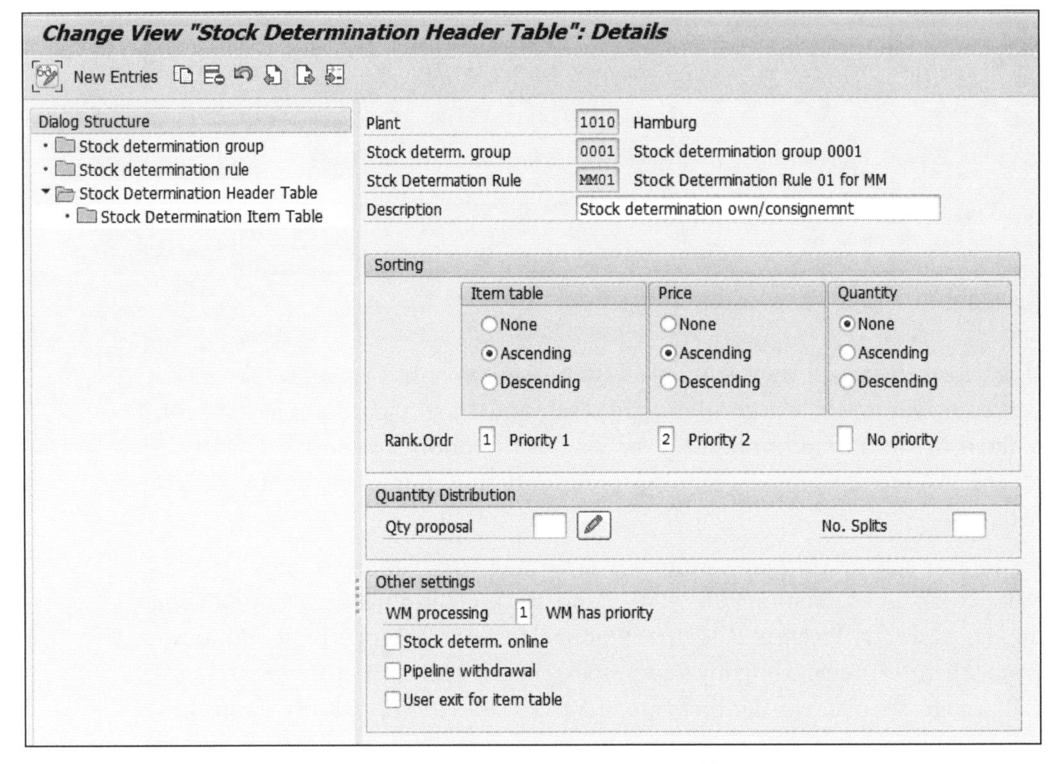

Figure 9.27 Stock Determination Fine-Tuning in the Header Table

Now, how can you get the system to apply this strategy? If you post a withdrawal to the cost center for our material TG11 using Transaction MIGO (the Post Goods Movement app), simply leave the **Storage Location** field empty and choose the **Stock Determination** button shown in Figure 9.28.

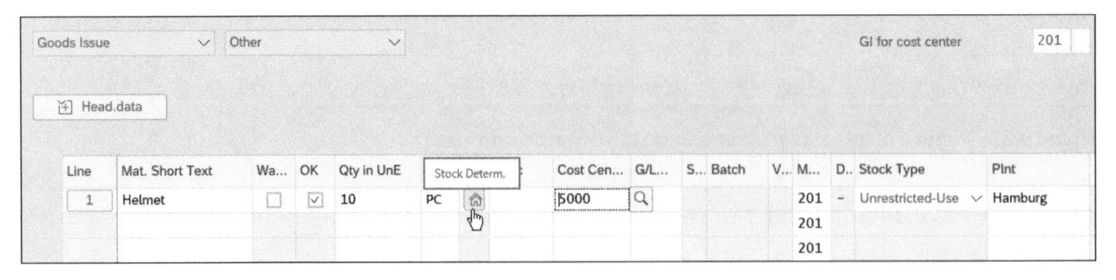

Figure 9.28 Stock Determination Button in the Post Goods Movement App (Transaction MIGO)

Conducting Physical Inventory

There are external and internal reasons for a company to carry out a physical inventory regularly:

- In many countries, it is a legal requirement that every company carry out a physical inventory of its material stocks at least once a year.
- Internally, it is important that the book inventory balances match the physical stock quantities. Incorrect inventory data leads to incorrect availability information.

The physical inventory is carried out based on *stock management units*. The following criteria are used in inventory management to uniquely identify a stock management unit:

- Material
- Plant
- Storage location
- Stock type
- Valuation type and batch
- Special stock

Each stock management unit of a material is counted separately, and the inventory differences are posted per stock management unit. For example, the material quantities must be counted and recorded separately for each stock type and storage location within a plant.

The physical inventory process consists of the following steps:

1. **Creating the physical inventory documents**
   Physical inventory documents can be created manually or in the background. A physical inventory document contains the stock management units to be counted. It consists of a header and items. The document header contains, among other things, a planned count date, the plant, and the storage location. If you want to count special stocks, enter the special stock indicator at the header level. This means that each physical inventory document contains a single plant/storage location, and you cannot have normal and special stock units within one physical inventory document. An item contains, for instance, the material number, the stock type, and if available, the batch or the valuation type. The physical inventory document also contains statuses at the header and item level that provide information on the progress of the physical inventory count.
2. **Entering the count**
   You enter the count results into the system. Here you enter the count date, which is used to determine the book inventory and the period for the inventory difference posting. The system calculates the stock differences. If some of the results are suspect, you can initiate a recount of the affected stocks.
3. **Assessing and posting stock differences**
   Inventory difference postings lead to a stock adjustment by quantity and value. Material and accounting documents are created.

You can block the materials of a physical inventory document for goods movements for a period during the physical inventory. You set the posting **Block** indicator either when you create the physical inventory document or later. The block is removed when the count is entered. For normal stock, the block is set at the storage location level. For special stocks, it is set at the level of the special stock segment (e.g., for consignment at the level of plant, storage location, special stock consignment, and then vendor).

If it is not possible to block the goods movements for organizational reasons, you can let the system determine the book inventory balance for the materials of a physical inventory document at any time (e.g., when you start counting) and transfer it to the physical inventory document. To do this, set the **Freeze Book Inventory** indicator in the physical inventory document. This prevents postings that are made for example between the count and the count entry from affecting the difference calculation.

In the material master record's storage location stock view you will find an indicator that informs you of the status of the physical inventory for the material in the storage location (see Figure 9.29):

- If you have set the goods movement block via the physical inventory document, the indicator is set to X.
- Otherwise, the flag has the value A and simply informs that the physical inventory is ongoing.
- As soon as the inventory is completed, the indicator is reset to blank.

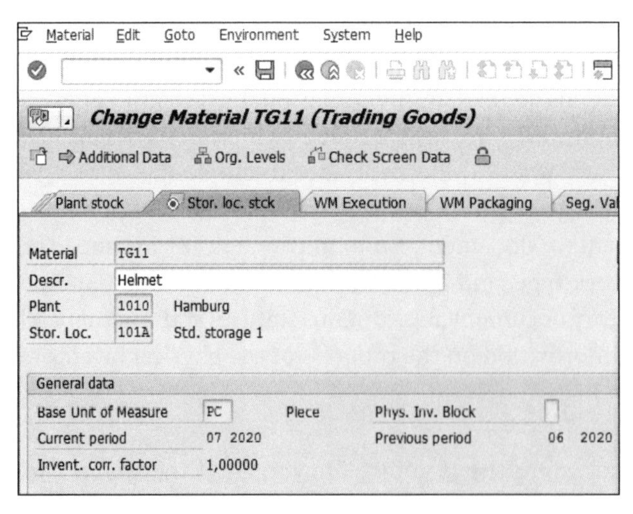

Figure 9.29 Physical Inventory Indicator in the Material Master Record

SAP S/4HANA provides tools that simplify the physical inventory:

- Merging of steps (e.g., you can create the physical inventory document and count in the system at once, which makes sense if you first created the physical inventory document in a subsystem)

- Collective processing through a batch input procedure
- The Create Physical Inventory Documents app to enable the mass creation of physical inventory documents

SAP S/4HANA provides the apps for carrying out a physical inventory but does not stipulate whether the inventory is to be carried out as part of an annual inventory or a continuous inventory. This is entirely the decision of the respective company.

SAP S/4HANA offers two further physical inventory procedures:

- The cycle counting procedure, in which inventory units are counted at regular intervals
- The physical inventory sampling procedure, in which only "selected" stock management units are physically counted

More detailed knowledge of these methods is not required for this certification.

Important Terminology

The following important terminology was used in this chapter:

- **Storage location**
  This is the organizational level where the inventory quantities are managed in inventory management.
- **Valuation area**
  This is the organizational level where the inventory values are managed.
- **Material document**
  A material document is created with every goods movement and contains all information about this goods movement.
- **Accounting document**
  A financial accounting document is generated for each valuated goods movement and contains all the information relevant to financial accounting.
- **Movement type**
  The movement type is a key in Customizing and controls the goods movements.
- **Reservation**
  This document contains the request to the warehouse to make a certain amount of material available for a certain purpose and a certain date.
- **Goods movement**
  Goods movements include goods receipts, goods issues, and stock transfers.
- **Stock determination strategy**
  These are automatic stock determination criteria that support the user in determining the storage location and the type of stock when posting a goods issue.

- **Stock determination group**
  The stock determination strategy is defined per plant, group, and rule. The group represents a material or a group of materials.
- **Stock determination rule**
  The stock determination strategy is defined per plant, group, and rule. The rule is defined differently for each application. In inventory management, it represents a movement type or a group of movement types.
- **Stock management unit**
  The stock management unit is the subdivision of the inventory that is to be counted.
- **Freeze Book Inventory indicator**
  This indicator in an inventory document item ensures that the current book inventory of the stock management unit is transferred to the inventory document.
- **Posting Block indicator**
  With this indicator in an inventory document, you ensure that no goods movement is possible for the material in the affected storage location until the result of the count is entered.

Practice Questions

These questions will help you evaluate your understanding of the topics covered in this chapter. They are similar in nature to those on the certification examination. Although none of these questions will be found in the exam itself, they will allow you to review your knowledge of the subject.

Select the correct answers, and then check the completeness of your answers in the next section. Remember that on the exam, you must select all correct answers—and only correct answers—to receive credit for the question.

1. True or false: Inventory management is only relevant until you are not using EWM. EWM then completely replaces inventory management—not only operationally but also in the background.

☐ **A.** True

☐ **B.** False

2. In inventory management, on which organizational unit are quantities of material stocks managed?

☐ **A.** Plant

☐ **B.** Warehouse number

☐ **C.** Storage location

☐ **D.** Bin location

3. True or false: It is possible for a material to be managed in stock in one plant but not in another plant.

☐ **A.** True

☐ **B.** False

4. With Transaction MIGO (or the Post Goods Movement app), what entries do you use to specify exactly which goods movement you want to book?

☐ **A.** Business transaction, reference, stock type

☐ **B.** Business transaction, reference, material type

☐ **C.** Business transaction, reference, movement type

☐ **D.** Business transaction, reference, reason for movement

5. Which of the following data can you enter differently for each item when creating a goods movement? (There are three correct answers.)

☐ **A.** Reference document

☐ **B.** Movement type

☐ **C.** Plant

☐ **D.** Company code

☐ **E.** Stock type

6. True or false: A single material document for unplanned withdrawals can contain both withdrawals for cost centers and withdrawals for projects or scrapping.

☐ **A.** True

☐ **B.** False

7. You want to make manual entry of the goods receipt value mandatory when posting an initial entry of stock. What is the best way to proceed?

☐ **A.** Adapt the field selection control for Transaction MIGO in Customizing by making the **External Amount** field a mandatory field.

☐ **B.** Create a customer-specific movement type by copying the standard movement type for the initial entry of inventory balances and set the **External Amount** field as a mandatory field.

☐ **C.** Adjust the field selection control of the standard movement type for the initial entry of inventory balances by making the **External Amount** field a mandatory field.

8. Through which of the following activities can reservations be automatically generated in the system? (There are two correct answers.)

☐ **A.** Through MRP

☐ **B.** By releasing a production order

☐ **C.** By placing a purchase order

☐ **D.** By creating a project

9. You unexpectedly require 200 pieces of a material for a cost center and 100 pieces for a project in a few days. You want to reserve the material in your plant to ensure that the required quantities are available when they are needed. How do you proceed?

☐ **A.** You manually create a reservation with two items.

☐ **B.** You manually create two reservations.

☐ **C.** You start an MRP run for the required material.

10. What are the features of the **Goods Movement Allowed** indicator in a reservation item? (There are two correct answers.)

☐ **A.** Reservations with this indicator are for information purposes only and are not passed on to MRP.

☐ **B.** It is automatically deleted by the system when the total reserved quantity is posted out.

☐ **C.** If not set, you can prevent reserved quantities from being withdrawn long before the requirements date.

☐ **D.** It can be set automatically as the request date approaches.

11. Which of the following statements apply to the two-step stock transfer posting between two plants?

☐ **A.** You can transfer stock quantities from any stock type in the issuing plant to any stock type in the receiving plant.

☐ **B.** You can post the goods issue from the supplying plant and the goods receipt to the receiving plant in one transaction.

☐ **C.** Once you have entered the goods issue in the issuing plant, the stock is posted to the stock in transfer of the receiving plant.

☐ **D.** When the goods receipt is posted in the receiving plant, an accounting document is created.

12. In your company, the valuation level is the plant. Which price is used to valuate the stock transfer quantity for stock transfers between two plants within a company code?

☐ **A.** The valuation price in the receiving plant

☐ **B.** The valuation price in the issuing plant

☐ **C.** The average of the valuation prices of the two plants

☐ **D.** No price; this type of stock transfer is not valued

13. True or false: For stock transfers between two plants, the stock in transit and the stock in transfer belong to the valuated stock of the receiving plant.

☐ **A.** True

☐ **B.** False

14. For plant internal stock transfer, an accounting document is only created in a two-step procedure.

☐ **A.** True

☐ **B.** False

15. For which objects can you define differentiated stock determination strategies in inventory management? (There are three correct answers.)

☐ **A.** Movement type

☐ **B.** Plant

☐ **C.** Storage location

☐ **D.** Material

☐ **E.** Business transaction

16. In inventory management, which of the following criteria are used to define stock management units that form the basis for the physical inventory count? (There are three correct answers.)

☐ **A.** Plant

☐ **B.** Warehouse number

☐ **C.** Bin location

☐ **D.** Stock type

☐ **E.** Storage location

17. What is the impact of setting the **Freeze Book Inventory** indicator in an inventory document?

☐ **A.** All materials in the physical inventory document are blocked for goods movements.

☐ **B.** Goods movements for materials in the physical inventory document are parked until the count is posted.

☐ **C.** The book inventory is determined for each material and fixed in the physical inventory document.

18. What controls whether a material may be kept in stock?

☐ **A.** The movement type

☐ **B.** The material type

☐ **C.** The stock type

☐ **D.** The valuation type

19. At what price are goods issues valuated?

☐ **A.** Valuation price of the material

☐ **B.** Average price from the info records

☐ **C.** Price entered manually by the user

☐ **D.** Price from the last goods receipt

Practice Question Answers and Explanations

1. Correct answer: **B**

   False. EWM builds on inventory management and is integrated with it. If an inventory management storage location is set as EWM-managed, the stocks of this storage location are managed operatively via EWM. Certain internal movements run entirely in EWM, but goods receipts, goods issues, or certain stock transfers result in material movements with material documents in inventory management as if the posting had taken place in inventory management. Inventory management is always there, either in the background or even operationally for storage locations that are not managed in EWM.

2. Correct answer: **C**

   In inventory management, material stocks are managed on a quantity basis at the storage location level, whereas value-based inventory management is usually carried out at the plant level (per SAP's recommendation). The warehouse number and storage bin are usually warehouse management units and specifically EWM units.

3. Correct answer: **A**

   True. Yes, it can happen that a material is managed in stock in one plant but not in another plant. This depends, on the one hand, on whether you have created the corresponding views (Accounting and, where necessary, Storage). It also depends on the material type with which you created the material. In Customizing for the material type, you can specify whether and how (by quantity and value or only by quantity) the materials of this material type are to be managed in the respective valuation areas.

4. Correct answer: **C**

   The movement type, together with the business transaction (**Goods Receipt**, **Goods Issue**, **Stock Transfer**, etc.) and the reference document, defines the type of movement to be posted. If the goods movement does not refer to a preceding document (unplanned goods movement), the reference field must still be filled with **Other**.

5. Correct answers: **B**, **C**, **E**

   Plant, stock type, and movement type are item data and can therefore be defined differently for each item. The reference (not the number of the reference document, but the category of the reference document) and the business transaction are entered once for each goods movement: they contribute to the determination of the material document type. Note that if you enter a goods receipt with reference to a purchase order, you can enter several purchase order numbers one after the other, but you cannot have a goods receipt with reference to a purchase order and a goods receipt with reference to a production order in the same document.

Note

In the standard SAP S/4HANA system, there is even a material type that only allows inventory management based on value (material type WERT). A value-only material represents a group of materials for which no inventory management is carried out for the material. The system allows only one such material type.

6. Correct answer: **A**

   True. When you post an unplanned withdrawal, you specify in the item—via the movement type—whether the item is to be scrapped and withdrawn for a cost center or for another account assignment. Thus you can have items with several movement types in a material document for unplanned withdrawals.

7. Correct answer: **B**

   Basically, you can modify the field selection control for the standard movement type 561, but we would recommend that you create a new movement type by copying the standard one.

8. Correct answers: A, B

   Stock transfer reservations can be generated by MRP. Dependent reservations are reservations of components of orders, projects, or networks and are generated when these objects are created.

9. Correct answer: B

   Since you have two different account assignments, you must create two reservations, each with the correct movement type. The movement type and account assignment object only are unique in a reservation. They cannot be entered at the item level. MRP cannot help here: Only stock transfer reservations can be created directly by MRP.

10. Correct answers: C, D

    If the indicator is not set, no goods movements can be posted with reference to the reservation item. In this way, you can prevent reserved quantities from being withdrawn well before the requirement date, which would confuse the entire planning process. You can schedule a report that runs regularly and sets the indicator shortly before the actual requirements date. A reservation item without the **Goods Movement Allowed** indicator is, of course, still passed on to MRP!

11. Correct answer: C

    A stock transfer posting in a two-step procedure is carried out in two stages. First, a goods issue is posted in the supplying plant. The quantity is posted to the stock in transfer of the receiving plant and an accounting document is created. Then the goods receipt is posted in the receiving plant. No accounting document is created at this point. The stock in transfer belongs to the receiving plant.

    A stock transfer between two plants in two steps can only be posted for a quantity in unrestricted-use stock.

12. Correct answer: B

    If the valuation level is the plant, a stock transfer between two plants in one company code is valuated. The stock transfer is valuated at the valuation price of the material in the issuing plant.

13. Correct answer: A

    True. When you transfer stock between two plants, the stock in transit is updated in the case of a stock transport order, and the stock in transfer is updated by a stock transfer posting in two steps. Both belong to the valuated stock of the receiving plant.

14. Correct answer: B

    False. An internal plant stock transfer is not valuated, and an accounting document is never created.

15. Correct answers: **A, B, D**
    Stock determination strategies are defined per plant, stock determination rule, and stock determination group. A stock determination rule represents one or more movement types, and a stock determination group represents one or more materials.
16. Correct answers: **A, D, E**
    The criteria for defining stock management units in inventory management are material, plant, storage location, stock type, batch, valuation type, and special stock. The warehouse number and storage bin are units from Warehouse Management (for example EWM). Note that the physical inventory is carried out in Warehouse Management if you are using a warehouse management system. Only the difference posting is transferred to inventory management. The definition of stock units in Warehouse Management is different and, of course, considers the storage bin.
17. Correct answer: **C**
    When the **Freeze Book Inventory** indicator is set, the current book inventory at that time is transferred to the physical inventory document. It will later be the basis for determining the inventory differences.
18. Correct answer: **B**
    In Customizing for the material type, you define whether materials of this material type are managed by value and/or quantity and, if so, whether they are managed in all or certain valuation areas (usually plants).
19. Correct answer: **A**
    The value of a goods issue posting is always *Withdrawal quantity × Current valuation price of the material in the plant*. For a stock transfer between two plants, the valuation price of the material in the issuing plant is taken.

Takeaway

You should now be familiar with the functions and tasks of inventory management. You should also be able to post planned and unplanned goods movements, and you should understand and be able to explain the effects of these postings. This chapter presented the reservation as a document for planning goods issues and stock transfers. You should be familiar with stock transfer options and be able to decide which one is best for which business situation. Finally, you should be able to implement a stock determination strategy to support users in their daily work and understand the stages of the physical inventory.

Summary

In this chapter, we explored which functions inventory management fulfills, how you can manage your stocks, and which organizational levels are involved in quantity and value-based inventory management.

We must be aware that the basic framework of warehouse management is always present, even if the operative business exists outside of the realm of warehouse management. Furthermore, inventory management is integrated with the overarching logistics. This basic knowledge, reviewed in this chapter, is therefore essential.

In the next chapter, we will again and more intensively examine stock values. We will discern how stock values are calculated and how the system is able to automatically determine the correct accounts in financial accounting when goods movements are posted.

Chapter 10
Valuation and Account Determination

Techniques You'll Master

- Master the principles of material valuation.
- Name the purpose and application areas of automatic account determination.
- List influencing factors for automatic account determination.
- Explain the possible settings for each influencing factor.
- Set up automatic account determination.
- Implement the split valuation.

In this chapter, we will briefly discuss the situations in which the system is required to automatically determine the accounts, and we will also summarize the basic principles of material stock valuation. The focus will then be on analyzing the various influencing factors and the available configuration options. Finally, we deal with the split valuation and its configuration.

Real-World Scenario

A chart of accounts is the individual account structure of a company, usually based on a selected account framework. Setting up one or more charts of accounts and assigning them to company codes is one of the first configuration activities when implementing an SAP S/4HANA system.

Many activities lead to a simultaneous update of accounts in financial accounting. For example, when raw materials are transferred from the warehouse to production, the system reduces the quantity in the warehouse and at the same time reduces the values on the stock accounts in the balance sheet. It would be unrealistic if the warehouse or production clerk who withdraws the goods also would have to enter the accounts in order to have the value updated. Therefore, account determination is performed in the background. Since the accounts to be determined are often customer-specific, the configuration must be adjusted in almost all implementation projects. Therefore, expertise in this area is one of the expected skills of a consultant. As a rule, account determination is set up in a collaborative effort involving both finance and materials management personnel.

Objectives of This Portion of the Test

The purpose of this part of the certification exam is to test your knowledge of the procurement of stock material. For the certification exam, you must have a good understanding of the following topics:

- Valuation fundamentals and valuation areas (recap from Chapter 2)
- Automatic account determination
- Account grouping code
- Valuation classes and account category references
- Transaction/event keys
- Account determination for special cases (e.g., for delivery costs, purchase orders assigned to an account, or taxes posted during invoice verification)
- Purpose and configuration of split valuation

Key Concept Refresher

In this chapter, we go through the basics of material stock valuation and analyze the automatic account determination. We want to learn how to influence the account determination and which levers are available to us. We will also discuss the separate valuation, its relevance, and its configuration.

Valuation Basics

Material stocks have a value. How this value is derived, how it changes, and which control options are available constitutes the topic of valuation.

The valuation of material stocks depends on the following factors:

- The valuation level
- The parameters in the material master record

Now let's take a closer look at these elements.

Valuation Level

In Chapter 2's discussion of organizational structure, we already introduced the *valuation area* as the organizational unit responsible for valuation.

The valuation area corresponds either to the company code or the plant. If the valuation area corresponds to the plant, you can valuate a material in different plants at different prices and/or using different procedures. If the valuation area corresponds to the company code, the valuation price of the material is the same in all plants in the company code.

Figure 10.1 and Figure 10.2 illustrate a setting where the valuation area corresponds to the plant.

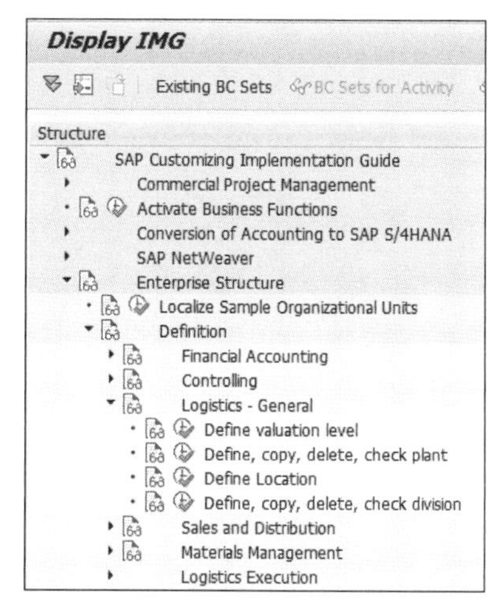

Figure 10.1 Customizing of the Valuation Area: Path

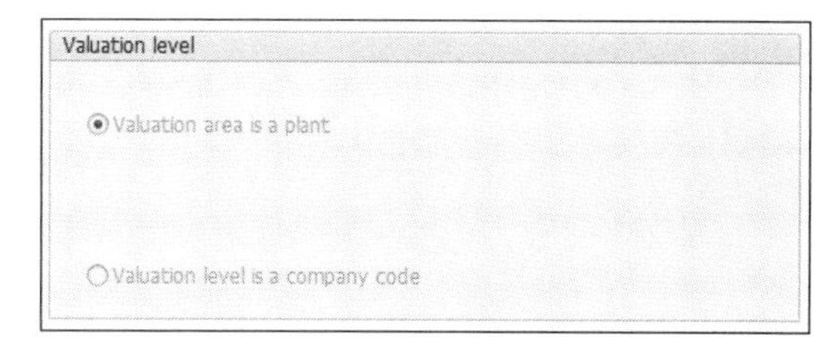

Figure 10.2 Customizing of the Valuation Area: Valuation Level

Note

SAP recommends valuation at the plant level. Valuation at the plant level is even mandatory if you want to use production planning and product cost controlling.

Once the valuation level is set, it should not be changed. Changing the valuation level can result in inconsistencies.

Once you have specified the valuation level, you can maintain valuation-relevant fields for each material and valuation area.

Parameters in the Material Master Record

Figure 10.3 and Figure 10.4 show a material master record with the fields relevant for valuation on the Accounting 1 view.

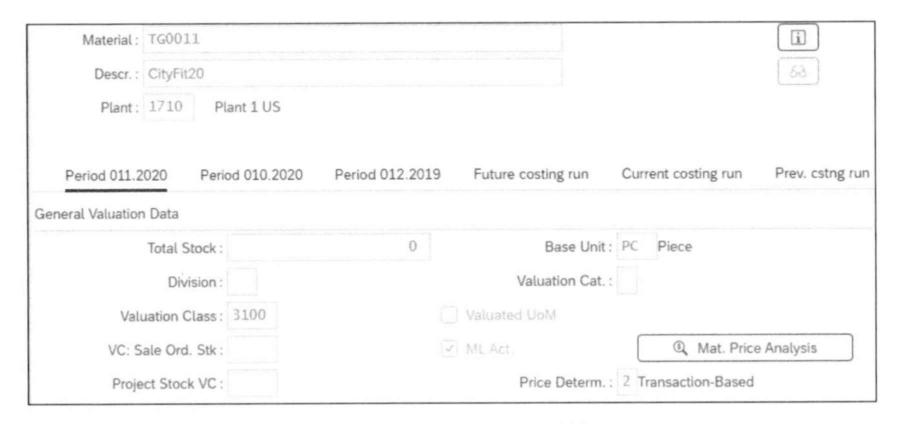

Figure 10.3 Accounting 1 View of the Material Master Record (I)

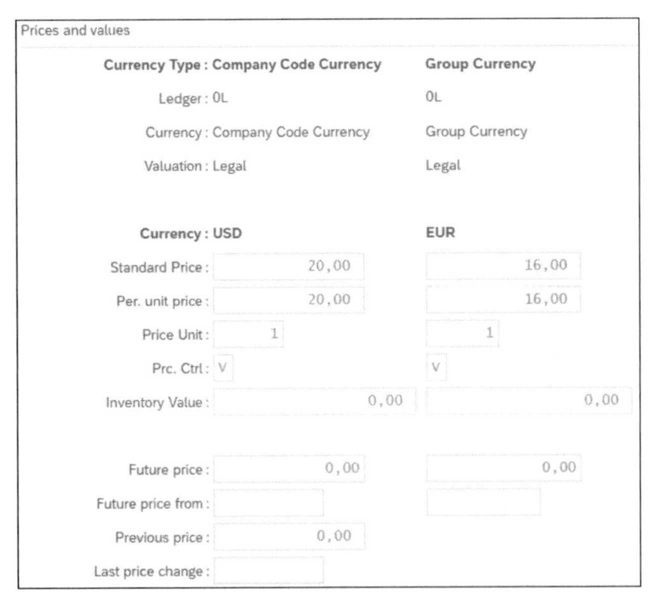

Figure 10.4 Accounting 1 View of the Material Master Record (II)

Let's walk through some of the material master fields that are relevant for valuation.

First, the **Price Control** and **Price** fields shown in Figure 10.4 determine the stock value. SAP S/4HANA offers two transaction-based valuation procedures:

- **The moving average price procedure**
  In the moving average price procedure, a new material price is recalculated after each goods receipt and after invoices that lead to a change in inventory value. This price is calculated by dividing the total stock value by the total quantity of the material in stock. In the moving average price method, the receipts are posted with the external receipt values and the issues are valuated with the current moving average price from the material master.

  In Chapter 3, we used examples to show in detail how the moving average price is formed and what the posting procedure for goods receipt and invoice receipt looks like for each price control. You can refer back to this chapter to refresh your knowledge, as needed.
- **The standard price procedure**
  If a material is valuated at the standard price, the standard price remains constant for at least one period. If you valuate a material at the standard price, all inventory postings are made with this constant standard price. The current inventory value is the product of the standard price and the current inventory quantity.

In addition to determining the inventory value using price control and price, you must also define the balance sheet accounts to which the inventory values are posted and the other accounts that are updated when goods movements occur.

The **Valuation Class** field maintained in the material master record at the valuation area level in Figure 10.3 is responsible for differentiated account determination per material. The valuation class in the material master record is therefore one of the factors influencing automatic account determination. Now let's examine the entire set of rules for account determination.

Automatic Account Determination

We have seen in previous chapters that goods movements lead to the simultaneous generation of accounting documents. In detail, the following postings can lead to a value update:

- Goods receipt
- Goods issue
- Stock transfers (except between storage locations within a plant)
- Price change
- Revaluation of a material quantity
- Incoming invoice
- GR/IR clearing account maintenance

When you enter a goods movement, you do not need to enter a general ledger account because SAP S/4HANA determines it automatically.

The SAP S/4HANA determines the general ledger accounts based on the following data:

- Organizational level
- Material
- Transaction

Organization Level

Automatic account determination is dependent on the company code and valuation area.

Company Code

The system uses the company code to determine the chart of accounts.

You may have already noticed, however, that you cannot enter a company code when you post a goods movement. Instead, you are required to enter a plant, which enables the system to determine the company code indirectly. Remember that a plant can be assigned to exactly one company code. A chart of accounts is assigned to each company code, as shown in Figure 10.5.

The chart of accounts is a key in the table for automatic account determination.

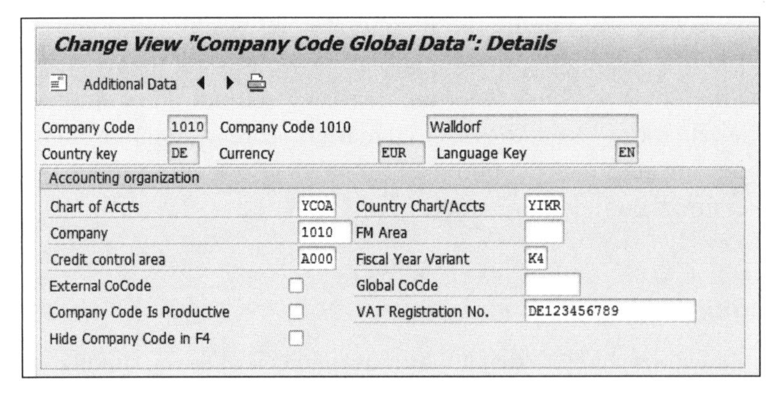

Figure 10.5 Customizing of Company Code 1010

Valuation Area

It is possible to set up account determination differently for each *valuation area*. It is also possible to group valuation areas and set a differentiated account determination for each group of valuation areas.

To achieve differentiated account determination per valuation area or per group of valuation areas, you must first activate valuation grouping in Customizing and then assign a *valuation grouping code* to each valuation area you want to group. This is shown in Figure 10.6.

The valuation grouping code is a key in the table for automatic account determination.

Note

- If you have not activated the valuation grouping code, all valuation areas are processed in the same way in terms of account assignment. Account determination cannot be configured to be dependent on the valuation area.
- If valuation grouping is activated but you do not assign a valuation grouping code to a valuation area, the system accesses the account determination table with the value blank.

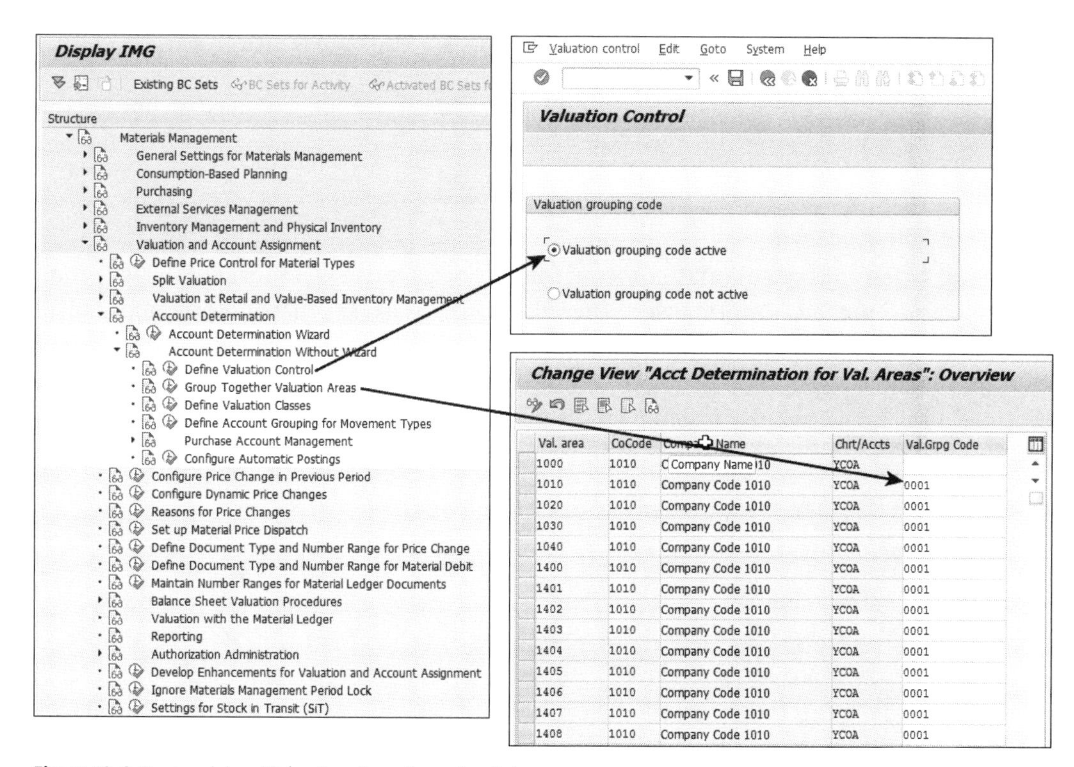

Figure 10.6 Customizing: Valuation Area Grouping Setup

Table 10.1 shows the structure of the account determination table. Note we have already explored two keys (influencing factors).

Client	Chart of Accounts	???	Valuation Grouping Code	???	??	General Ledger Account Debit	General Ledger Account Credit
400	YCOA	BSX	0001	--	3000	13100000	13100000
400	YCOA	GBB	0001	VBR	3000	51100000	51100000
400	YCOA	BSX	0001	--	3050	13500000	13500000
400	YCOA	GBB	0001	VBR	3050	51500000	51500000

Table 10.1 Logical Structure of the Account Determination Table (with Chart of Accounts and Valuation Grouping Code)

In addition to the organizational units, automatic account determination depends on the material type and the material master record.

Material Master Data

We will now investigate the relationship between the material master data and the account determination process.

Valuation Class

When you create a master record for a stock material, you must enter a *valuation class* in the Accounting view. The valuation class is a key that allows you to control account determination, depending on the material. The valuation classes also allow you to group materials; materials with the same valuation class are treated the same from the account determination view.

The valuation class is a further key in the account determination table in Table 10.2.

Client	Chart of Accounts	???	Valuation Grouping Code	???	Valuation Class	General Ledger Account Debit	General Ledger Account Credit
400	YCOA	BSX	0001	--	3000	13100000	13100000
400	YCOA	GBB	0001	VBR	3000	51100000	51100000
400	YCOA	BSX	0001	--	3050	13500000	13500000
400	YCOA	GBB	0001	VBR	3050	51500000	51500000

Table 10.2 Logical Structure of the Account Determination Table (with Valuation Class)

When creating/maintaining a material, you select the valuation class from a value list. Figure 10.7 shows that the selection options are not the same for every material:

we see that for material TG11, only one valuation class is available for selection, whereas for material RM124, seven are proposed.

This restriction depends on the material type. In Figure 10.7, the material TG11 is a trading good, while RM124 is a raw material.

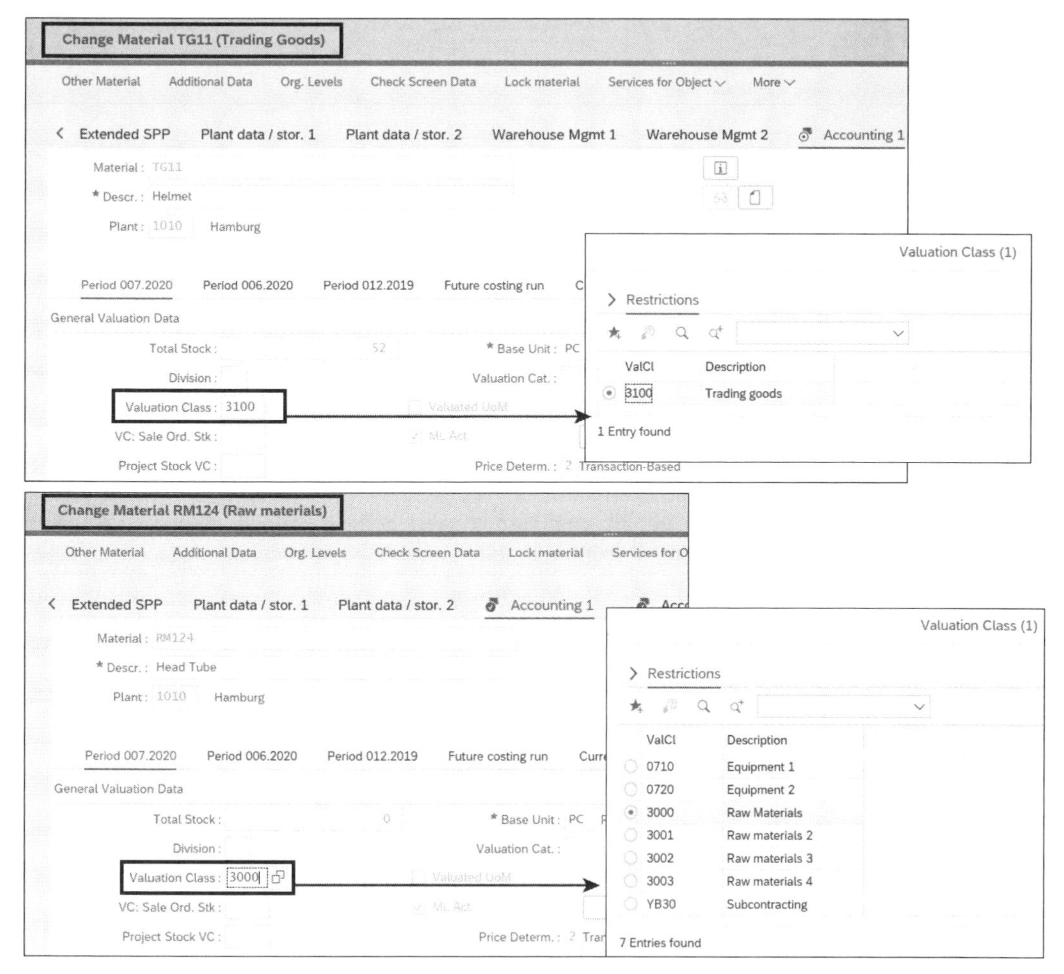

Figure 10.7 Selection Options for the Valuation Class, Depending on the Material Type

Material Type

The valuation classes allowed depend on the *material type*. You establish the relationship between valuation classes and material types using the *account category reference*. These are assigned to the material types. You can only assign one account category reference per material type.

Figure 10.8 shows the account category references respectively assigned to material type HAWA (trading goods) and material type ROH (raw materials).

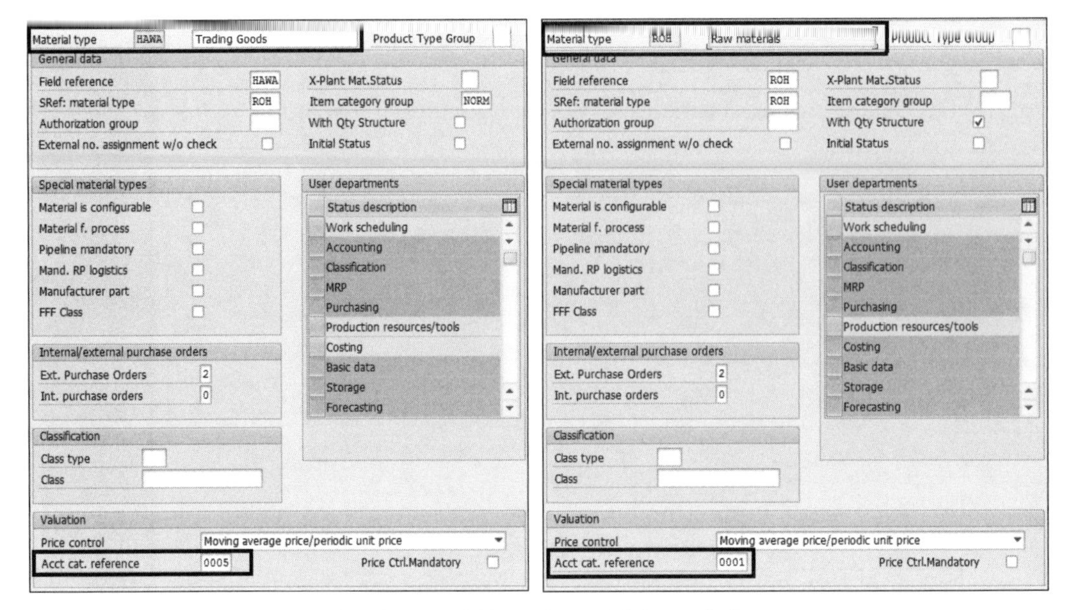

Figure 10.8 Customizing the Material Type: Account Category Reference

While you can only assign one account category reference to a material type, several material types can have the same account category reference. Figure 10.9 illustrates the possible relationships between material type, account category reference, and valuation class.

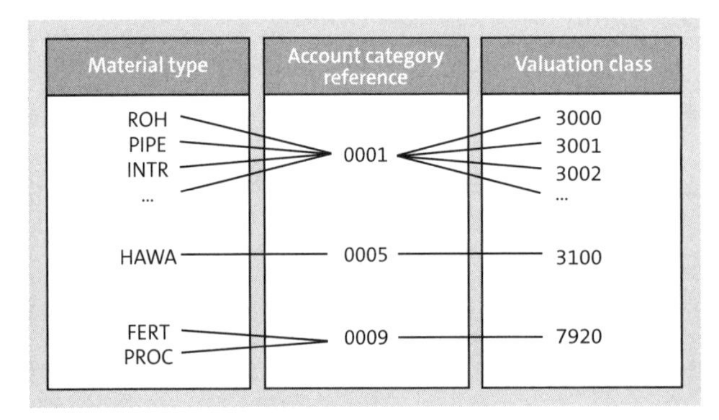

Figure 10.9 Possible Relationships between Material Type, Account Category Reference, and Valuation Class

Figure 10.10 shows the Customizing settings. Here, the account category references are created first. Then you assign the allowed valuation classes to each account class reference and a single account class reference to each material type. (Of course, it is not necessary to assign an account category reference to material types that are intended for non-stocked or non-valuated materials.)

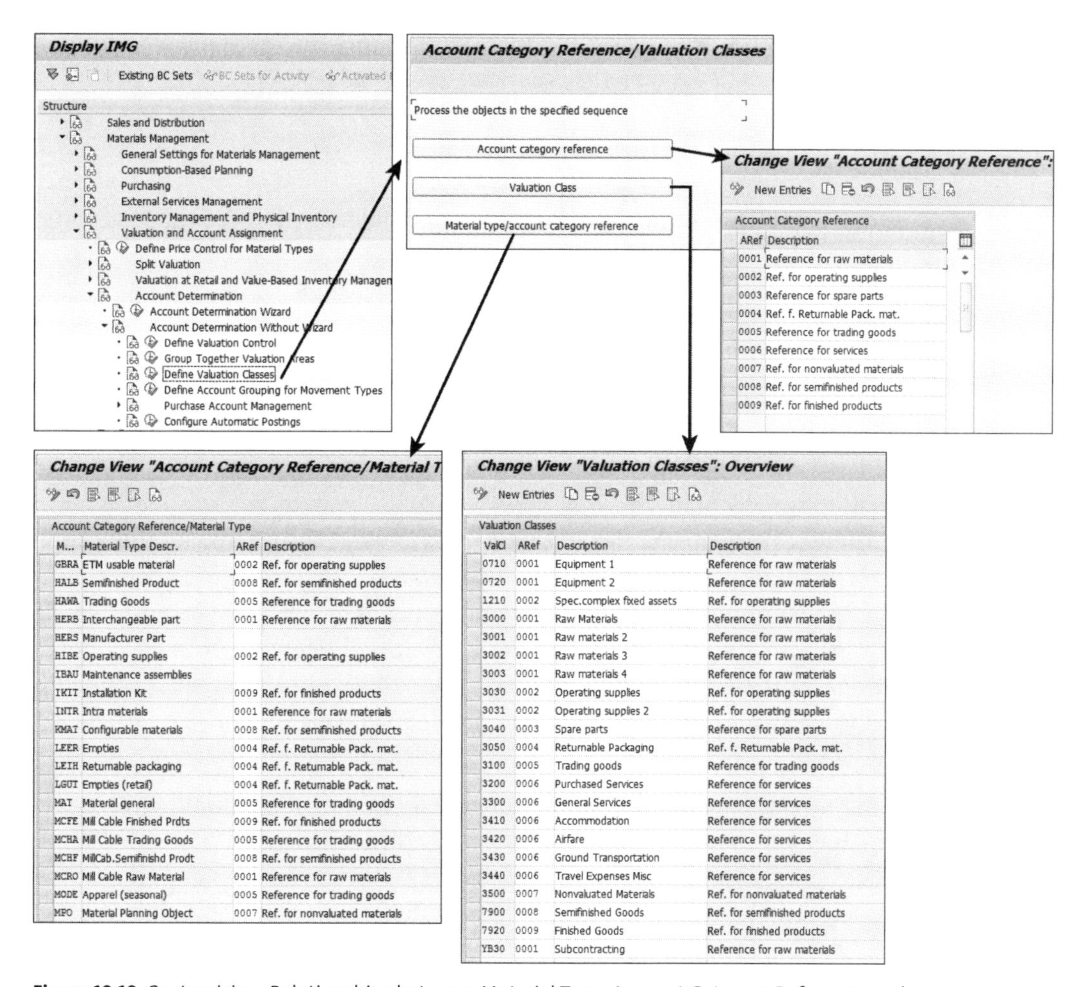

Figure 10.10 Customizing: Relationships between Material Type, Account Category Reference, and Valuation Class

The reason why a differentiation and preselection of the possible valuation class is made via the material type is to avoid, for example, a raw material ending up in the same account as a finished product.

So far in this chapter we have learned that the organizational unit (valuation area) and the material can lead to a differentiated account determination. For example, when you post a goods receipt or goods issue, you can update to a different stock account, depending on the valuation area and material.

Now let's look at the next influencing factor. Consider the example of goods issue. If you withdraw a quantity of a material for a cost center or scrap it, the stock account of the material in the valuation area is credited in each case. However, the offsetting entry is not made to the same account in both cases: when withdrawing to the cost center, the value is posted to a consumption account, whereas when scrapping, the value is posted to a scrapping account.

With this example, we understand that the accounts that are updated during a goods movement depend on the type of transaction itself.

Transaction

We now concretely look at how the transaction influences the account determination.

Value Strings and Transaction Keys

When you post a goods movement, an invoice, or one of the previously mentioned accounting-relevant logistics transactions in SAP S/4HANA, the system determines a so-called *value string* internally.

The value string contains information about all the maximum possible postings that can take place during this business transaction. The system determines the value string, depending on your entries. For example, in the case of a goods movement, the determination of the value string depends on whether you post a goods receipt, goods issue, or stock transfer, with or without reference, with what movement type, with or without a special stock key, and so forth. You cannot configure which value strings exist and how they are defined. This is defined internally by the system.

You may recall that in Chapter 3 we analyzed the postings that occur at goods receipt for a purchase order, depending on whether the material is valuated at the standard price or the moving average price. We noticed that, in addition to the stock account, the GR/IR clearing account (and, in the case of the standard price, often the price difference account) is also posted to. This is due to the value string determined in the background.

The value string WE01 is used for goods receipts into the warehouse with reference to a standard purchase order. Figure 10.11 shows the maximum possible postings that the value string WE01 allows. Each posting line is represented by a *transaction key*: BSX for the stock posting, WRX for the posting to the GR/IR clearing account, PRD for the posting to price differences, and so on.

Note

Not all transaction keys are used each time you post a goods receipt for a warehouse order. For example, FRL and FRN are only used for subcontracting orders, EIN and EKG are only in certain countries such as France or Belgium, and so forth.

The second BSX key is used when you post the goods receipt back in the previous period.

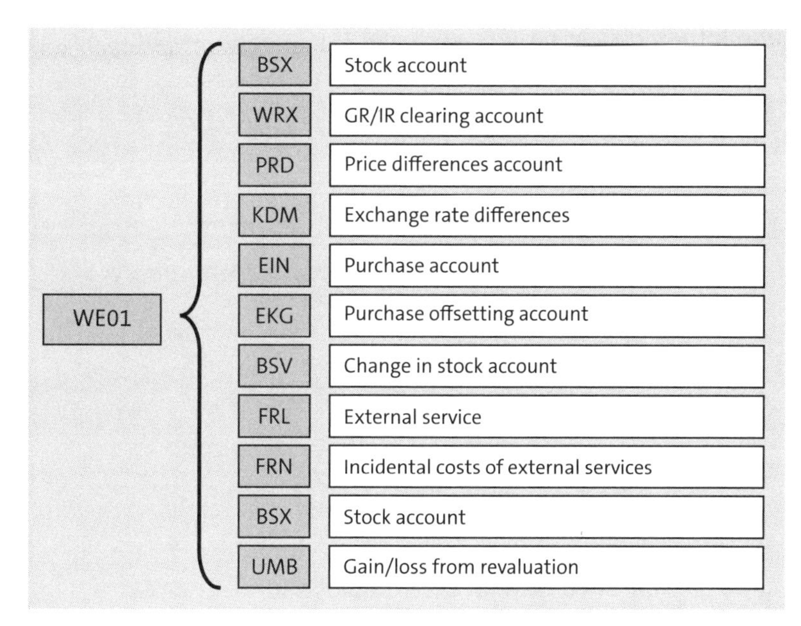

Figure 10.11 Example of Value String WE01

The transaction key is a further key in the account determination table in Table 10.3.

Client	Chart of Accounts	Transaction Key	Valuation Grouping Code	???	Valuation Class	General Ledger Account Debit	General Ledger Account Credit
400	YCOA	BSX	0001	--	3000	13100000	13100000
400	YCOA	GBB	0001	VBR	3000	51100000	51100000
400	YCOA	BSX	0001	--	3050	13500000	13500000
400	YCOA	GBB	0001	VBR	3050	51500000	51500000

Table 10.3 Logical Structure of the Account Determination Table (with Transaction Key)

You will likely ask yourself how you can find out the possible individual posting lines if, for example, you need to check and set up account determination for goods movement postings. The simulation function enables you to check the settings for automatic account determination.

Simulation Function

To access the simulation function in Customizing, go to **Materials Management • Valuation and Account Assignment • Account Determination • Account Determination Without Wizard • Configure Automatic Postings**. Then choose the **Simulation** button. Choose a material, a plant, and, for example, a movement type for which you want to check the possible posting lines and the account assignments, as shown in Figure 10.12.

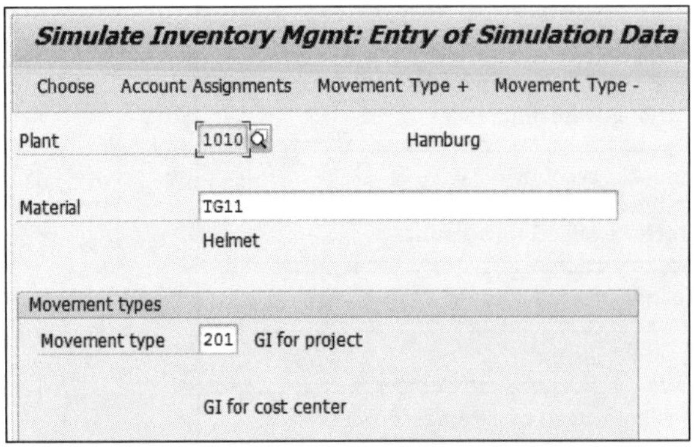

Simulate Inventory Mgmt: Entry of Simulation Data

Choose | Account Assignments | Movement Type + | Movement Type -

Plant 1010 Hamburg

Material TG11
Helmet

Movement types
Movement type 201 GI for project

GI for cost center

Figure 10.12 Initial Screen of the Simulation Function

Figure 10.13 shows the simulated account determination for the movement type withdrawal to the cost center (201) for a trading good. The posting lines displayed in the lower block represent the transaction keys that belong to the corresponding value string WA01. You recognize the structure of the account determination table.

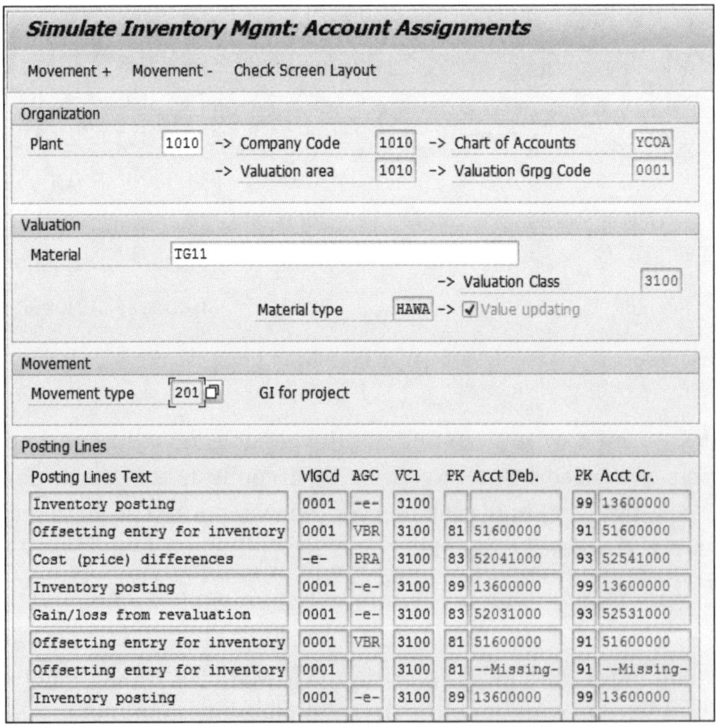

Simulate Inventory Mgmt: Account Assignments

Movement + | Movement - | Check Screen Layout

Organization
Plant 1010 -> Company Code 1010 -> Chart of Accounts YCOA
-> Valuation area 1010 -> Valuation Grpg Code 0001

Valuation
Material TG11
-> Valuation Class 3100
Material type HAWA -> Value updating

Movement
Movement type 201 GI for project

Posting Lines

Posting Lines Text	VlGCd	AGC	VC1	PK	Acct Deb.	PK	Acct Cr.
Inventory posting	0001	-e-	3100			99	13600000
Offsetting entry for inventory	0001	VBR	3100	81	51600000	91	51600000
Cost (price) differences	-e-	PRA	3100	83	52041000	93	52541000
Inventory posting	0001	-e-	3100	89	13600000	99	13600000
Gain/loss from revaluation	0001	-e-	3100	83	52031000	93	52531000
Offsetting entry for inventory	0001	VBR	3100	81	51600000	91	51600000
Offsetting entry for inventory	0001		3100	81	--Missing-	91	--Missing-
Inventory posting	0001	-e-	3100	89	13600000	99	13600000

Figure 10.13 Posting Lines in the Simulation Function

If you take a closer look at Figure 10.13 (simulation for movement type 201), you will see the various influencing factors we have discussed so far:

- The **Organization** section
  - Company code and chart of accounts transaction keys (posting lines)
  - Valuation area and valuation grouping code
- The **Material** section
  - Material, material type, and valuation class
- The **Posting Lines** section
  - Posting lines text corresponding to the transaction key
  - Valuation grouping code (the **VlGCd** column)
  - Valuation class (the **VCl** column)

You may also note that three columns have not been discussed yet: the two **PK** columns and the **AGC** column. PK stands for *posting key*. The posting key is a key for financial accounting and controls, among other things, whether the posting is a debit or credit posting. You can use the posting key to define different accounts, depending on whether it is a debit or credit posting. This can be useful for some postings (e.g., price difference postings).

AGC stands *account grouping code*, and we will explore its relevance in the next section.

Account Grouping Code

We continue looking at the example of the value string WA01. It is assigned internally to various goods issues, such as withdrawal for a cost center or scrapping.

In all cases, the value of the withdrawal is posted from the stock account. For material TG11 (see Figure 10.13), this would be account 13600000, as you can see in the **Inventory Posting** line, and corresponds to transaction key BSX.

The offsetting entry, however, depends on the type of withdrawal—that is, on the movement type. If the goods are scrapped, a scrapping account is debited. If the goods are withdrawn to a cost center, for example, a consumption account is posted to.

However, according to value string WA01 in Figure 10.14, there is only one entry for the offsetting entry with transaction key GBB. This means that further differentiation is necessary.

Value strings contain the transaction keys for the postings that lead to general ledger account updates in financial accounting. If a transaction key is too imprecise for your requirements (e.g., GBB), you can activate the general modification for it. You can then define account grouping codes to differentiate the postings.

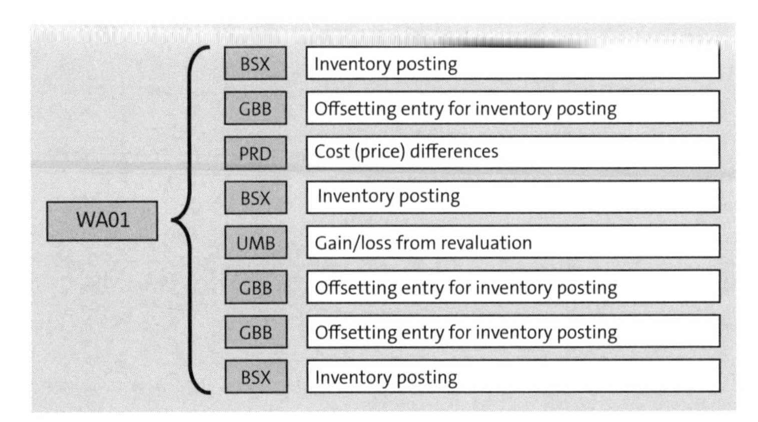

Figure 10.14 Example of Value String WA01

In Figure 10.14, which represents the simulation for movement type 201, you can see that for the line **Offsetting Entry for Inventory**—transaction key GBB—the system uses the account grouping code VBR to determine the consumption account. For the goods movement postings, you can assign account grouping codes, depending on the movement type and other indicators (e.g., special stock indicator).

In Table 10.4, you can see that we have now examined and explained all keys for automatic account determination.

Client	Chart of Accounts	Transaction Key	Valuation Grouping Code	Account Grouping	Valuation Class	General Ledger Account Debit	General Ledger Account Credit
400	YCOA	BSX	0001	--	3000	13100000	13100000
400	YCOA	GBB	0001	VBR	3000	51100000	51100000
400	YCOA	BSX	0001	--	3050	13500000	13500000
400	YCOA	GBB	0001	VBR	3050	51500000	51500000

Table 10.4 Logical Structure of the Account Determination Table (with Account Grouping)

Customizing Automatic Account Determination

Now that we know the influencing factors, we go through the Customizing steps and learn the remaining details that play a role in automatic account determination.

Figure 10.15 shows the Customizing path and the list of transaction keys. We take a closer look at BSX (**Inventory Posting**) and GBB (**Offsetting Entry for Inventory Posting**).

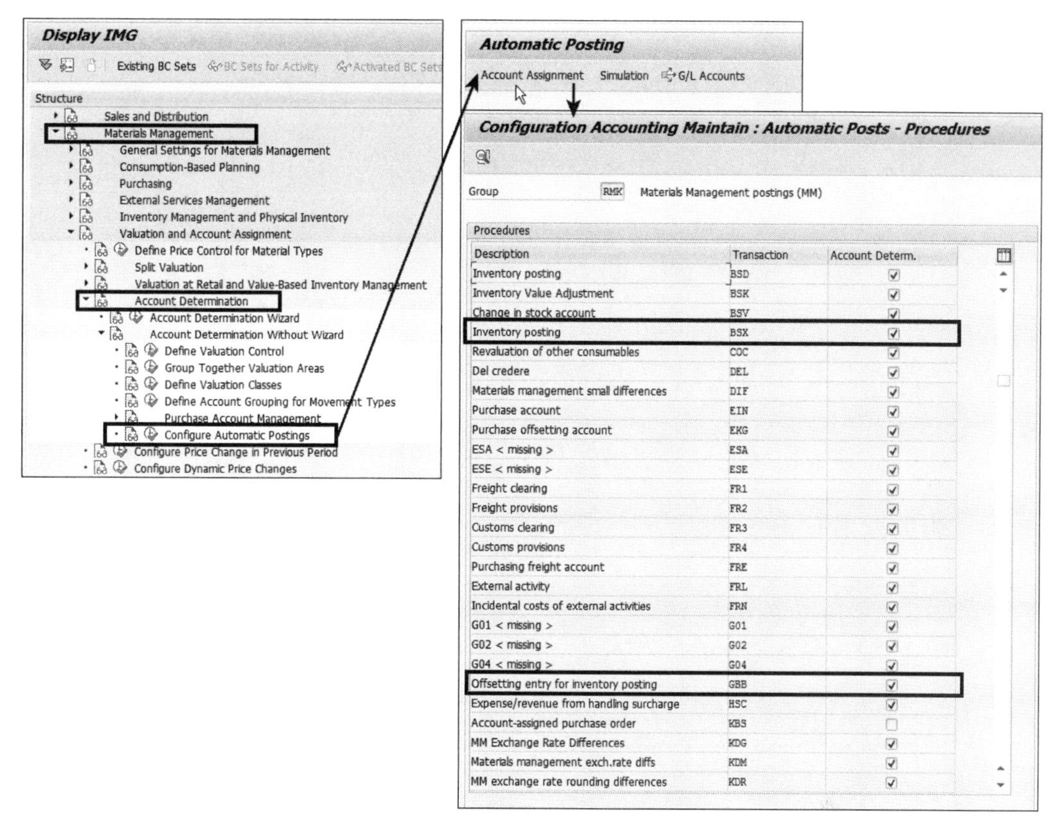

Figure 10.15 Customizing for Account Determination: Path and Initial Steps

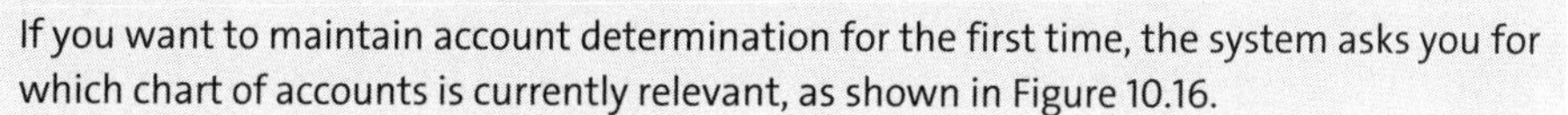

Note

If you want to maintain account determination for the first time, the system asks you for which chart of accounts is currently relevant, as shown in Figure 10.16.

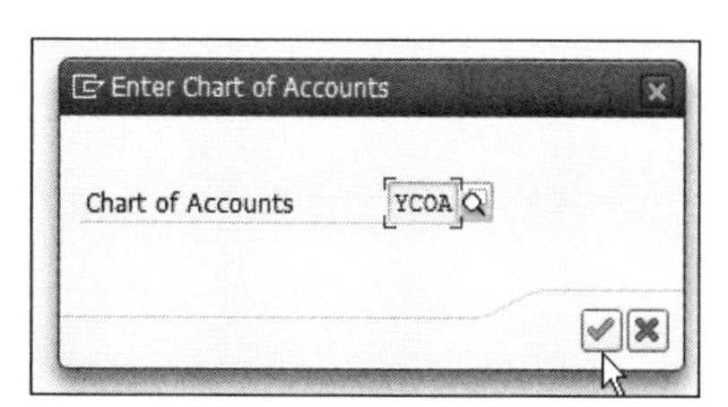

Figure 10.16 Customizing for Account Determination: Define the Chart of Accounts

Figure 10.17 shows that account determination for the inventory posting (BSX) depends on the valuation grouping code (see the **Valuation Modif.** column) and the valuation class. If we select the **Rules** button, we find that we can influence these dependencies to a certain extent, as shown in Figure 10.18.

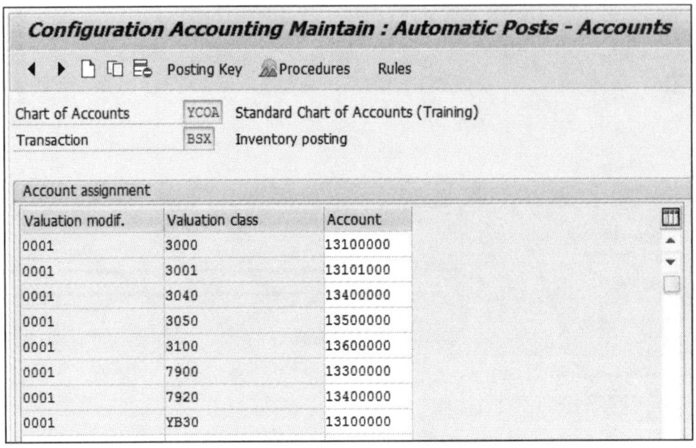

Configuration Accounting Maintain : Automatic Posts - Accounts

Posting Key Procedures Rules

Chart of Accounts YCOA Standard Chart of Accounts (Training)
Transaction BSX Inventory posting

Account assignment

Valuation modif.	Valuation class	Account
0001	3000	13100000
0001	3001	13101000
0001	3040	13400000
0001	3050	13500000
0001	3100	13600000
0001	7900	13300000
0001	7920	13400000
0001	YB30	13100000

Figure 10.17 Customizing for Account Determination: Transaction Key BSX (Inventory Posting)

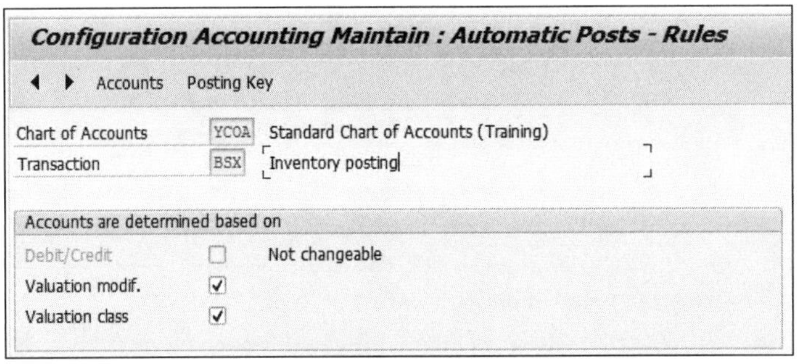

Configuration Accounting Maintain : Automatic Posts - Rules

Accounts Posting Key

Chart of Accounts YCOA Standard Chart of Accounts (Training)
Transaction BSX Inventory posting

Accounts are determined based on

Debit/Credit ☐ Not changeable
Valuation modif. ☑
Valuation class ☑

Figure 10.18 Customizing for Account Determination: Rules for Transaction Key BSX

Figure 10.19 shows that account determination for the offsetting entry for inventory posting (GBB) depends on the valuation grouping code (the **Valuation Modif.** column), the account grouping code (the **General Modification** column), and the valuation class. If we select the **Rules** button, we notice that the general modification is activated, which means that operation GBB can be further subdivided using the account grouping code, as shown in Figure 10.20.

Up until now, we have dealt with the general automatic account determination for goods movements and invoice posting. However, there are also special forms of account determination, which we will now discuss.

Configuration Accounting Maintain : Automatic Posts - Accounts

Posting Key Procedures Rules

Chart of Accounts YCOA Standard Chart of Accounts (Training)

Transaction GBB Offsetting entry for inventory posting

Account assignment

Valuation modif.	General modification	Valuation class	Debit	Credit
0001	VAY	7900	54083000	54083000
0001	VAY	7920	54083000	54083000
0001	VBO	3000	51950000	51950000
0001	VBO	3001	51950000	51950000
0001	VBO	3050	51900000	51900000
0001	VBO	3100	51950000	51950000
0001	VBO	7900	51950000	51950000
0001	VBO	7920	50300000	50300000
0001	VBO	YB30	51950000	51950000
0001	VBR		51100000	51100000
0001	VBR	3000	51100000	51100000
0001	VBR	3001	51100000	51100000
0001	VBR	3040	54400000	54400000
0001	VBR	3050	51500000	51500000

Figure 10.19 Customizing for Account Determination: Transaction Key GBB (Offsetting Entry for Inventory Posting)

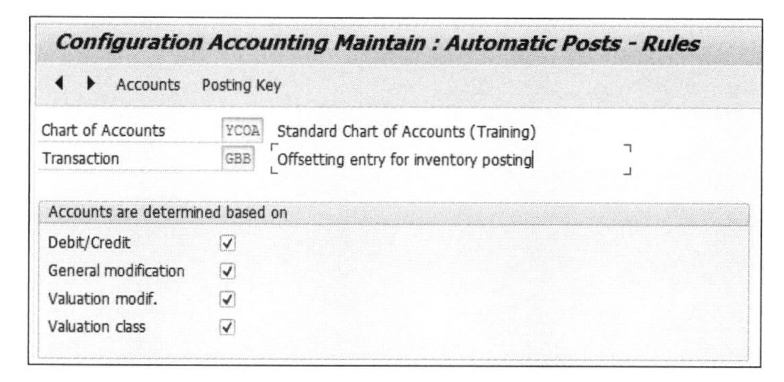

Figure 10.20 Customizing for Account Determination: Rules for Transaction Key GBB (Offsetting Entry for Inventory Posting)

Account Determination for Special Cases

In this section, the following special features are presented:

- Account determination for planned delivery costs
- Automatic determination of an account for account assigned purchase order items (or purchase requisitions items)
- Determination of tax accounts when posting invoices

Let's review each one.

Planned Delivery Costs

You can enter delivery costs in a purchase order item using specific condition types. If delivery costs are already entered in the order item, we call them *planned delivery costs*. Planned delivery costs are part of the purchase price of a material. At goods receipt (we assume that the goods receipt is posted before the invoice receipt), the total purchase value (*Goods value + Delivery costs*) is posted to the stock account at the moving average price. The offsetting entry is posted to the GR/IR clearing account in the amount of the goods value, while the delivery costs are posted to a freight clearing or provision account.

This freight clearing account is determined using a transaction key that does not come from the value string but from the calculation schema. You can enter a transaction key in the pricing procedure of the purchasing document, in the row corresponding to the condition type for freight (see the **Accruals** column in Figure 10.21). These transaction keys for freight postings can be freely defined in Customizing.

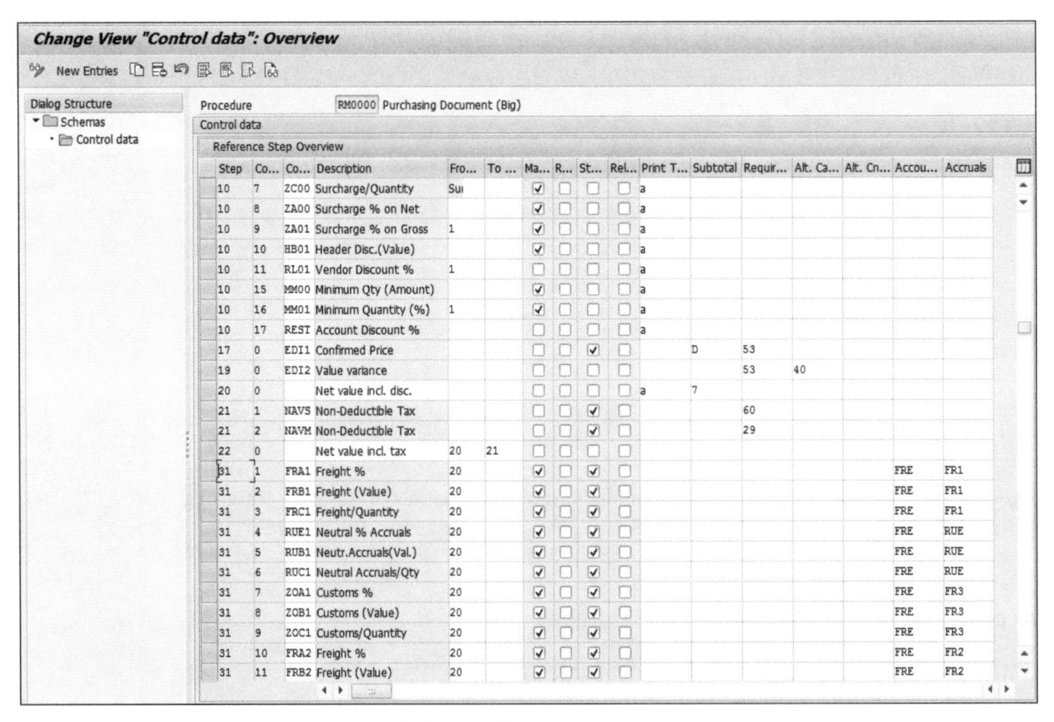

Step	Co...	Co...	Description	Fro...	To ...	Ma...	R...	St...	Rel...	Print T...	Subtotal	Requir...	Alt. Ca...	Alt. Cn...	Accou...	Accruals
10	7	ZC00	Surcharge/Quantity	Sui		☑	☐	☐	☐	a						
10	8	ZA00	Surcharge % on Net			☑	☐	☐	☐	a						
10	9	ZA01	Surcharge % on Gross	1		☑	☐	☐	☐	a						
10	10	HB01	Header Disc.(Value)			☑	☐	☐	☐	a						
10	11	RL01	Vendor Discount %	1		☐	☐	☐	☐	a						
10	15	MM00	Minimum Qty (Amount)			☑	☐	☐	☐	a						
10	16	MM01	Minimum Quantity (%)	1		☑	☐	☐	☐	a						
10	17	REST	Account Discount %			☐	☐	☐	☐	a						
17	0	EDI1	Confirmed Price			☐	☐	☑	☐		D	53				
19	0	EDI2	Value variance			☐	☐	☐	☐			53	40			
20	0		Net value incl. disc.			☐	☐	☐	☐	a	7					
21	1	NAVS	Non-Deductible Tax			☐	☐	☑	☐			60				
21	2	NAVM	Non-Deductible Tax			☐	☐	☑	☐			29				
22	0		Net value incl. tax	20	21	☐	☐	☐	☐							
31	1	FRA1	Freight %	20		☑	☐	☑	☐						FRE	FR1
31	2	FRB1	Freight (Value)	20		☑	☐	☑	☐						FRE	FR1
31	3	FRC1	Freight/Quantity	20		☑	☐	☑	☐						FRE	FR1
31	4	RUE1	Neutral % Accruals	20		☑	☐	☑	☐						FRE	RUE
31	5	RUB1	Neutr.Accruals(Val.)	20		☑	☐	☑	☐						FRE	RUE
31	6	RUC1	Neutral Accruals/Qty	20		☑	☐	☑	☐						FRE	RUE
31	7	ZOA1	Customs %	20		☑	☐	☑	☐						FRE	FR3
31	8	ZOB1	Customs (Value)	20		☑	☐	☑	☐						FRE	FR3
31	9	ZOC1	Customs/Quantity	20		☑	☐	☑	☐						FRE	FR3
31	10	FRA2	Freight %	20		☑	☐	☑	☐						FRE	FR2
31	11	FRB2	Freight (Value)	20		☑	☐	☑	☐						FRE	FR2

Figure 10.21 Customizing: Calculation Schema RM0000

> **Note**
> Unlike planned delivery costs, unplanned delivery costs are not known at the time of ordering. In Customizing for logistics invoice verification, you can define how these costs are to be posted. In the standard system, the costs are distributed among the invoice items.

However, you can opt for the unplanned delivery costs to be posted to a separate account. In this case, you must assign a general ledger account for the transaction key UPF in the account determination table.

Account Assigned Purchase Order/Purchase Requisition Items

When you create a purchase order item with account assignment, the system can propose an account. To do this, you must define an account grouping code for the account assignment category in Customizing.

The system determines and proposes an account by using (depending on the rules defined) the valuation grouping code of the valuation area, the valuation class of the material, the transaction key GBB, and the account grouping code of the account assignment category. Figure 10.22 shows the account grouping code (account modification) VBR for account assignment category K.

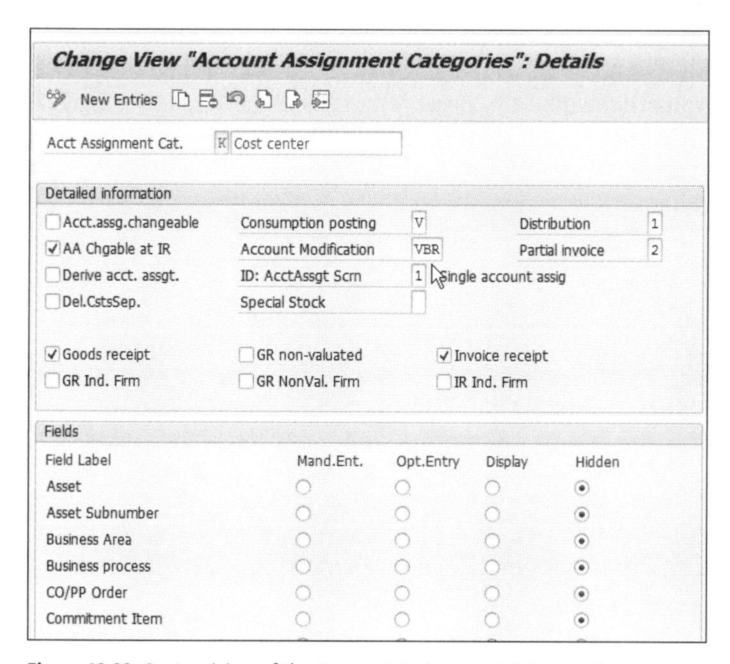

Figure 10.22 Customizing of the Account Assignment Category K

Note

You can define valuation classes for material groups. This enables an account proposal to be made for purchase order items with account assignment without material.

You assign valuation classes to material groups in Customizing for purchasing under **Entry Aid for Items without Material Master**.

Tax Accounts When Posting Invoices

In Customizing for financial accounting, a tax schema (or tax determination procedure) is assigned to the country in which your company code is located. This schema provides various condition types for calculating the tax amounts and transaction keys for determining the tax accounts.

You assign the accounts for tax postings in Customizing under **Financial Accounting • Financial Accounting Global Settings • Tax on Sales/Purchases • Posting • Define Tax Accounts**. The standard transaction key for input tax is VST.

Split Valuation

Sometimes the valuation area is not enough for the stock valuation of some materials. Consider the example of a tool that you manage in stock. You manage this material as both a new material and a repaired material and want new and repaired materials to be valued at different prices.

Here is another example: a material is both externally procured and produced in-house. You want to valuate the quantity procured externally at the moving average price, while the quantities produced in-house are valuated at the standard price.

Split valuation enables a differentiated valuation of stock quantities of a material within a valuation area.

Setup

Split valuation must be first set up in Customizing, so let's walk through it.

Activate Split Valuation

If you want to valuate the quantity of some materials differently according to certain criteria, you must first make sure that split valuation is activated in the system. Figure 10.23 shows the Customizing path for configuring split valuation.

In Figure 10.24, we see that split valuation is activated in the system.

Warning

The fact that you have activated split valuation does not mean that all materials must be valuated using split valuation. It merely means that you can use split valuation for materials whenever you deem it necessary.

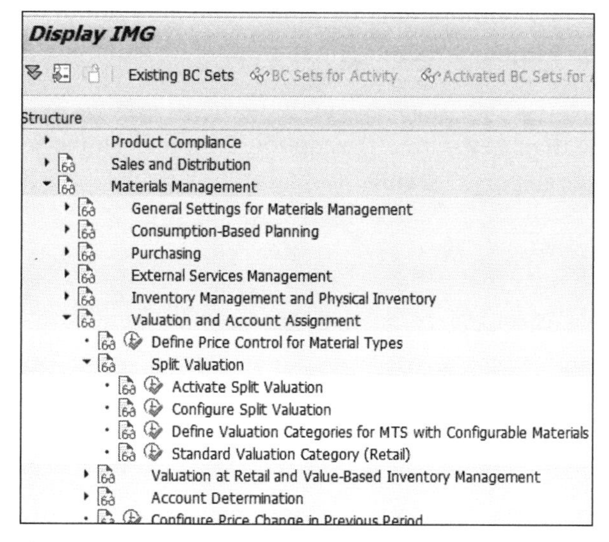

Figure 10.23 Customizing Path for Split Valuation

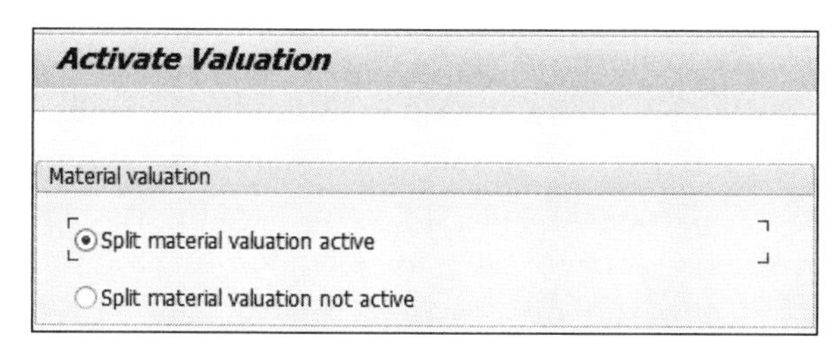

Figure 10.24 Customizing: Activate Split Valuation

Define Valuation Categories

You must then define the criteria according to which materials are to be split-valuated. These criteria can be, for example, the procurement type (procured externally or produced in-house), the country of origin (USA, Germany, France, China, etc.), or the state of the product (new, used, etc.). The possible criteria are represented in the system with *valuation categories*.

Define Valuation Types

Once the possible differentiation criteria have been established, you must define the attributes of the partial stocks. For the valuation category (criterion) procurement type, for example, the attributes can be *procured externally* or *produced in-house*. For the valuation category state, the attributes can be *new*, *used*, or *repaired*. The possible values of a criterion are represented in the system by *valuation types*.

Procedure in Detail

For setting up split valuation, you proceed as follows:

1. Define the global valuation categories, independent of the valuation area. The valuation category is a one-digit key with a description. In our example in Figure 10.25, this is valuation category **C, Status**.

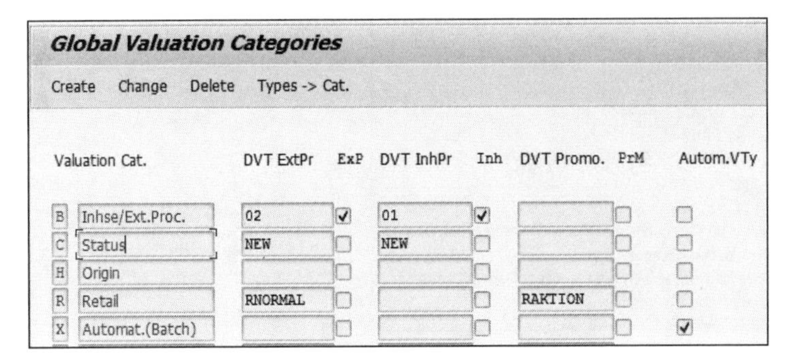

Figure 10.25 Customizing: Overview of the Valuation Categories

2. Define global valuation types, independent of the valuation area. A valuation type consists of a key with attributes, as shown in Figure 10.26. For individual valuation types, you can specify whether they can be used in an internal and/or external procurement process. With the valuation type, you can set the account determination by specifying different account category references for each valuation type. For example, the stock of the "new" material can be managed in a different stock account than the stock of the "repaired" material.

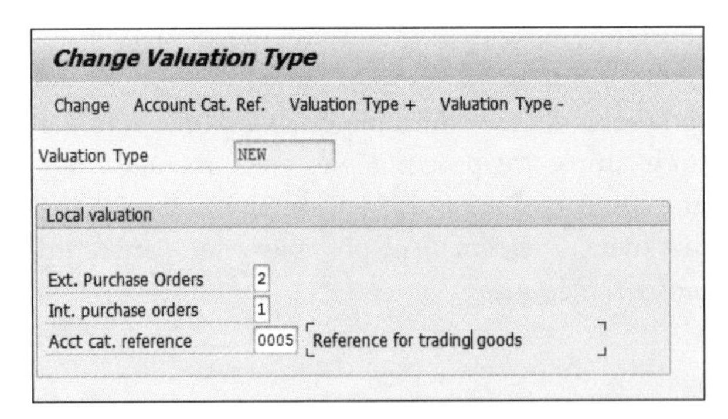

Figure 10.26 Customizing: Example of a Valuation Type

Note

The account category reference assigned to the valuation type plays the same role as the one assigned to the material type: restriction of the possible valuation classes when maintaining a material master record.

3. Assign valuation types to valuation categories, independent of the valuation area. After you have assigned valuation types to the valuation categories, you

can define for each valuation category a default valuation type for external procurement and a default valuation type for internal procurement. You can also specify whether these default values are binding.

4. Assign globally defined valuation categories to valuation areas in which they can be used; these are called *local definitions*. Define the valuation categories, valuation types, and the assignment of types to categories in Customizing for materials management under **Valuation and Account Assignment • Split Valuation • Configure Split Valuation**.

Maintaining Materials That Are Subject to Split Valuation

After you have activated split valuation and created valuation categories and valuation types, you can create certain materials so that they are subject to split valuation.

The first step is to enter a valuation category in the Accounting view of the material on the valuation area level (usually plant; see Figure 10.27). This causes the material to be split-valuated in this valuation area.

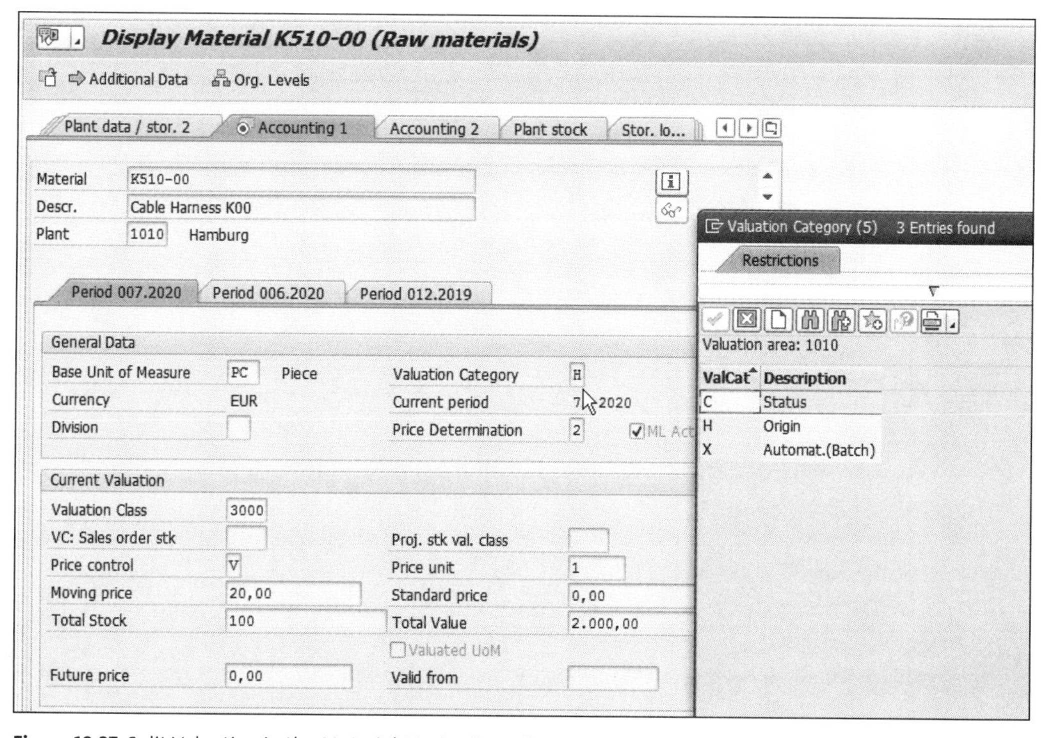

Figure 10.27 Split Valuation in the Material Master Record

Warning

For a material subject to split valuation, you can only use the moving average price method at the valuation area level.

Next, you must also create accounting data for all possible valuation types for the material, as shown in Figure 10.28 and Figure 10.29. At the valuation area–valuation type level, you can choose between price control using the standard price or the moving average price and between different valuation classes. The number of valuation classes available depends on the account category reference assigned to the valuation type.

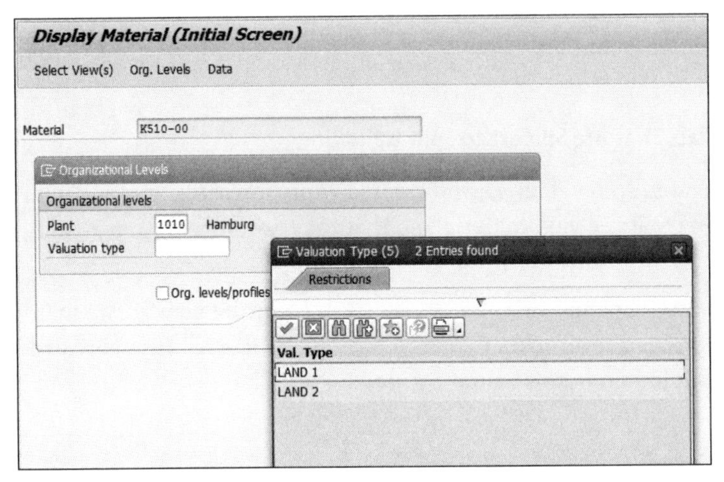

Figure 10.28 Selecting from Available Valuation Types in the Material Master Record

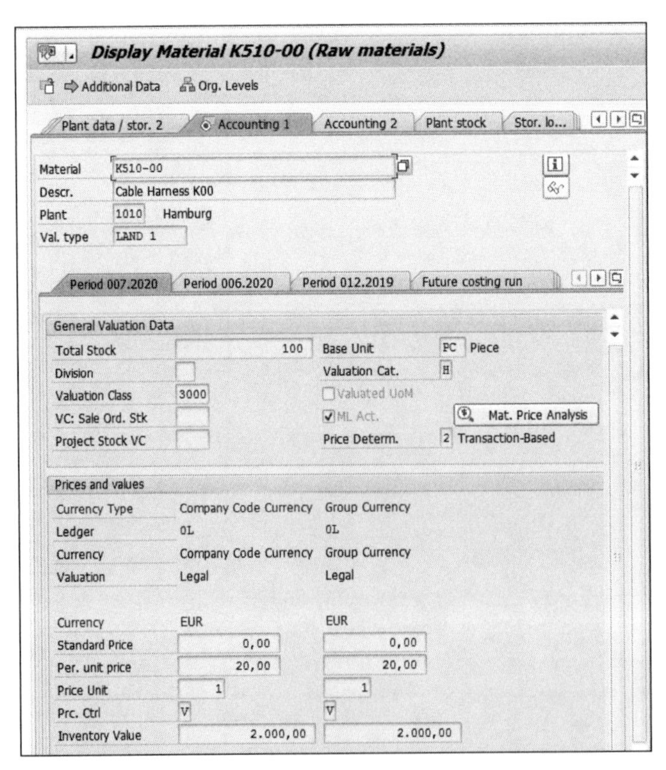

Figure 10.29 Accounting Data on Valuation Area (Plant) and Valuation Type Level in the Material Master Record

For every transaction involving a material that is subject to split valuation and is relevant for valuation (goods receipt, goods issue, invoice receipt, physical inventory, etc.), you must always specify which partial stock (valuation type) is affected. The change only applies to the partial stock affected; other partial stocks are not touched. In addition to the affected partial stock, the total stock is also updated. The total stock is calculated from the sum of the stock values and the stock quantities of the individual substocks.

Important Terminology

The following important terminology was used in this chapter:

- **Chart of account**
  This is a complete register of general ledger accounts required to cover the business needs of the company code it is assigned to.
- **Valuation area**
  This is the valuation level—either the plant (recommended) or the company code.
- **Valuation grouping code**
  This key enables you to group valuation areas that should be treated as one from the account determination point of view.
- **Account category reference**
  This key is assigned to a material type or a valuation type (for split valuation). The account category references, in turn, are assigned valuation classes. In this way, you can ensure that only certain valuation classes are allowed for certain material types or valuation types.
- **Valuation class**
  This is a key in the material master record at the valuation area level and, in the case of split valuation, at the valuation type level, that directly influences automatic account determination for a material.
- **Value string**
  This is determined internally by the system and contains all possible posting lines (transaction keys) for the currently executed transaction.
- **Transaction key**
  Each posting line of a value string is represented by a transaction key: BSX, WRX, PRD, etc. The transaction key directly influences automatic account determination.
- **Account grouping code**
  This key enables further differentiation of the posting if a transaction key is too imprecise (such as GBB).
- **Split valuation**
  This is the option of separately valuating parts of the stock of a material within the plant.

- **Valuation category**
  This is the criterion according to which partial stocks of a material are to be valuated separately.
- **Valuation types**
  These are characteristics of the valuation category: if the valuation category (criterion for a separate valuation) is the country of origin, the valuation type can be China, Canada, Italy, and so on.

Practice Questions

These questions will help you evaluate your understanding of the topics covered in this chapter. They are similar in nature to those on the certification examination. Although none of these questions will be found in the exam itself, they will allow you to review your knowledge of the subject.

Select the correct answers, and then check the completeness of your answers in the next section. Remember that on the exam, you must select all correct answers—and only correct answers—to receive credit for the question.

1. Which of the following fields in the material master record govern stock valuation? (There are two correct answers.)

☐ **A.** Valuation class

☐ **B.** Material price group

☐ **C.** Purchasing value key

☐ **D.** Price control indicator

2. True or false: When you configure the organizational structure of your company, you must create and assign valuation areas.

☐ **A.** True

☐ **B.** False

3. True or false: The company code is not relevant for account determination if the valuation level corresponds to the plant.

☐ **A.** True

☐ **B.** False

4. You want to valuate your materials according to different procedures for each plant, so you have decided to set valuation at the plant level. However, you want to have a common account determination across all plants. How can you achieve this? (There are two correct answers.)

☐ **A.** You do not activate the valuation grouping code. In this way, you switch off the account determination dependency on the valuation area.

☐ **B.** You activate the valuation grouping code and assign the same valuation grouping code to all plants.

☐ **C.** You cannot achieve this. Since the valuation area is a key of the account determination table, you must set the account determination for each plant.

5. What is the name of the field with which you can restrict the possible valuation classes per material type or group of material types?

☐ **A.** Valuation grouping code

☐ **B.** Account grouping code

☐ **C.** Account category reference

☐ **D.** Account assignment category

☐ **E.** Valuation area

6. Which of the following factors can influence automatic account determination? (There are three correct answers.)

☐ **A.** Transaction via transaction key

☐ **B.** Material via valuation class

☐ **C.** Vendor via account group

☐ **D.** Valuation area via valuation grouping code

☐ **E.** Purchase order via purchase order type

7. Which of the following statements apply to the value string? (There are two correct answers.)

☐ **A.** You can configure the value string in Customizing.

☐ **B.** The value string is part of the key of the account determination table.

☐ **C.** The value string contains transaction keys.

☐ **D.** The system automatically determines the value string that is assigned to a posting.

8. Which of the following postings are relevant for automatic account determination? (There are two correct answers.)

☐ **A.** Transfer posting between two storage locations within a plant

☐ **B.** Transfer posting from unrestricted use stock to blocked stock within a storage location

☐ **C.** Transfer posting between two plants of one company code

☐ **D.** Transfer posting from consignment stock to own stock

9. At which levels can you assign a valuation class to a material? (There are two correct answers.)

☐ **A.** Valuation area

☐ **B.** Valuation category

☐ **C.** Material type

☐ **D.** Valuation type

10. True or false: The general modification is not activated for the procedure BSX (inventory posting) by default but could be activated if required so that a separate account grouping code could be used to determine different stock accounts depending on the movement type.

☐ **A.** True

☐ **B.** False

11. You work with planned delivery costs. These are part of the purchase value and debit the stock account and possibly the price difference account at goods receipt. But where is the offsetting entry being posted?

☐ **A.** On the GR/IR clearing account

☐ **B.** On an account determined via a specific transaction key contained in value string WE01

☐ **C.** On an account determined via a transaction key from the calculation schema

☐ **D.** On an account determined via an account grouping code of transaction key GBB

12. True or false: When you create a purchase order item with account assignment, the system only determines a default account if you have specified a material number.

☐ **A.** True

☐ **B.** False

13. Which transaction key is used to determine the default account for purchase order items with account assignment?

☐ **A.** GBB

☐ **B.** BSX

☐ **C.** VBR

☐ **D.** WRX

14. To which objects is an account category reference assigned? (There are two correct answers.)

☐ **A.** Material type

☐ **B.** Valuation area

☐ **C.** Material class

☐ **D.** Valuation type

15. Which key determines the criteria for dividing the stocks of a material by value within a valuation area?

☐ **A.** Valuation grouping code

☐ **B.** Valuation class

☐ **C.** Valuation category

☐ **D.** Valuation type

16. True or false: All materials that are subject to split valuation can only be split-valuated according to the same criterion.

☐ **A.** True

☐ **B.** False

17. True or false: It is possible to have some accounts determined per valuation area (valuation area grouping code), but other accounts are determined independently of the valuation area.

☐ **A.** True

☐ **B.** False

18. Which of the following statements apply to the split valuation? (There are three correct answers.)

☐ **A.** Split valuation cannot be used for materials managed in batches.

☐ **B.** The valuation type must be specified for each goods movement of a split-valuated material.

☐ **C.** The stock value of a material subject to split valuation can be posted to a different stock account for each valuation type.

☐ **D.** The moving average price is the only price control allowed for partial stocks of a split-valuated material.

☐ **E.** You must specify which valuation categories can be used in a valuation area.

Practice Question Answers and Explanations

1. Correct answers: **A, D**

   The **Price Control** field determines whether the material is valuated according to the standard or the moving average price. The valuation class influences the account determination. The **Material Price Group** is a field relevant for sales, and the purchasing value key contains default values for the purchase order.

2. Correct answer: **B**

   False. You do not need to create additional organizational units for the valuation area. You simply decide at which level (company code or plant) the valuation shall take place; then the valuation area corresponds either to the plant or the company code.

3. Correct answer: **B**

   False. The company code always plays a role in account determination because it contains the chart of accounts. If the valuation level corresponds to the company code and valuation grouping is activated, the company code plays a double role: once for the chart of accounts and once via the valuation grouping code.

4. Correct answers: **A, B**

   You can make account determination independent of the valuation area (in this case, the plant) by deactivating the valuation grouping code. You achieve the same effect if you activate the valuation grouping code and assign the same valuation grouping code to all plants. We recommend this second option, as it requires less effort if at some point you decide otherwise and want to determine the accounts differently for each plant.

   Note that the valuation area is an influencing factor for account determination, but only if the valuation grouping code is activated. It is not itself a key in the account determination table but rather the valuation grouping code.

5. Correct answer: **C**

   With the valuation grouping code, you group valuation areas. The account grouping code corresponds to a more detailed breakdown of a transaction. The account category reference is assigned to one or more material types and allows you to restrict the possible valuation classes for these material types. The account assignment category has nothing to do with account determination: it determines whether a purchase order item is intended for consumption and which account assignment element is to be entered. The valuation area is simply the organizational unit where material valuation takes place: either the plant or the company code.

6. Correct answers: **A, B, D**

   The transaction key (representing a posting line of a posting), the valuation class (for the material), and the valuation grouping code (for the valuation area) are part of the key of the account determination table.

7. Correct answers: **C, D**

   The system automatically determines the value string assigned to a particular posting, depending on the parameters you enter manually, and parameters derived internally by the system. The value string contains the maximum possible transaction keys for a posting. The value string cannot be customized. The transaction key, not the value string, is part of the key of the account determination table.

8. Correct answers: **C, D**

   Transfer postings between two plants (provided that the plant is the valuation level) and transfer postings from vendor consignment stock (non-valuated) to your own stock are relevant because they create accounting documents. Transfer postings between two storage locations within a plant and transfer postings from unrestricted-use stock to blocked stock within a plant or storage location do not create accounting documents. Therefore, automatic account determination is not necessary.

9. Correct answers: **A, D**

   You maintain the valuation classes in the material master at the valuation area level (usually at the plant level). If a material is subject to split valuation, you also maintain the required valuation class for the valuation type.

10. Correct answer: **B**

    False. The general modification cannot be activated for the BSX (inventory posting) procedure. The option is not available in the rules for BSX. This option is available for certain transaction keys—for example, for GBB (offsetting entry for inventory posting) or PRD (cost [price] differences)—but not for all. The determination of the stock account cannot be further differentiated by movement type, for example.

11. Correct answer: **C**

    Freight costs are entered in the purchase order item using condition types (FRA, FRB, etc.). Transaction keys are defined for these condition types in the pricing procedure (also called *calculation schema*), which enable the freight accounts to be determined.

12. Correct answer: **B**

    False. For items with a material number, the system determines the default account with the valuation class from the material master record (among other influencing factors). For materials without a valuation class (e.g., materials with material type UNBW), the system uses the valuation class "blank." You can also define a valuation class for material groups. If you create a purchase order without a material, the system then uses the valuation class of the material group during account determination. If there is no valuation class for the material group, the system uses the valuation class "blank."

13. Correct answer: **A**

    The default account for purchase order items with account assignment is determined—among other factors—by the transaction key GBB and the account grouping code defined in Customizing for the account assignment category.

14. Correct answers: **A, D**

    Account category references are assigned to material types and valuation types and restrict the selection of valuation classes for the materials of the respective material types or valuation types.

15. Correct answer: **C**

    The valuation category is defined in Customizing and assigned to materials. It represents a criterion according to which materials can be valuated separately.

16. Correct answer: **B**

    False. For each material that is to be split-valuated, you specify the criterion (valuation category) individually at the valuation area level. A prerequisite is that the valuation category exists in Customizing.

17. Correct answer: **A**

    True. Thanks to the rules in Customizing for account determination at the transaction key level, it is possible to activate the valuation grouping code as an account determination factor for some transaction keys (such as BSX) while disabling it for other transaction keys (such as WRX).

18. Correct answers: **B, C, E**

    The valuation type must be specified for each goods movement of a split-valuated material. The stock value of a material subject to split valuation can be posted to a different stock account, depending on the valuation type, thanks to the valuation class that you enter in the material data for each valuation type. You must first define valuation categories in Customizing and then specify which valuation categories can be used in which valuation area. Batch-managed materials can be subject to split valuation. Batches can be regarded as valuation types and can be valuated separately. The valuation category that corresponds to split valuation by batch is "X". Partial stocks of split-valuated materials (valuation type level) can be valuated with both the standard price and the moving average price. At the valuation area level, however, you can only use the moving average price.

Takeaway

This chapter provided an overview of material stock valuation. You should now be able to define the valuation level and know how to maintain a material from a valuation point of view. The goal of this chapter was to help you understand automatic account determination and know when it comes into effect. At this point, you should be able to name and use the various influencing factors to implement customer-specific account determination. You should be familiar with the technical terminology used herein. You should also understand that automatic account determination can be used in different ways and in different applications, such as in purchasing, where it can provide account proposals for purchase order items with account assignments. This chapter outlined the various options offered by split valuation. You should be able to set up and create materials that are subject to split valuation.

Summary

The logic of the automatic account determination is proven and has been mostly carried over from the SAP ERP system. Readers who have experience with SAP ERP likely found many familiar things in this chapter.

Account determination is an important part of the integration between logistics and finance and is usually performed as a joint task in implementation projects. Financial accounting will certainly determine which accounts are available and in which processes they are to be posted to. However, as a logistics consultant, you know the materials and processes best and must be able to adapt account determination to customer requirements.

In the next chapter, we will continue our exploration of integrative topics relating to finance, with a focus on invoice verification, where account determination is used again.

Chapter 11
Logistics Invoice Verification

Techniques You'll Master

- Enter a supplier invoice with reference to a procurement process, including taxes, cash discounts, and foreign currency.
- Explain the difference between purchase-order based and goods-receipt based invoice verification.
- Explain the invoice parking functions and their advantages.
- Outline the options for posting invoices with variances.
- Set tolerances for deviations, and name all possible blocking reasons for an invoice.
- Know how to release blocked invoices.
- Enter a subsequent debit, subsequent credit, and credit memo, or cancel an invoice.
- Enter planned and unplanned delivery costs.
- Explain and perform Evaluated Receipt Settlement (ERS).
- Maintain the goods receipt/invoice receipt (GR/IR) clearing account.

In this chapter, we refresh your knowledge about the functions of logistics invoice verification: we enter or park an invoice with a reference, recording delivery costs, taxes, and discounts. We also enter credit notes, subsequent debits, or credits, careful to mention the effects of our postings, as well as the configuration options. We recapitulate the options we have for posting an invoice with differences, making note of the blocking reasons, the configurable tolerances, and the options for releasing blocked invoices. Finally, we look at the Evaluated Receipt Settlement (ERS) procedure, which can significantly streamline the procurement process.

Real-World Scenario

The logistic invoice verification with invoice posting and payment release completes the procurement process from the perspective of a consultant for sourcing and procurement. Of course, the invoices must then be paid, but this is the task of the payment program of the financial accounting department.

In practice, more and more companies are trying to minimize or even eliminate the tedious process of entering incoming invoices. For example, invoices from suppliers can be received electronically in the SAP Ariba network and transferred to SAP S/4HANA. Invoices can also be transmitted by EDI and posted automatically using an IDoc. There are also many third-party providers on the market that offer, for example, scanning and transferring invoices to SAP S/4HANA.

Then you will ask why we attach so much importance to this topic during certification. The reason is simple: No matter how an invoice is posted, whether manually or electronically, the internal system checks and postings are the same. And you, as a consultant, need to understand which checks are running in the background and which options are available to you to control the system behavior.

Besides entering and posting invoices, there are other activities that take place directly in SAP S/4HANA, such as releasing blocked invoices or processing parked invoices.

Objectives of This Portion of the Test

The purpose of this part of the certification exam is to test your general knowledge of logistics invoice verification. The certification exam expects you to have a solid understanding of the following topics:

- Possible features of a supplier invoice and the posting options
- Dealing with deviations

- Configuration options for the invoice parking and posting
- Configuration options for invoice blocking
- Posting of credit memos, subsequent debits, or subsequent credits
- Prerequisites and execution of Evaluated Receipt Settlement (ERS)

Key Concept Refresher

SAP S/4HANA offers various options for entering and checking a supplier invoice, such as:

- Online with the Create Supplier Invoice—Advanced app (Transaction MIGO) or with the Create Supplier Invoice app
- In the background
- Via EDI with IDocs or service interfaces (API)

In this chapter, we focus on the online option and especially on the entry information, checks, and postings that are basically the same, however you post an invoice.

We start with the entry of incoming invoices and learn how to deal with deviations. We will also learn how to configure invoice blocking and how to release invoices. We will create credit notes, subsequent credit notes, or debits, and analyze the consequences of our posting. A special topic will be delivery costs and, finally, we will discuss how to manage the GR/IR clearing account.

Processing Incoming Invoices

When an invoice refers to a purchase order, most of the data can be transferred directly from the purchase order to the invoice document.

To do this, simply enter the purchase order number as a reference and the system will automatically generate the invoice lines from the purchase order items. You can confirm the entries as they are from the purchase order (if correct) or change them to match the actual invoice (if there are differences).

Tip

Two apps are available for the manual entry of an invoice. While the Transaction MIRO (Create Supplier Invoice—Advanced) offers more options and is therefore a bit more complex, the Create Supplier Invoice app is relatively straightforward and intuitive to use. We recommend that you gain experience with both apps. It would go beyond the scope of this book to list every single function of Transaction MIRO. Invoice verification is a little theory and a lot of practice. Here we refresh the theory and some important functions. We recommend that you practice booking an invoice with both the Transaction MIRO and the SAP Fiori app for each of the following topics (taxes, foreign currency etc.) to better familiarize yourself with them.

Figure 11.1 shows how to enter a supplier invoice using the Create Supplier Invoice app, as follows:

❶ Enter basic data like company code, gross invoice amount, invoice date, and reference.

❷ Enter the reference purchase order and generate invoice items.

❸ Check the item details and post the invoice.

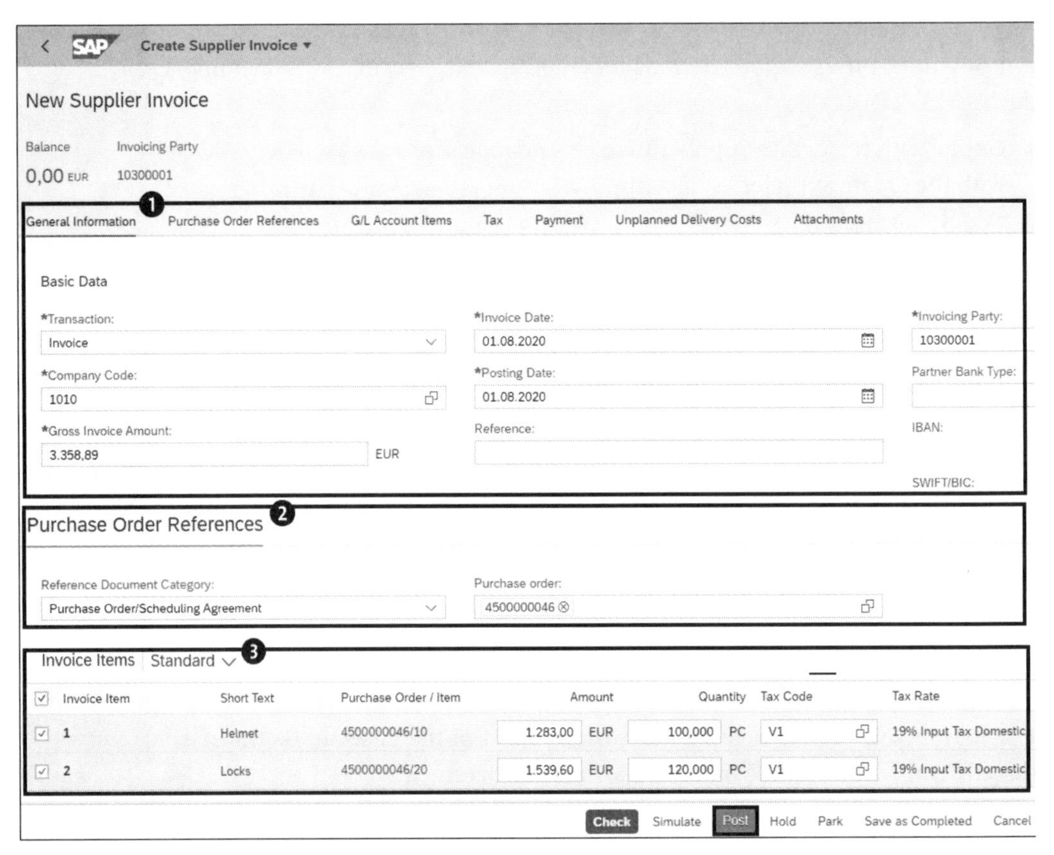

Figure 11.1 Create Supplier Invoice App

Figure 11.2 shows how to enter a supplier invoice using the Create Supplier Invoice—Advanced app (Transaction MIRO), as follows:

❶ Enter basic data like transaction, amount, tax amount invoice date, and reference.

❷ Enter the reference purchase order, and generate invoice items.

❸ Check the item details, and post the invoice.

You do not have to specify the supplier; the system can determine the vendor through the reference to the purchase order (or scheduling agreement). If you are working with Transaction MIRO, you will see information on the supplier in the header area of the screen after you enter the purchase order: name, address, and bank details of the supplier. From here, you can branch to the display of the business partner master record.

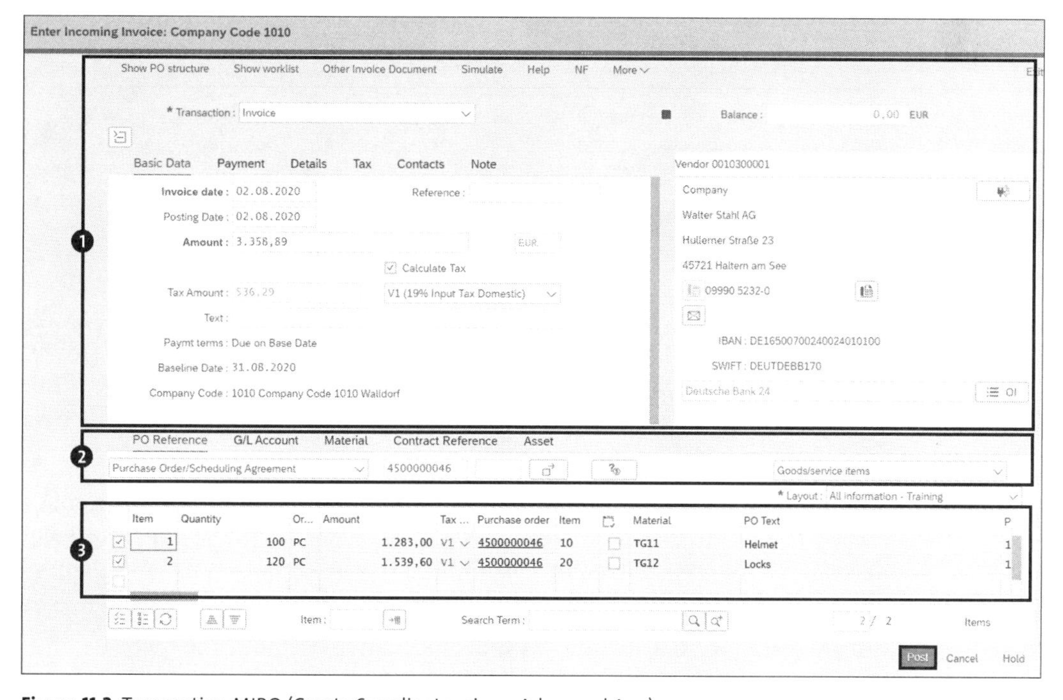

Figure 11.2 Transaction MIRO (Create Supplier Invoice—Advanced App)

Assigning the Incoming Invoice

As already mentioned, when you enter the incoming invoice, it is important to assign the incoming invoice to a purchasing process so that you obtain the default values that come from the purchase order data and the purchase order history, and so that the purchase order history can be updated again when the invoice is posted.

You determine the items to be invoiced by entering a reference (see Figure 11.3). Goods or service items can be selected with the following reference documents:

- Service entry sheet
- Delivery note
- Purchase order or scheduling agreement

For planned delivery costs, you can use the following reference documents:

- Purchase order or scheduling agreement
- Bill of lading

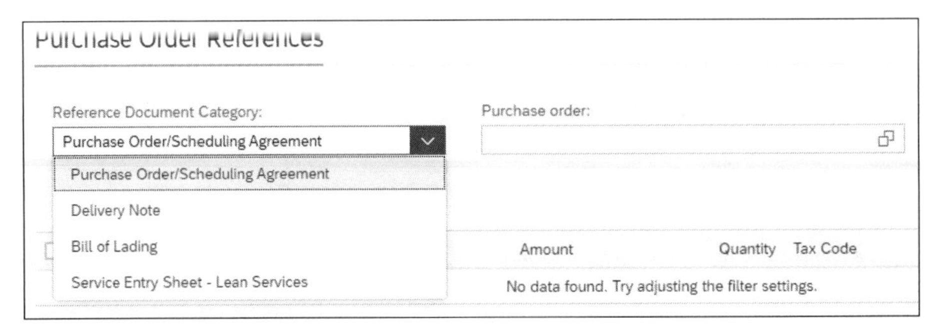

Figure 11.3 Reference Documents

Verifying and Posting an Invoice

Once you have entered a reference document (e.g., a purchase order), the system lists all items in this purchase order, as shown in Figure 11.4. However, there is a difference between the items for which an invoice is expected and those for which no invoice is yet expected.

- A purchase order item where an invoice is expected is an item where the goods receipt has been partially or completely posted but no invoice has yet been entered. The quantity delivered but not yet invoiced and its corresponding value are proposed by the system. The system also automatically flags the selection indicator for such an item, as shown with invoice item 1 and invoice item 2 in Figure 11.4.
- A purchase order item for which no invoice is expected is a purchase order item for which no goods receipt has yet been posted. Even though the system lists these items, neither quantities nor values are proposed, and the selection indicator is not selected, as shown with invoice item 3 in Figure 11.4.

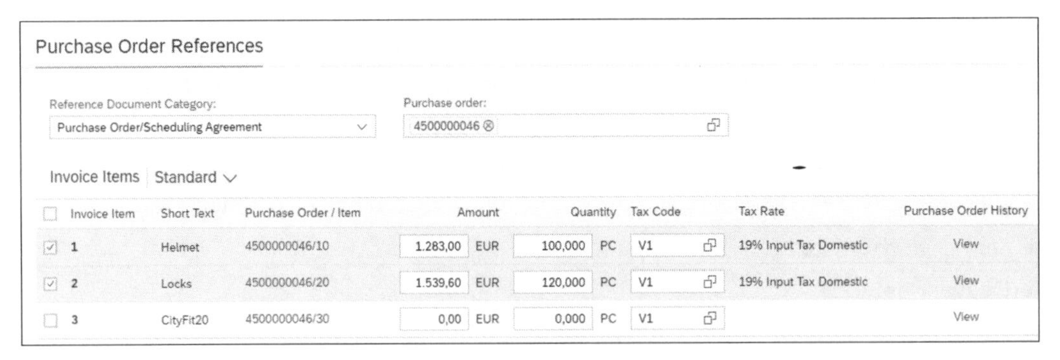

Figure 11.4 Create Supplier Invoice App: Proposed Invoice Items

Compare the proposed invoice items with the items in the vendor invoice and correct the proposed values for amount, quantity, and tax code, if necessary. The selection indicator must be flagged for the items to be posted.

When you choose the **Simulate** or **Post** function, the system compares the quantity and amount data of each supplier invoice item with the data of the corresponding purchase order item. You can use the **Simulate** function to display the account movements before posting the document.

As a result of the invoice posting, an invoice document and an accounting document are posted, and the purchase order history is updated.

Handling Invoices for Purchase Order Items with Account Assignment

In a purchase order item assigned to an account, account assignment elements such as cost center, order, and so on are entered, depending on the account assignment category. Also, an account is either entered manually or defaulted by the system.

Depending on the account assignment category, it may be possible to change the account assignment data at invoice entry, provided no goods receipt is posted or the goods receipt is posted non-valuated.

Note

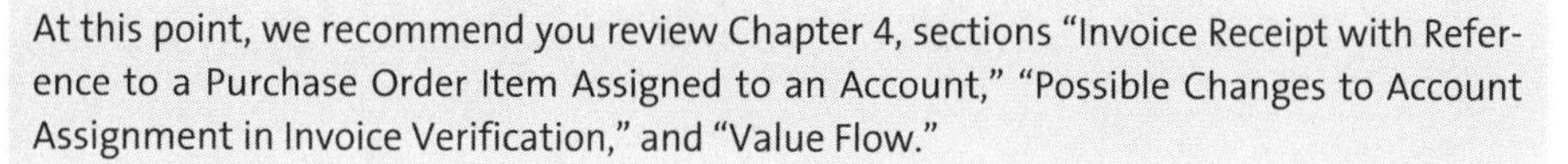

At this point, we recommend you review Chapter 4, sections "Invoice Receipt with Reference to a Purchase Order Item Assigned to an Account," "Possible Changes to Account Assignment in Invoice Verification," and "Value Flow."

Entering Taxes

A tax code is assigned to each invoice item, and the system calculates the tax base for each tax rate using the values and tax codes of the individual items. The item tax codes are displayed in the item list of the invoice and can be changed there, if necessary.

The tax code for each item is determined from the purchase order. The system checks the tax data when you simulate or post an invoice. You enter the tax codes and amounts specified in the incoming invoice differently, depending on whether you use the Create Supplier Invoice app or Transaction MIRO.

- **Create Supplier Invoice app**
  The system automatically creates tax lines based on the tax codes from the items. You only need to intervene if there is a discrepancy compared to the invoice. The upcoming figures illustrate this system behavior: the two selected positions have different tax rates (Figure 11.5). The system has automatically generated two corresponding tax lines (Figure 11.6 and Figure 11.7).

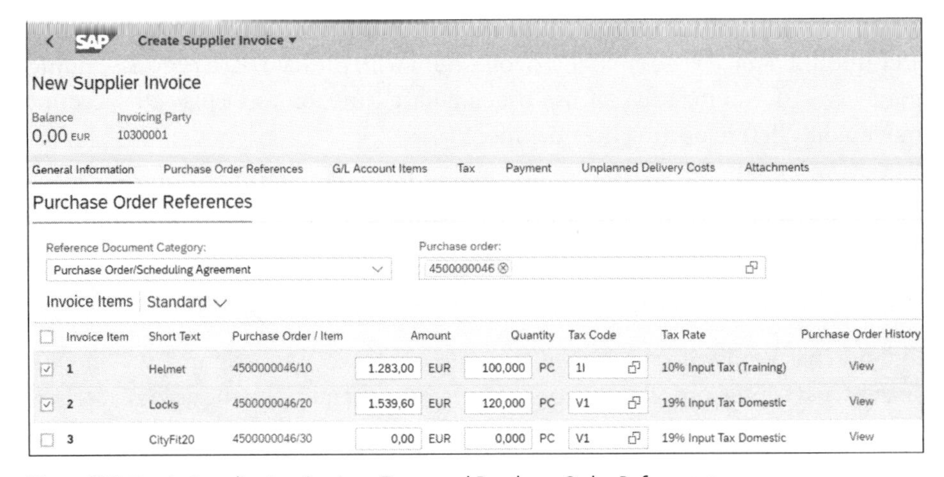

Figure 11.5 Create Supplier Invoice App: Taxes and Purchase Order References

Create Supplier Invoice

New Supplier Invoice

Balance: 0,00 EUR
Invoicing Party: 10300001

General Information | Purchase Order References | G/L Account Items | Tax | Payment | Unplanned Delivery Costs | Attachments

Tax Code	Tax Rate	Tax Amount
1I	10% Input Tax (Training)	128,30 EUR
V1	19% Input Tax Domestic	292,52 EUR

Figure 11.6 Create Supplier Invoice App: Taxes

Simulation

Simulation Results

Post | Change Simulation Currency

Company Code: 1010
Simulation Currency: EUR

Overview | Details

Debit		Credit	
21120000 (Goods Received/Invoice Received)	1.283,00 EUR	21100000 (Walter Stahl AG / D-45721 Haltern am See)	3.243,42 EUR
21120000 (Goods Received/Invoice Received)	1.539,60 EUR		
12600000 (Input Tax (VST))	128,30 EUR		
12600000 (Input Tax (VST))	292,52 EUR		
Total	3.243,42 EUR	Total	3.243,42 EUR

Details

Standard

Item	Account type	G/L Account	Name	Amount	Tax Code	Company Code
1	K	**21100000**	Walter Stahl AG / D-45721 Haltern am See	-3.243,42 EUR	**	1010
2	S	**21120000**	Goods Received/Invoice Received	1.283,00 EUR	1I	1010
3	S	**21120000**	Goods Received/Invoice Received	1.539,60 EUR	V1	1010
4	S	**12600000**	Input Tax (VST)	128,30 EUR	1I	1010
5	S	**12600000**	Input Tax (VST)	292,52 EUR	V1	1010

Figure 11.7 Create Supplier Invoice App: Taxes—Simulation

- **Create Supplier Invoice—Advanced app (Transaction MIRO)**
  There are two places where you can enter the tax code and tax amount:
  - The **Basic Data** tab shown in Figure 11.8 is suitable if the invoice contains a single tax rate. You can either enter the tax amount yourself or have the system calculate it automatically, based on the tax codes from the items. For this, you must set the indicator **Calculate Tax**.

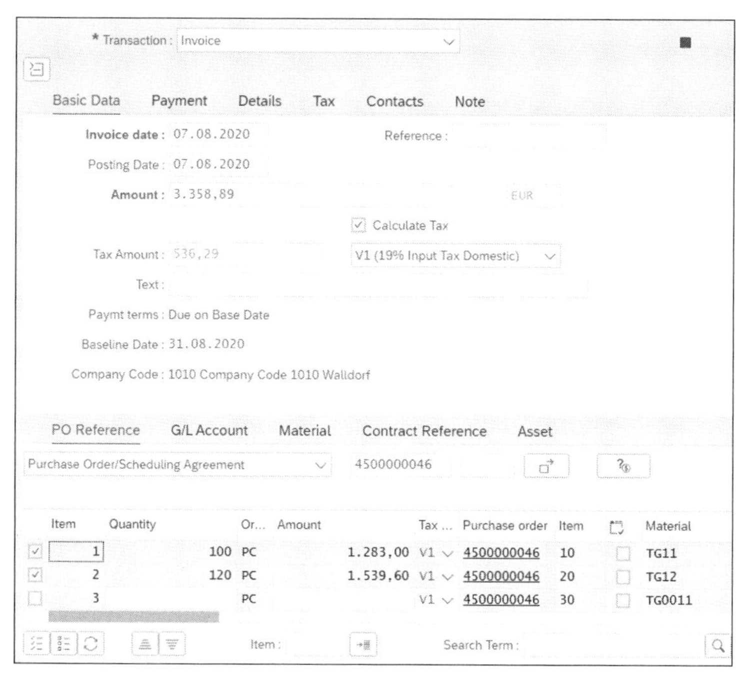

Figure 11.8 Transaction MIRO: Tax on the Basic Data Tab

Note
The system proposes a tax code on the **Basic Data** tab page: this default value comes from Customizing and can be set per company code.

  - The **Tax** tab shown in Figure 11.9 is intended for use when the vendor invoice contains more than one tax rate. As with the procedure described previously for a single tax rate on the **Basic Data** tab page, you can either enter the tax amounts yourself or have the system calculate them automatically from the items based on the tax codes. To do this, you must set the **Calculate Tax** indicator.

Note
You can select the **Calculate Tax** indicator either on the **Basic Data** tab page or on the **Tax** tab page for the same effect.

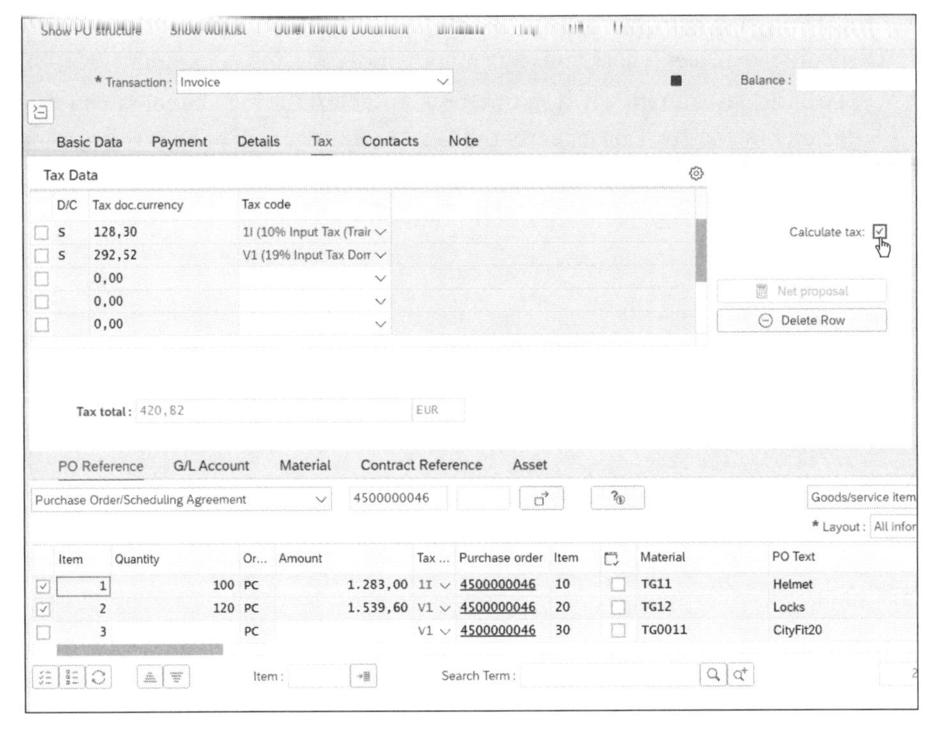

Figure 11.9 Transaction MIRO: Taxes on the Tax Tab

Note

You can configure the tax codes, including the tax rate, in Customizing for Financial Accounting.

Posting Cash Discount

We first look at how and where terms of payment can be maintained and stored, and then how granted cash discount can be posted.

Maintaining Terms of Payment

You agree to terms of payment with the supplier and store these in the supplier (business partner) master record. There are two places in the business partner master record where you can store terms of payment:

- At the company code level for the business partner role "FI vendor" (see Figure 11.10). These apply to invoices without reference to a purchasing document.
- At the purchasing organization level for the role "Supplier" (see Figure 11.11). These apply to purchase orders and other purchasing documents.

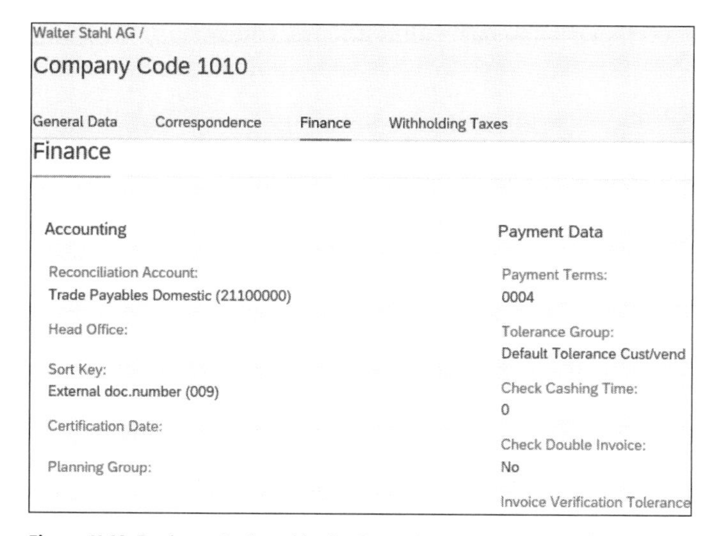

Figure 11.10 Business Partner Master Data: Payment Terms at the Company Code Level

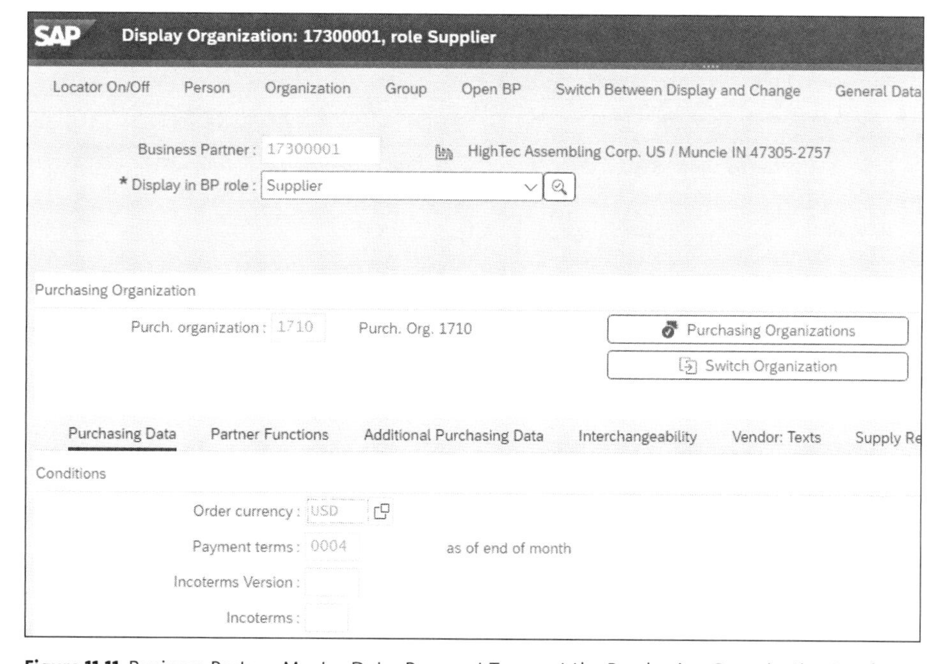

Figure 11.11 Business Partner Master Data: Payment Terms at the Purchasing Organization Level

Note

The terms of payment at the company code level and those for purchasing order documents may differ.

Terms of payment are defined and configured in Customizing for Financial Accounting using a four-digit alphanumeric code. The terms of payment can have up to three levels—for example, if you pay within 10 days, you get a 3% discount; within 20 days, you get a 2% discount; and within 30 days, you pay the net price.

The terms of payment from the supplier master (purchasing organization level) are automatically adopted in purchase orders for this supplier and can be adjusted there, if required. The terms of payment from the purchase order are inherited into the invoice, where they can also be adjusted according to the real incoming invoice, as shown in Figure 11.12.

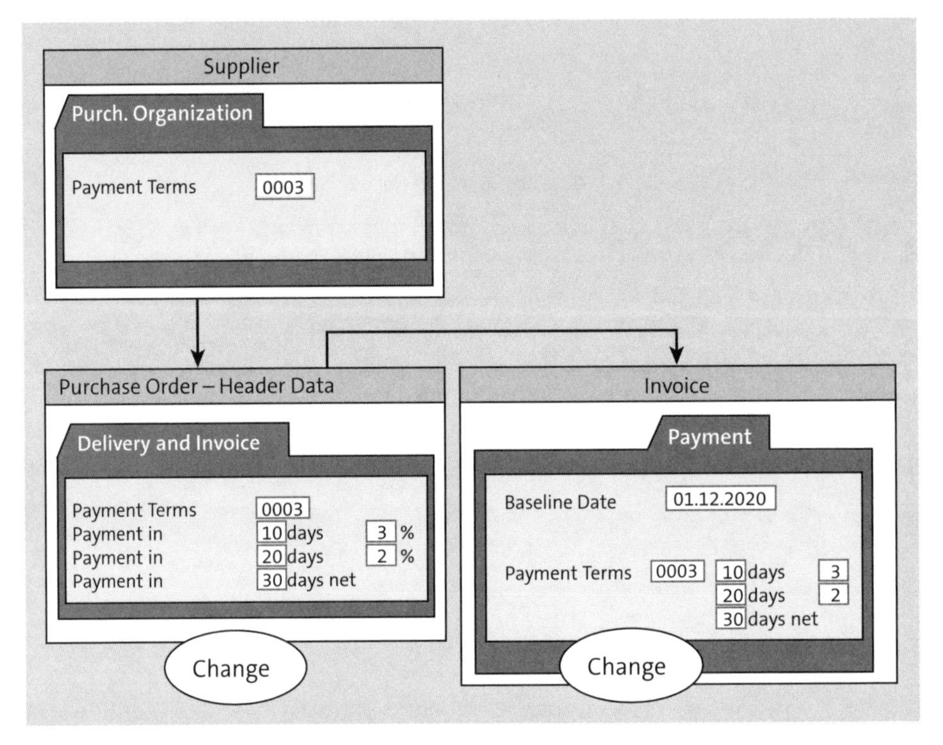

Figure 11.12 Maintaining Terms of Payment

Note

If you enter an invoice with reference to more than one purchase order, the system proposes the terms of payment from the first referred purchase order.

Gross and Net Posting

There are two options for posting the cash discount:

- **Posting gross**
  When the invoice is posted, the cash discount amount is not considered; it is only posted to a cash discount account when the payment is made. Therefore, the cash discount amount is not credited to the stock or cost account (Figure 11.13).

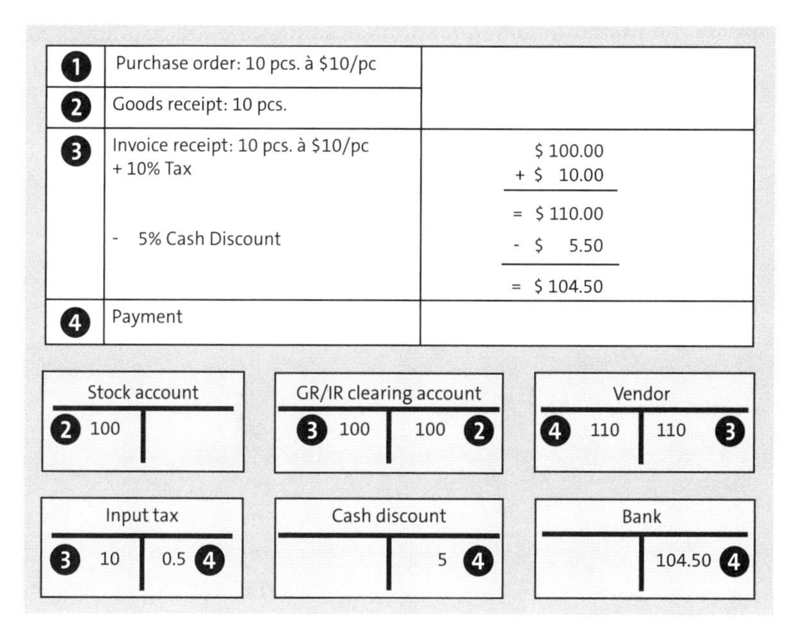

Figure 11.13 Cash Discount Posting Gross

- **Posting net**
  When you post the incoming invoice, the cash discount amount is credited to the stock or cost account and the offsetting entry is made to a cash discount clearing account. The cash discount clearing account is cleared at the time of payment (Figure 11.14).

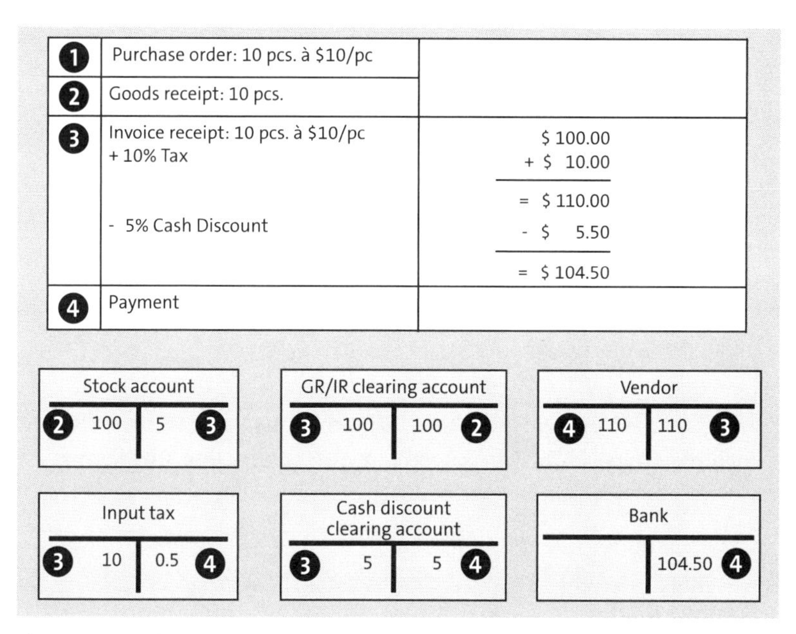

Figure 11.14 Cash Discount Posting Net: Example of Postings for a Stock Item—Material Valuated at the Moving Average Price

Warning

In net posting, the cash discount amount is treated as a price difference for posting to the stock account or cost account:

- For a purchase order item with account assignment, the cash discount amount is credited to the cost account.
- For a stock order item, it depends on the valuation method of the material.
  - If the material is valuated at the moving average price, the system credits the stock account for the material by the amount of the cash discount granted for the item, if there is enough stock coverage. This changes the moving average price.
  - If the material is valuated at the standard price, the cash discount amount is posted to a price difference account.

Refer to Chapter 3 for further information about posting logic for goods receipt and invoice receipt, depending on the valuation procedure.

Whether you post gross or net depends on the document type. If you choose gross posting, select document type **RE**. As you can see in Figure 11.15, the **Net document type** indicator is not selected in Customizing for document type RE.

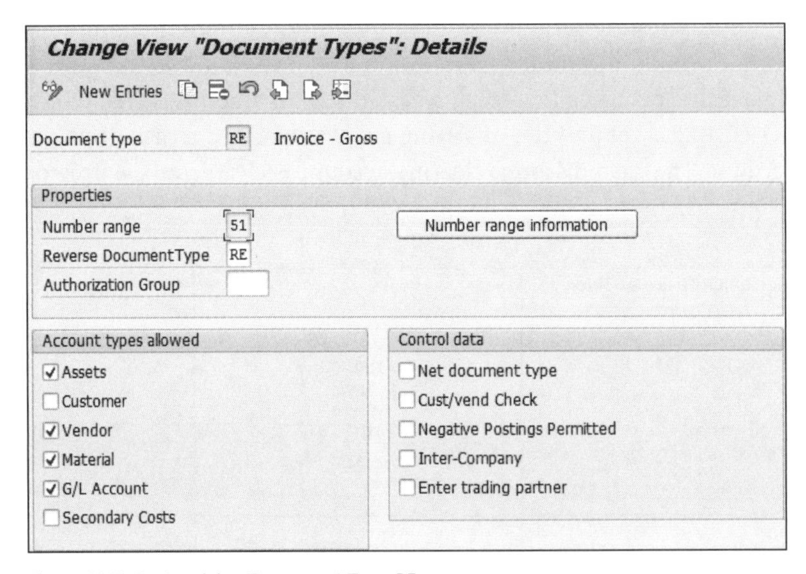

Figure 11.15 Customizing Document Type RE

If you choose the gross posting, select document type **RN**. As you can see in Figure 11.16, the **Net document type** indicator is selected in Customizing for document type RN.

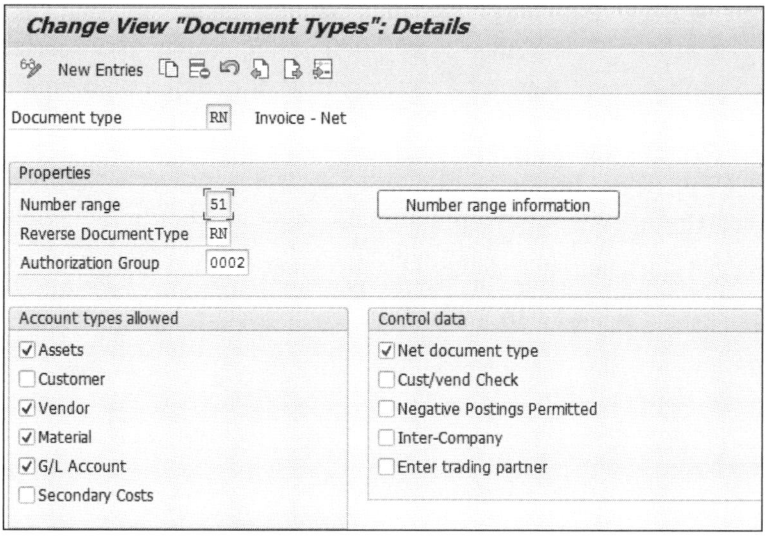

Figure 11.16 Customizing Document Type RN

Entering Invoices in Foreign Currency

An invoice in a foreign currency is an invoice where the amount is specified in a currency that differs from the local currency set in the company code.

The purchase order itself can be in a foreign currency. When ordering in a foreign currency, you decide whether the exchange rate is fixed or not.

- *If the exchange rate is fixed* in the purchase order, then the system uses the exchange rate from the purchase order to convert the foreign currency into the local currency at goods receipt and invoice receipt. As a result, there are no exchange rate differences between the goods receipt and invoice receipt.
- *If the exchange rate is not fixed*, then the system uses the valid exchange rate when posting a goods receipt. At invoice entry, the system proposes the current valid exchange rate, but you can also enter the exchange rate of your choice manually. Therefore, exchange rate differences can occur between the goods receipt and invoice receipt. Exchange rate differences are posted in different ways:
  - For purchase order items with account assignment, the exchange rate difference is credited or debited to the consumption account.
  - For a material with moving average price and stock coverage, the stock account is credited or debited with the amount of the exchange rate difference.
  - For a material with standard price control or a material with moving average price control but no stock coverage, the method for posting the exchange

rate difference depends on the Customizing settings. You can decide in Customizing whether the exchange rate difference is posted to a special exchange rate difference account or to the normal price difference account for price variances.

Note

You can override the rules we have just described: if you set the indicator **Treatment of Exchange Rate Diffs. in Company Code Currency** per company code in Customizing, then exchange rate differences are always posted to an exchange rate difference account, even for purchase order items assigned to an account and materials with a moving average price.

Posting Delivery Costs

Delivery costs can be planned in the purchase order or unplanned.

Planned Delivery Costs

Planned delivery costs are entered in the purchase order and are part of the procurement price.

When the goods receipt is posted, it is included in the valuation of the ordered material in the case of a stock item or is debited from the cost elements in the case of an item assigned to an account. The offsetting entry is made to a freight clearing account.

The freight clearing account is cleared when the invoice is received.

If no price differences occur between the planned delivery costs from the purchase order and the requested delivery costs in the invoice, posting the invoice has no effect on the valuation of the material. If price differences occur, they are treated as price variances.

Warning

If you use Transaction MIRO (Create Supplier Invoice—Advanced app), do not forget to choose the correct item type (see Figure 11.17) so that the system can propose the planned delivery costs.

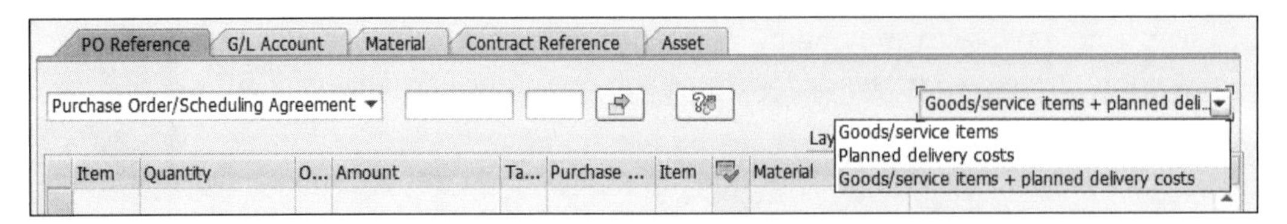

Figure 11.17 Transaction MIRO (Create Supplier Invoice—Advanced): Item Types

Unplanned Delivery Costs

Unplanned delivery costs are not recorded in the purchase order and only appear in the invoice, either as an additional line in the invoice for goods or services or in a separate invoice.

Figure 11.18 and Figure 11.19 show how you can enter unplanned delivery costs in an invoice, depending on whether you work with Transaction MIRO or with the Create Supplier Invoice app.

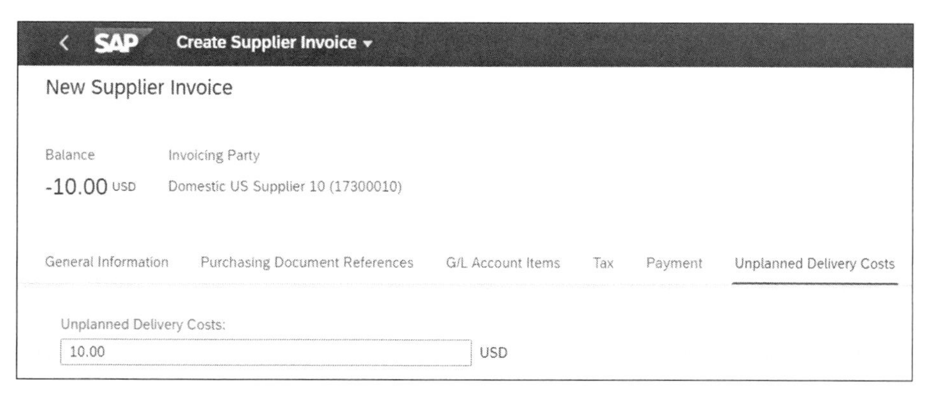

Figure 11.18 Create Supplier Invoice App: Enter Unplanned Delivery Costs

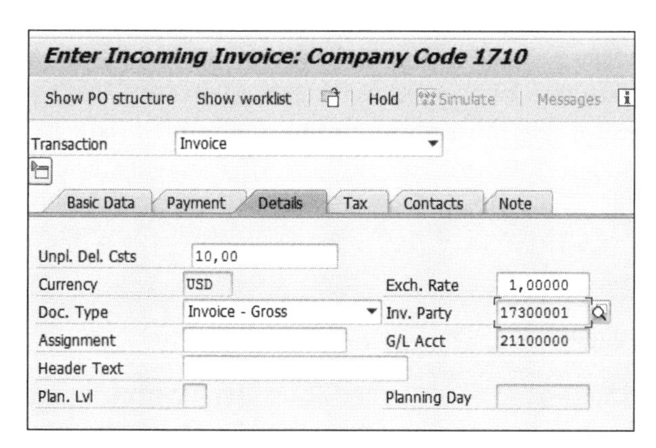

Figure 11.19 Transaction MIRO: Enter Unplanned Delivery Costs

You configure in Customizing, per company code, whether the system automatically distributes the unplanned delivery costs to the invoice items or posts the costs to a separate general ledger (G/L) account. Follow **Materials Management • Logistics Invoice Verification • Incoming Invoice • Configure How Unplanned Delivery Costs Are Posted**. Figure 11.20 shows the configuration for company code 1010: delivery costs are distributed among invoice items.

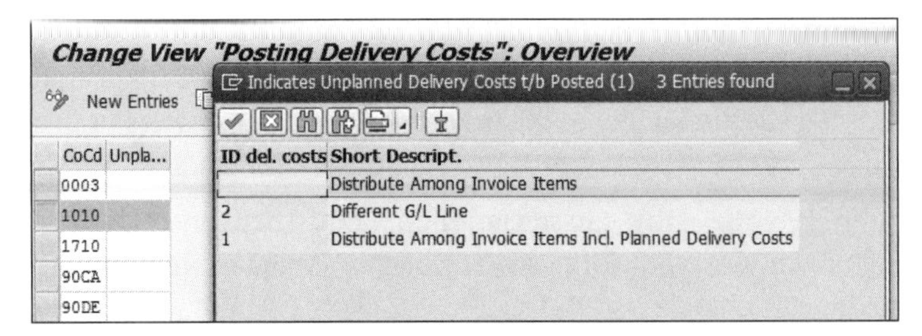

Figure 11.20 Customizing: How to Post Unplanned Delivery Costs

For distribution to the invoice items, the system distributes the unplanned delivery costs among the items in proportion to the total value calculated so far and the values in the current invoice. This means that if partial invoices have already been posted previously, they are considered when distributing the delivery costs!

When the delivery costs are distributed among the invoice items, the amounts in the invoice items are automatically increased by the delivery costs portion. When the invoice is posted, the unplanned delivery costs are treated as price variances.

Note

The posting of price differences is described in detail in Chapters 3 and 4.

In Customizing, if you have chosen to post unplanned delivery costs to a separate G/L account, you must also define the G/L account to be automatically posted to. To do this, assign a G/L account for the transaction key UPF (Unplanned Delivery Costs) in the automatic account determination (see Chapter 10). The total amount of unplanned delivery costs is then posted to this G/L account when the invoice is posted.

Entering Invoice Items without Reference to a Purchase Order

You can enter invoice items (or entire invoices—but it makes less sense!) without reference to a purchase order by directly adding a G/L account line. With Transaction MIRO, it is even possible to debit or credit stock accounts by entering the material number (stock accounts cannot be posted directly from Materials Management). You only need to ensure that direct posting is activated in Customizing for Invoice Verification.

Subsequent Debits and Credits

A subsequent debit is posted if items that have already been invoiced are subsequently debited with additional unplanned costs or if the invoice amount was erroneously too low and is hereby corrected.

A subsequent credit memo is posted if, for example, a discount is subsequently granted on items that have already been invoiced or if the amount already invoiced was erroneously too high and is therefore corrected.

To enter a subsequent debit or credit memo, an invoice must therefore already have been posted for the items concerned. You post a subsequent credit or debit by choosing the correct transaction (Figure 11.21 and Figure 11.22). Then you refer to the purchase order or delivery note as for an invoice. Note that you should *not* refer to the previous invoice.

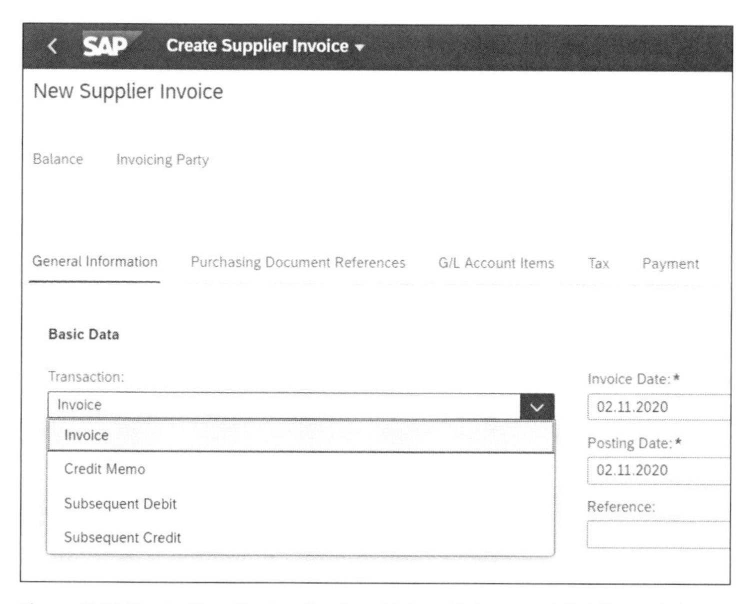

Figure 11.21 Create Supplier Invoice App: Enter a Subsequent Debit or a Subsequent Credit

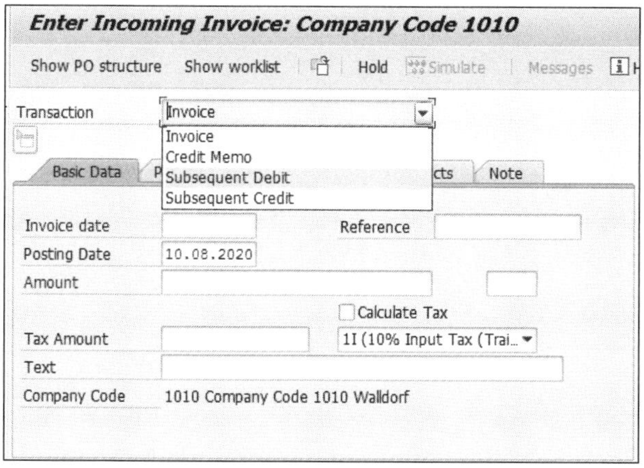

Figure 11.22 Transaction MIRO: Enter a Subsequent Debit or a Subsequent Credit

When you post the subsequent debit or credit memo, the offsetting entry to the vendor account depends on whether the corresponding quantity has already been received or not:

- When the goods receipt is posted, the offsetting entry is made to the stock account or price difference account for a stock item, depending on the price control of the material, or to the cost account for an item assigned to an account, as shown in Figure 11.23:

  ❶ Purchase order: 10 pieces at €10/piece

  ❷ Goods receipt: 10 pieces

  ❸ Invoice receipt: 10 pieces at €10/piece

  ❹ Subsequent credit: 10 pieces at -€1/piece

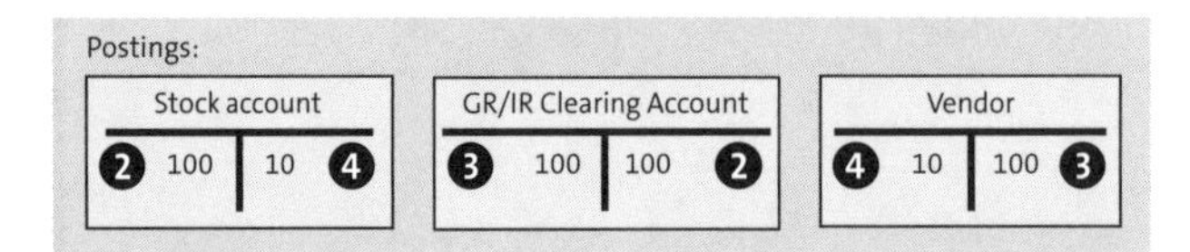

Figure 11.23 Posting a Subsequent Credit when the Corresponding Quantity Has Been Delivered

- When the goods receipt has not yet been posted, the offsetting entry is made to the GR/IR clearing account, as shown in Figure 11.24.

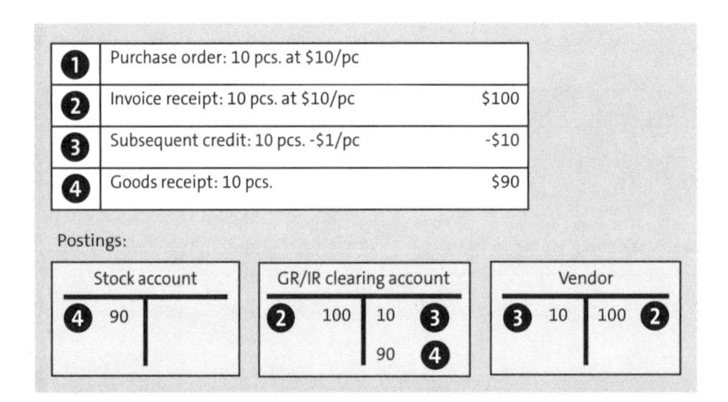

Figure 11.24 Posting a Subsequent Credit before the Corresponding Quantity Has Been Delivered

Note

This prompts us to recall a golden posting rule regarding the GR/IR clearing account: Goods receipt before invoice receipt means posting at purchase order price. Invoice receipt before goods receipt means posting at invoice price!

Posting a Credit Memo

In contrast to a subsequent credit memo, in which a quantity that has already been invoiced is only credited by value, a credit memo reduces the invoiced quantity.

- You receive and post a credit memo if the *quantity* invoiced is too high due to an error or a return delivery.
- You post a subsequent credit if the *price* invoiced was too high or if you receive a discount for a certain quantity that has already been invoiced.

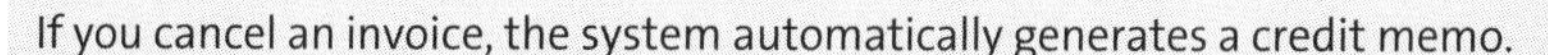

Note

If you cancel an invoice, the system automatically generates a credit memo.

Holding Invoices

You can interrupt the entry of an invoice at any time and save the current entry version. It may be necessary if you are interrupted in the middle of entering data and do not want to start all over again later. The *hold* function is available for this purpose.

Holding a document has the following implications:

- An invoice document (not an accounting document) with the *held* status is created.
- No information is yet passed on to financial accounting.
- The purchase order history is not updated.

You can further process a held invoice:

- You can delete it.
- You can change it and continue to hold it.
- You can post it.
- You can park or save it as complete.

Note

If you hold an invoice in Transaction MIRO or the Create Supplier Invoice app, the document has the status **Entered and Held (D)**. A held document in Transaction MIR7 (Park Invoice) has the status **Parked and Held (C)**.

Note

When you start to enter an invoice with the Create Supplier Invoice app but are interrupted and leave the app, the system automatically saves a draft, and when you call the app again, the system asks if you want to continue with the data from the previous invoice entry.

Parking an Invoice or Saving an Invoice as Complete

You can *park* a document if, for example, the information required to post the document is missing and you do not want to enter the data again. This park function also can be used by your organization to process the entry and posting of invoices according to the principle of dual control: a first person enters the invoice; a second person posts it.

Unlike holding, you must enter a minimum of data before you can park an invoice. Parking an invoice has the following effects:

- An invoice document (not an accounting document) with the *parked* status is created.
- The purchase order history is updated.
- An accounting document number is assigned. The document is listed as a parked open item.
- The balance may or may not be zero.

Saving as completed is like parking, except that the invoice must also be ready for posting (zero balance and no error messages).

If you use document parking to operate according to the principle of dual control, you can use *workflows* to automate the exchange of documents between accounting clerks.

SAP provides the following standard workflows related to document parking:

- **Workflow for completing parked documents**
  If invoice documents are to be parked and then completed by different employees, you can use the document completion workflow. When a user parks an invoice, users who are assigned as processors of the task *Complete parked invoice* of the workflow receive a work item in their inbox and are to process parked documents further: save as complete, delete, or post.
- **Workflow for releasing and posting saved as complete documents**
  If an invoice is to be entered first and then released and posted by two different users, you can use the release and post workflow. This means that a user can save an invoice document completely but cannot post it. The person responsible for release receives a work item and decides whether the invoice document is to be released and posted or rejected.

Warning
Do not confuse this with the payment release, which concerns invoices blocked for payment and is dealt with in one of the next sections.

Prepayment

You can prepay held, parked, or saved as complete invoices, which means you can pay invoices that have not yet been posted. You can opt for prepayment to prevent the supplier from being affected by your internal organizational processes (e.g., the "four-eye principle") and to enable you to benefit from the payment terms.

The following prerequisites must be fulfilled in the system for prepayment:

- The prepayment must be permitted for the company code in Customizing.
- In Customizing for account determination, a G/L account (prepayment clearing account) must be assigned to the transaction key PPX (prepayment).
- In the vendor master record, prepayment must be allowed by setting the **Prepayment** indicator in the company code data.

Warning

- Prepayment is not possible via the Create Supplier Invoice app.
- If an invoice is parked (or held or saved as complete) with prepayment, accounting documents are generated when parking.

Verifying Invoices on a Purchase Order or Goods Receipt Base

We distinguish between purchase-order-based (PO-based) and goods-receipt-based (GR-based) invoice verification:

- In PO-related invoice verification, the system generates an invoice item for each purchase order item in the item list of Transaction MIRO or the Create Supplier Invoice app. The system proposes the quantity to be invoiced as the difference between the total quantity already delivered and the total quantity already invoiced for the purchase order item. This means that if partial deliveries exist for a purchase order item, they are not proposed as individual invoice items. It does not matter whether you make the assignment using a purchase order or a delivery note. If you choose a delivery note as a reference document, the system determines the relevant PO items belonging to it, with their total quantities to be settled in the same way as when you reference a purchase order. The system proposes not only the quantity from the assigned delivery note but, as noted previously, the total open quantity for the purchase order item, considering all goods receipt and all invoices related to this item.
- In GR-based invoice verification, a single invoice item is generated for each partial delivery quantity. If you make the assignment with the purchase order, the system proposes one invoice item for each partial delivery. If you make the assignment using the delivery note, the system proposes one invoice item corresponding to the quantities posted in this goods receipt and not yet invoiced; each invoice item can be uniquely assigned to one goods receipt item, and this assignment is reflected in the purchase order history. In GR-based invoice verification, the invoice quantity cannot be greater than the goods receipt quantity.

We illustrate the differences between PO-based and GR-based invoice verification using two examples. First, Figure 11.25 shows a purchase order item 4500000049/10 for which PO-based invoice verification is specified.

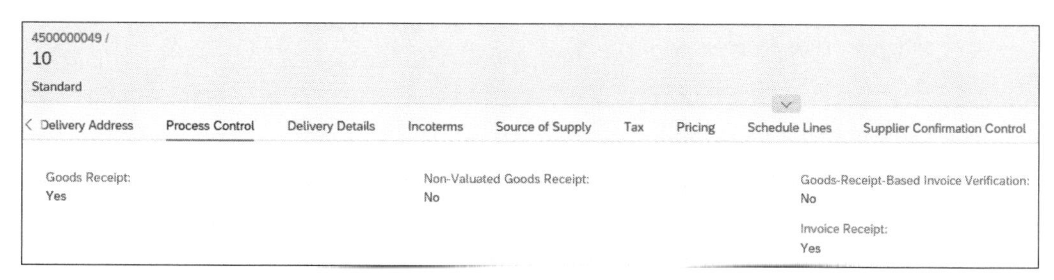

Figure 11.25 Manage Purchase Orders App: PO Item—Control Fields for PO-Based Invoice Verification

We assume that three partial deliveries (goods receipts) have been posted for the purchase order item: 30, 15, and 10 pieces. So far, no invoices have been posted for these purchase order items. Figure 11.26 shows the corresponding process flow.

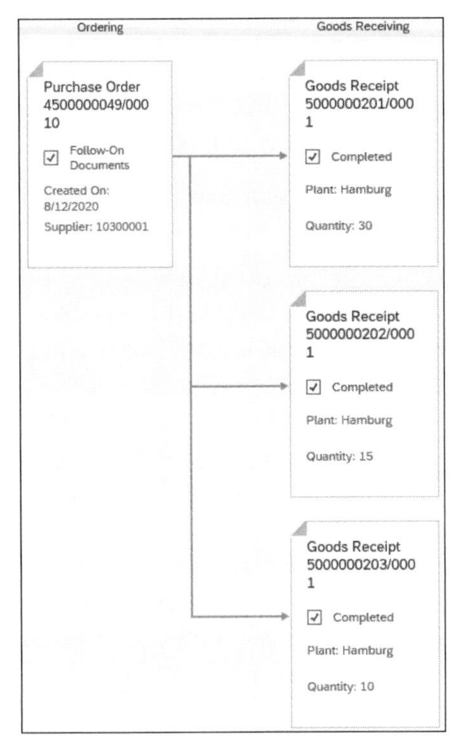

Figure 11.26 Manage Purchase Orders App: Process Flow

When you now enter an invoice for the purchase order item, the system proposes a single invoice item with the total of the goods receipt quantities, as shown in Figure 11.27.

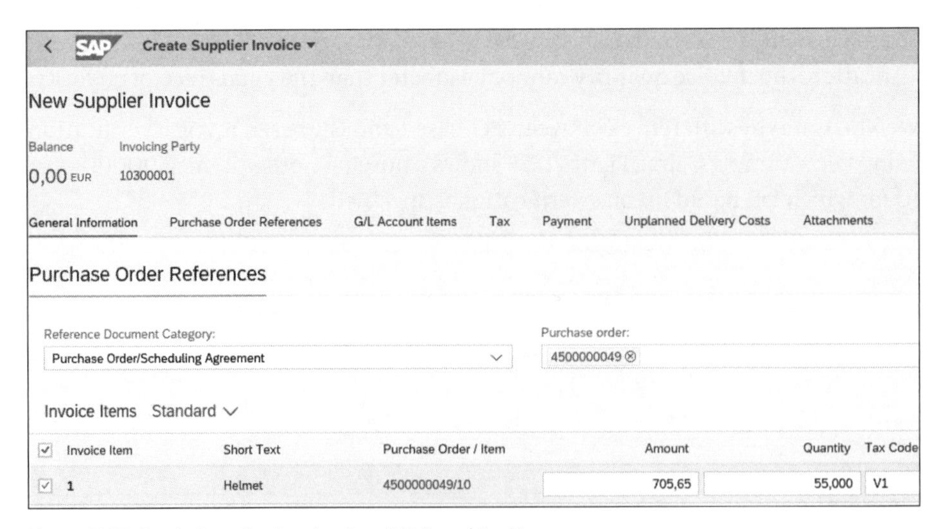

Figure 11.27 Create Supplier Invoice App: PO-Based Posting

In contrast, Figure 11.28 shows a purchase order item 4500000050/10 for which GR-based invoice verification is specified.

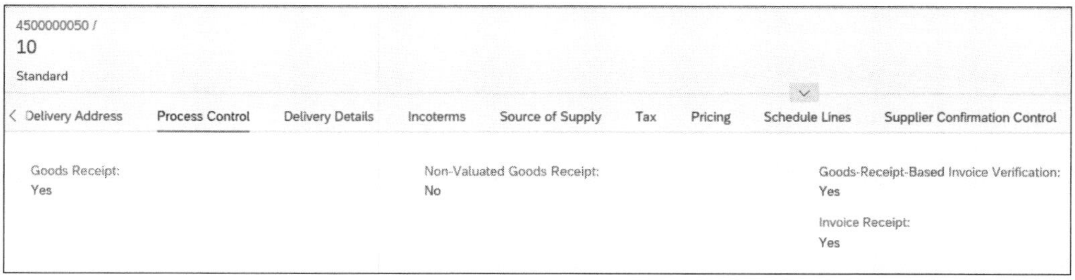

4500000050 /
10
Standard

< Delivery Address | Process Control | Delivery Details | Incoterms | Source of Supply | Tax | Pricing | Schedule Lines | Supplier Confirmation Control

Goods Receipt: Yes
Non-Valuated Goods Receipt: No
Goods-Receipt-Based Invoice Verification: Yes
Invoice Receipt: Yes

Figure 11.28 Manage Purchase Orders App: PO Item—Control Fields for GR-Based Invoice Verification

We assume that three partial deliveries (goods receipts) have been posted for the purchase order item: 30, 15, and 10 pieces. So far, no invoices have been posted for these purchase order items. Figure 11.29 shows the corresponding process flow.

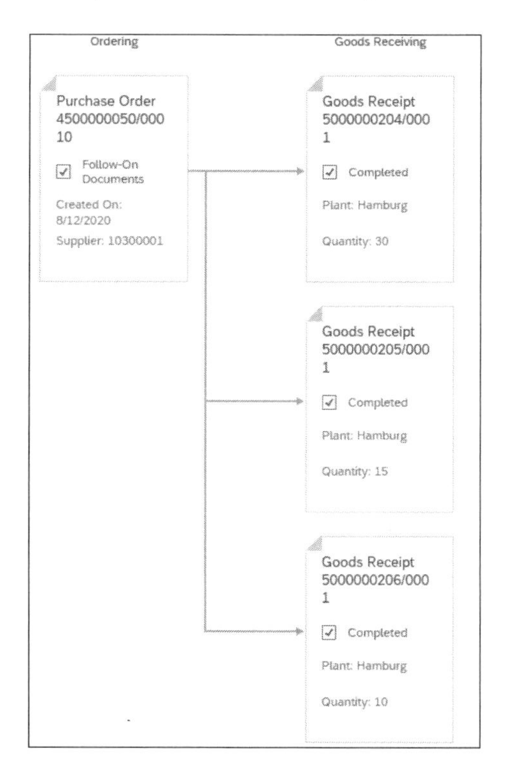

Figure 11.29 Manage Purchase Orders App: Process Flow

When you now enter an invoice for the purchase order item, the system proposes one invoice item per goods receipt, as shown in Figure 11.30.

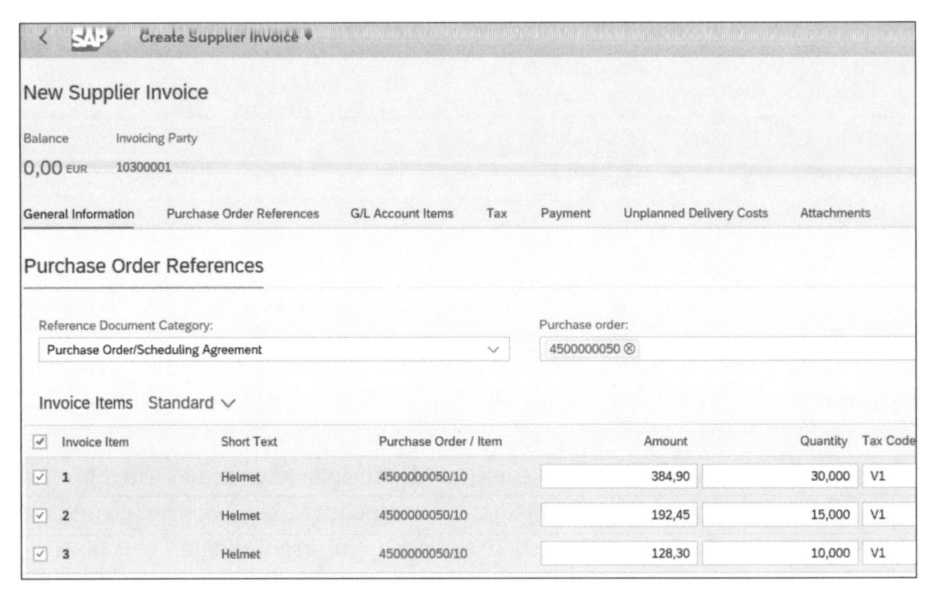

Figure 11.30 Create Supplier Invoice App: Posting GR-Based

Dealing with Deviations

When you enter an invoice with reference to a purchase order, the system proposes a quantity and amount per item. These default values are derived from the purchase order history: as a rule, the system proposes the quantity that has already been delivered but not yet invoiced. If an incoming invoice item differs from these values, you have two options: First, you can overwrite the default values and post the invoice with these new values. The system can block the invoice for payment if the differences exceed defined tolerance limits. Alternatively, you can choose to not accept these differences and "reduce" the invoice. Let's examine these two options and their consequences in detail.

Overwriting Default Values and Posting the Invoice with the New Values

The first option in the case of deviations is to overwrite the default values.

In the case of minor deviations, it is not worth carrying out extensive investigations. For this reason, you can set up tolerance limits for each type of variance in Customizing. If you detect a variance in an invoice item and overwrite the default values, the system checks whether the variances are within the tolerance limits in Customizing.

Although invoices with variances can always be posted, the system behaves differently, depending on whether the difference is within or outside the tolerance intervals:

- If the difference is *within the tolerance limits*, the system does not issue a message. The invoice is posted and is released for payment.

- If the difference is outside the tolerance interval but in your favor, that is, if the value falls *below a lower tolerance limit*, the system simply issues a warning message for your information. The invoice is posted and is released for payment.
- If an *upper tolerance limit is exceeded*, the system also issues a warning message. This time, however, the invoice is automatically blocked for payment. If only one item is affected by a difference that exceeds the tolerated limit, the entire invoice is blocked. The item in question is marked with a blocking reason. You must release a blocked invoice before it can be released for payment.

For an invoice item, you can define tolerance limits for the following variances: quantity, price, purchase order price quantity, and date. Let's look at each.

Quantity Differences

Say you have ordered 10 pieces of a material: 6 have been delivered, and 10 are invoiced.

For a purchase order, the open quantity to be invoiced is the difference between the quantity delivered and the quantity already invoiced. A quantity variance exists if the quantity invoiced does not correspond to the open quantity. Let's examine how the system checks these differences and what the posting looks like in case of quantity differences.

The quantity difference with +/- sign is multiplied by the price and compared with the tolerance limits. This ensures that invoice items with a high price tolerate less quantity variance than invoice items with a lower price. You can also define percentage limits for checking the quantity variance where the system does not take the price into account.

In Customizing, follow **Materials Management • Logistics Invoice Verification • Invoice Block • Set Tolerance Limits**.

Figure 11.31 shows the available tolerance key for company code 1010.

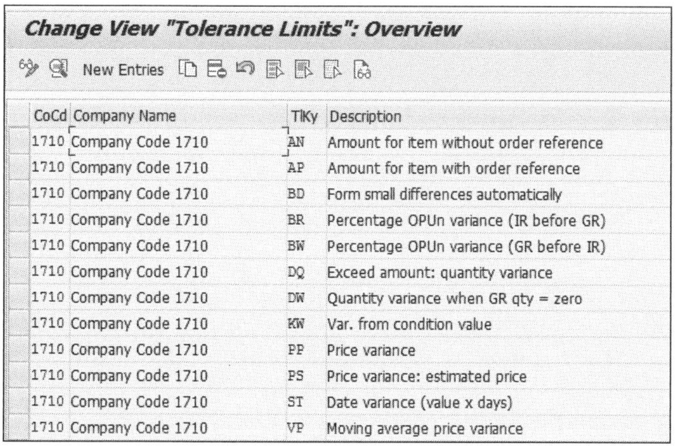

Change View "Tolerance Limits": Overview

New Entries

CoCd	Company Name	TlKy	Description
1710	Company Code 1710	AN	Amount for item without order reference
1710	Company Code 1710	AP	Amount for item with order reference
1710	Company Code 1710	BD	Form small differences automatically
1710	Company Code 1710	BR	Percentage OPUn variance (IR before GR)
1710	Company Code 1710	BW	Percentage OPUn variance (GR before IR)
1710	Company Code 1710	DQ	Exceed amount: quantity variance
1710	Company Code 1710	DW	Quantity variance when GR qty = zero
1710	Company Code 1710	KW	Var. from condition value
1710	Company Code 1710	PP	Price variance
1710	Company Code 1710	PS	Price variance: estimated price
1710	Company Code 1710	ST	Date variance (value x days)
1710	Company Code 1710	VP	Moving average price variance

Figure 11.31 Customizing: Tolerance Keys for Variances

Figure 11.32 shows the settings for the tolerance key for quantity variances.

Change View "Tolerance Limits": Details

New Entries

Tolerance Key DQ Exceed amount: quantity variance
Company Code 1010 Company Code 1010
Amounts in EUR Euro (EMU currency as of 01/01/1999)

Lower Limit
Absolute
(•) Do Not Check
( ) Check Limit
Value

Percentage
(•) Do Not Check
( ) Check Limit
Tolerance Limit %

Upper Limit
Absolute
( ) Do Not Check
(•) Check Limit
Value 100,00

Percentage
(•) Do Not Check
( ) Check Limit
Tolerance Limit %

Figure 11.32 Customizing: Tolerance Limits for Quantity Variances

Now let's review postings with quantity differences. In the example shown in Figure 11.33, the invoiced quantity is greater than the goods receipt quantity. When you post the invoice, a balance is generated on the GR/IR clearing account. A further goods receipt is expected. When you post this goods receipt, the GR/IR clearing account is balanced. Here you can see:

❶ Purchase order: 10 pieces at €12/piece

❷ Goods receipt: 6 pieces at €12/piece

❸ Invoice receipt: 10 pieces at €12/piece (quantity variance)

❹ Goods receipt: 4 pieces at €12/piece

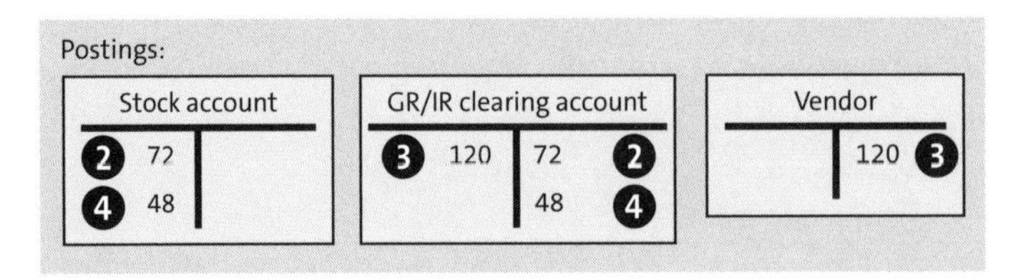

Figure 11.33 Postings for Quantity Variances

In the opposite case (when the invoice quantity is less than the goods receipt quantity), the GR/IR clearing account is not cleared until the total goods receipt quantity is invoiced.

Price Differences

The next type of deviation is a price difference—if, for example, you have agreed to an order price of 12 EUR per piece but the invoice price is 11 EUR. A price variance occurs if the invoice price (invoice amount divided by the invoice quantity) does not correspond to the net price of the purchase order.

Figure 11.34 shows you how to check against tolerance limits for price differences. Here you can define absolute and/or percentage tolerance values.

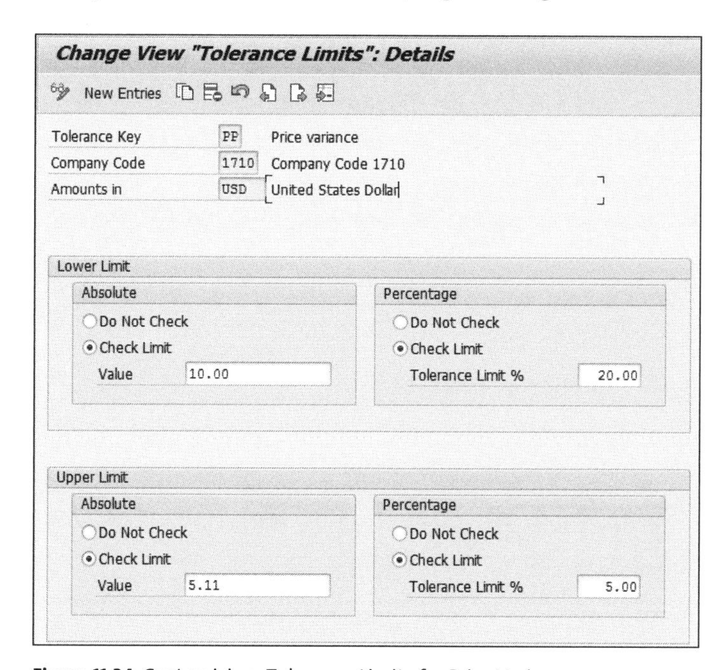

Figure 11.34 Customizing: Tolerance Limits for Price Variances

In the case of price variances, the account postings depend on the valuation of the invoiced material or whether we are dealing with an order item assigned to an account.

We assume that the goods receipt was before the invoice receipt (i.e., that the goods receipt was posted at the purchase order price).

If the material is valuated at a standard price, the system posts the price variance to a price difference account, as shown in Figure 11.35. Assume that the material has a standard price of €10/piece. Take the following values:

❶ Purchase order: 10 pieces at €12/piece

❷ Goods receipt: 10 pieces at €12/piece

❸ Invoice receipt with price variance: 10 pieces at €11/piece

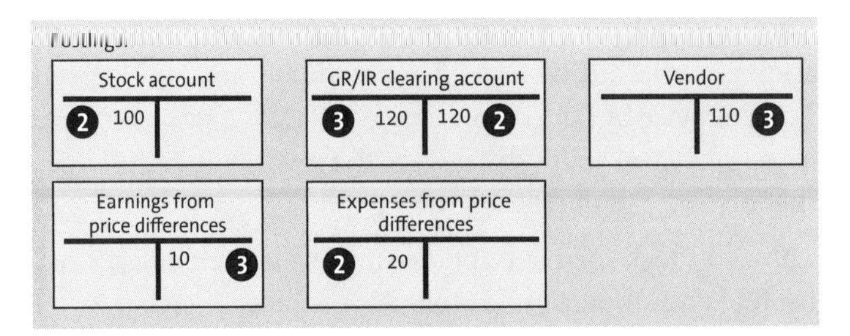

Figure 11.35 Postings in Case of a Price Difference in Your Favor for a Material with a Standard Price

If the material is valuated to the moving average price and there is enough stock coverage, the system posts the price variance to the stock account (as shown in Figure 11.36), and the moving average price is updated. If there is no or insufficient stock coverage, the system posts the difference to a price difference account for the quantity not in stock:

❶ Purchase order: 10 pieces at €12/piece

❷ Goods receipt: 10 pieces at €12/piece

❸ Invoice receipt with price variance 10 pieces at €11/piece

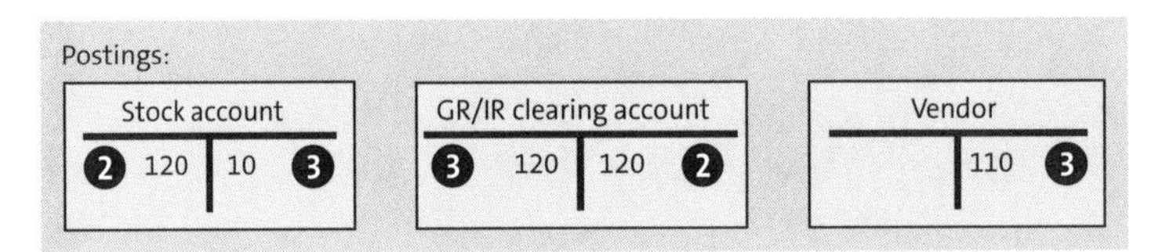

Figure 11.36 Postings in Case of a Price Difference in Your Favor for a Material with a Moving Average Price

For a purchase order item with account assignment, the system posts the difference to the cost account.

Purchase Order Price Quantity Variances

This next type of variance can occur if you order in a certain unit of measure (e.g., a piece) but the price unit is different (e.g., per kilogram).

Let's consider an example to illustrate this: Say you order 10 cauliflower heads. However, the price for cauliflower is not per piece (head) but per kilogram, at a rate of 2 EUR per kilogram. A cauliflower normally weighs 1.2 kg. When you specify an order price unit (kg) for an order item that differs from the order unit (pc), you must enter the quantity in both units at goods receipt and at invoice entry. You can have the system automatically block an invoice for payment if the relationship between the order unit and the order price unit differs from that at goods receipt (or from the purchase order, if no goods receipt is made) and exceeds a defined percentage tolerance limit (for example, 20%). In our example, if the ordered cauliflowers no longer weigh around 1.2 kg but 1.5 kg, the system would block the invoice for payment.

Date Variances

A date variance exists if the invoice receipt date is before the delivery date specified in the purchase order. The system performs the following calculation for each item: *Item value × (Planned delivery date – Invoice receipt date).*

The system then compares the result with the absolute upper limit that you defined in Customizing, as shown in Figure 11.37. The value of the invoice item thus plays a role in determining the date variance.

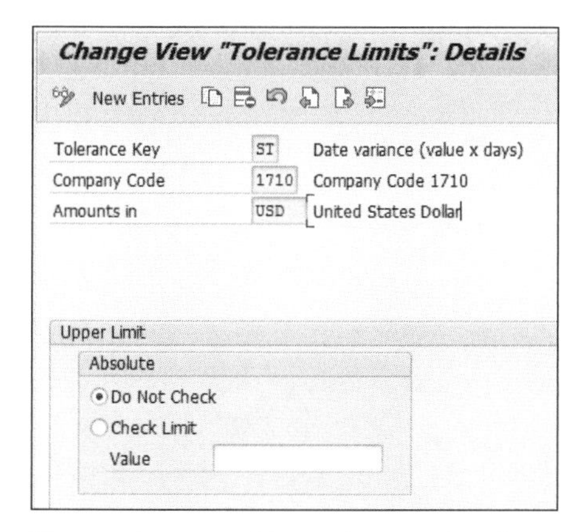

Figure 11.37 Customizing: Tolerance Limit for Date Variances

Reducing Invoices

If you receive a vendor invoice with a difference either in the amount or in the quantity (or both), you can refuse this difference and choose to "reduce" the invoice.

Using Transaction MIRO, you select the correction indicator **Supplier error: reduce invoice** on the item line, as shown in Figure 11.38. Using the Create Supplier Invoice app shown in Figure 11.39, select the **Reduce Invoice** button at the item level. In addition to the default values, which correspond to the expected values, enter the data (quantity and amount) from the vendor invoice.

The system then simultaneously posts an invoice for the values specified on the vendor invoice and a credit memo for the difference from the expected values. The difference value is posted in both documents to a clearing account for invoice reduction.

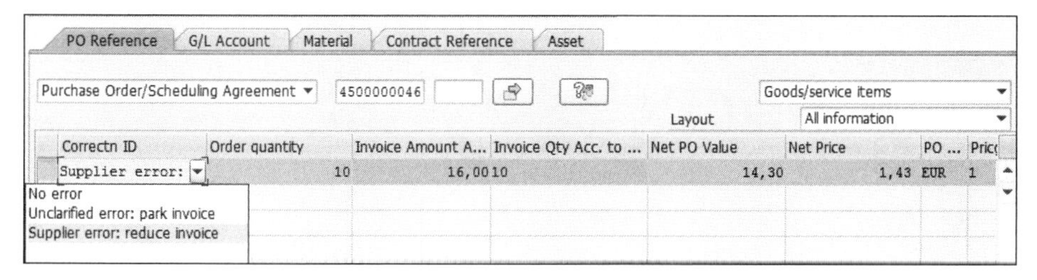

Figure 11.38 Reduce an Invoice with Transaction MIRO

Amount: * 1,283.00 EUR
Amount from Supplier: 2,000.00 EUR
Quantity: * 100.000 PC
Quantity from Supplr: 100.000 PC
OK Cancel

Figure 11.39 Reduce an Invoice with the Create Supplier Invoice App

Note

The differences we have discussed so far have always related to a specific position. However, it can also happen that you enter an invoice and find a balance between the total invoice value and the sum of the items, without being able to know exactly where this difference comes from. You can, of course, take the trouble to investigate, but you can also have the system automatically post this difference up to a certain amount, either as a small difference or as a reduction.

To do this, create tolerance groups in Customizing (as shown in Figure 11.40), in which you define when a small difference posting line and when a reduction is to be posted. These tolerance groups are then assigned to vendors in the vendor master records.

Change View "Supplier-Specific Tolerances": Details
New Entries
Company Code 1710 Company Code 1710
Tolerance Grp 0001
Currency USD

Automatic Acceptance of Negative Differences
Absolute Lower Limit: Check Limit ✓; NegAccLowerLimt 200.00-
Percentage Lower Limit: Check Limit ✓; PercLowerAccLim 4.00 %
Negative Small Difference: Check Limit ✓; Small Diff. 10.00-

Automatic Acceptance of Positive Differences
Positive small difference: Check Limit ✓; Small Diff. 5.00
Absolute Upper Limit: Check Limit ✓; PosAccUpperLimt 50.00
Percentage Upper Limit: Check Limit ✓; PercUpperAccLim 2.00 %

Automatic Invoice Reduction
Absolute Upper Limit: Check Limit ✓; Inv.Red.Tol.Amn 40.00
Percentage Upper Limit: Check Limit ✓; Per.Inv.Red.Tol 5.00 %

Figure 11.40 Customizing: Supplier-Specific Tolerances

Releasing Blocked Invoices

We have learned that the system can automatically block for payment invoices with variances, depending on the tolerance limits you have defined in Customizing. In addition to the automatic blocking of invoices due to deviations, there are other possibilities to block invoices for payment.

Invoices can be blocked not only because of deviations but also for the following reasons:

- The user can decide to enter a payment block manually in the header data or to block an item manually.
- The system can automatically block an invoice item because the amount exceeds a predefined value. This is the case when you have:
  - activated the amount check in Customizing per company code;
  - defined in Customizing for which PO item category and goods receipt indicator this amount check should apply when posting the invoice; and
  - defined the amount from which an invoice item is to be blocked. (Note there are two entries: one for invoice items with reference to a purchase order item and one for invoice items without reference to a purchase order item.)
- In addition to the automatic blocking of invoices due to deviations or due to the amount, it is also possible to randomly block invoices for payment. You can use the stochastic block for this type of random check. Stochastic blocks apply to the entire invoice, not at the item level. To automatically block invoices, you must:
  - Activate the stochastic block per company code in Customizing.
  - Set a threshold value and a percentage. This determines the probability of a block. If the invoice value is greater than or equal to the threshold value, the probability that the invoice will be blocked is equal to the specified percentage. If the invoice value is below the threshold value, the probability decreases proportionally.

Follow in Customizing **Materials Management • Logistics Invoice Verification • Invoice Block** to activate and set the amount block and the stochastic block.

Note

If you use Quality Management, an invoice can also be blocked for quality reasons.

Blocked invoice items due to deviations, amount, quality, or user decision are marked with a so-called blocking reason indicator. The blocking flag is also set at the header level.

Blocked invoices must be released for payment in a separate step after the reasons for the block have been clarified and any follow-up actions decided, if necessary.

The Release Blocked Invoices app (Transaction MRBR; Figure 11.41) is available for this purpose. Proceed as follows:

❶ Select the invoices to be processed.

❷ Specify the processing mode.

❸ Select the blocking procedure.

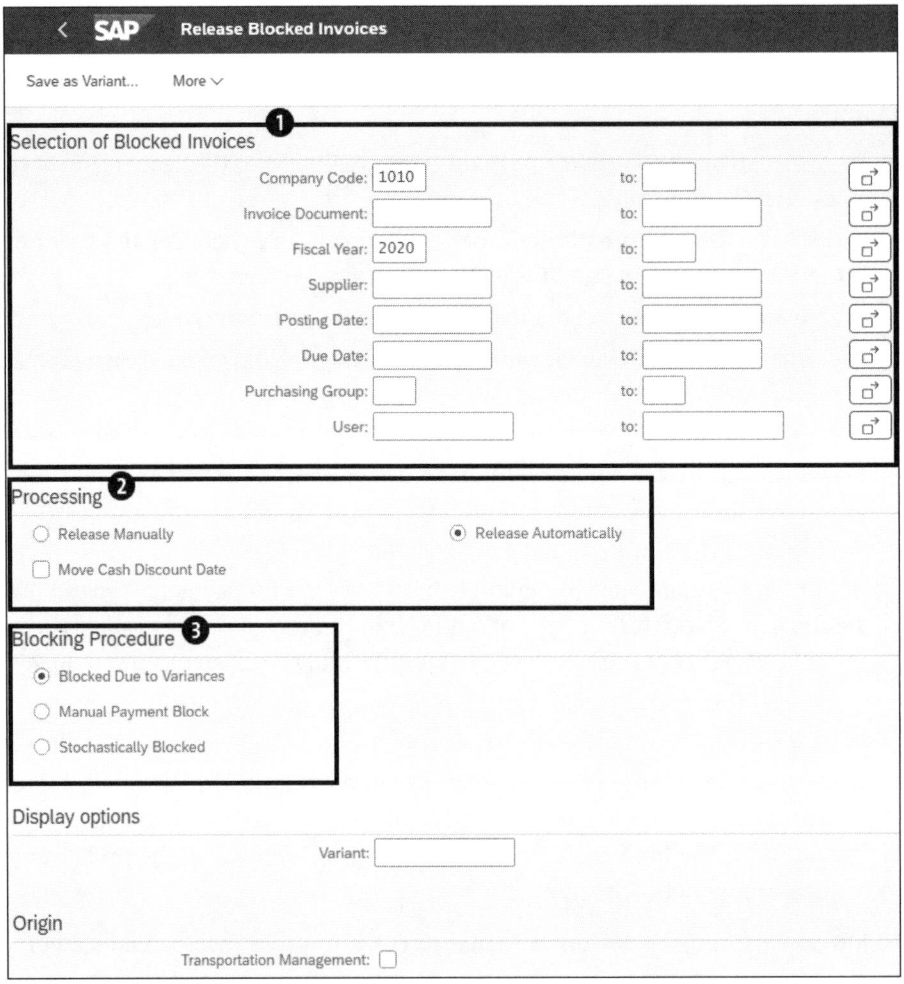

Figure 11.41 Release Blocked Invoices App (Transaction MRBR)

Automatic Release

The Transaction MRBR can be scheduled as a regular job so that the invoices for which the blocking reasons are no longer valid can be released automatically.

In which cases can a reason for blocking be no longer valid?

- If you receive a further delivery or a credit memo, the block due to quantity variance (blocking reason M) is no longer valid.
- If you change the purchase order price or receive a subsequent credit, the block due to price variance (blocking reason P) may no longer apply.

- Blocking reason Q (quality variance) becomes invalid due to a positive usage decision in Quality Management.
- The blocking reason due to the date (blocking reason D) becomes invalid as soon as the delivery date is reached.

When you start Transaction MRBR for an automatic release (either online or in the background), the system checks, item by item according to the selection criteria, whether the blocking reasons have become obsolete. If they have, they are removed, and if no items have a blocking reason, the entire invoice is released for payment. If blocking reasons still exist because they could not be removed automatically, the invoice must be released manually.

Manual Release

If you want to release invoices manually, make your selection both in the **Selection of Blocked Invoices** area and in the **Blocking Procedure** area. The system lists the blocked invoices according to your selection. There are two possibilities here:

- You can delete blocking reasons in individual invoice items. When you delete the last blocking reason in an invoice and save your changes, the system releases the invoice for payment.
- You can also release an invoice as a whole; in this case, the payment block in the vendor line is deleted, but the blocking reasons in the items are not deleted.

Note

You can use a workflow for invoices that are blocked due to price variances: the buyer responsible receives a work item, can adjust the price in the purchase order, or release the invoice directly. He or she can also set the status *Cannot be clarified*, which means that the invoice verification clerk is assigned the work item again.

Automating Invoice Posting: Evaluated Receipt Settlement (ERS)

In agreement with your supplier, you can decide that he or she no longer needs to send you invoices, but that you automatically generate invoice documents based on deliveries.

The prerequisites are as follows:

- The ERS indicator must be set in the supplier master record in the purchasing organization data.
- The ERS indicator must be set in the purchase order item. It is copied from the vendor master.
- The indicator for GR-based invoice verification must be set in the purchase order item.
- The purchase order item must contain a tax code.

- The purchase order header must contain a terms of payment key
- The price in the purchase order item is not an estimated price.

To automatically settle goods receipts, choose the Create Evaluated Receipt Settlement app (Transaction MRRL). On the SAP Fiori launchpad, use the Schedule Supplier Invoice Jobs—Advanced app.

Figure 11.42 shows the first step when scheduling a job using the Schedule Supplier Invoice Jobs—Advanced app: the job template is preselected, and you can specify a job name.

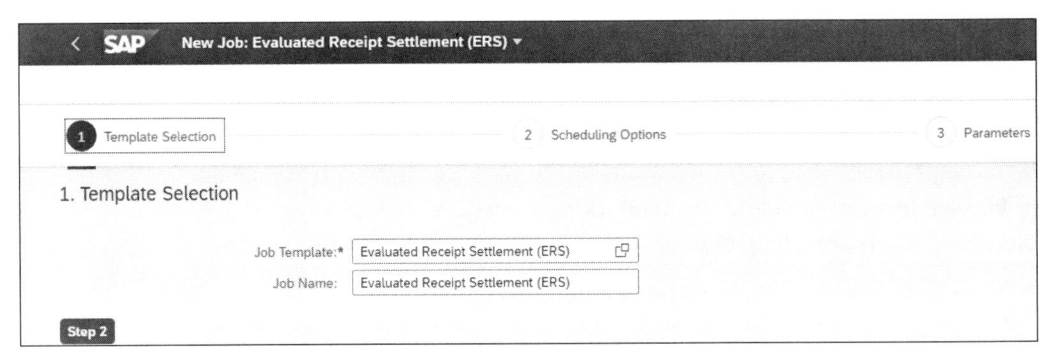

Figure 11.42 Schedule Supplier Invoice Jobs—Advanced App: Step 1

Figure 11.43 shows the second step when scheduling a job using the Schedule Supplier Invoice Jobs—Advanced app: setting the scheduling and recurrence options.

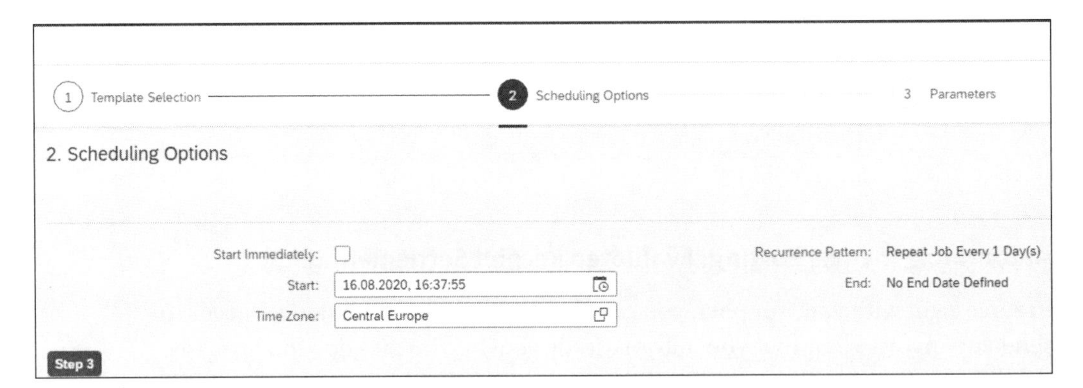

Figure 11.43 Schedule Supplier Invoice Jobs—Advanced App: Step 2

Figure 11.44 shows the third step when scheduling a job using the Schedule Supplier Invoice Jobs—Advanced app: specifying the parameter for selecting the documents.

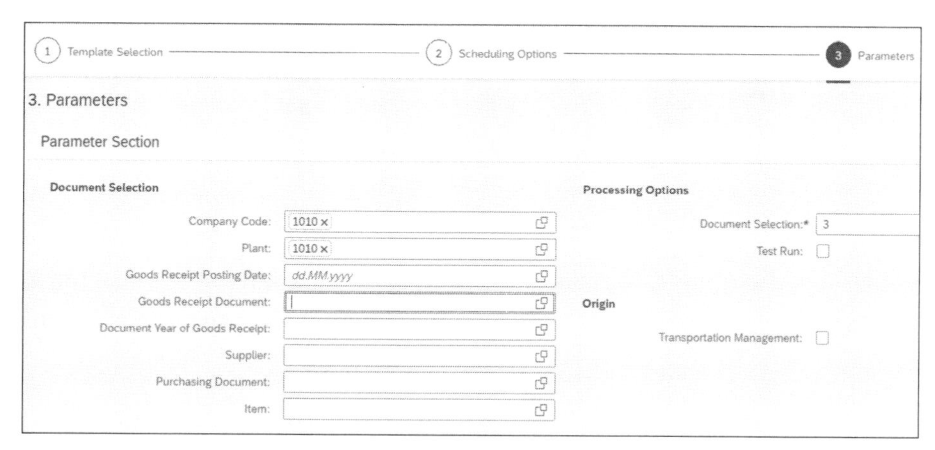

Figure 11.44 Schedule Supplier Invoice Jobs—Advanced App: Step 3

Maintaining GR/IR Clearing Account

The GR/IR clearing account is posted when a goods receipt or an invoice is posted with reference to the purchase order. We have already discussed the posting logic in Chapters 3 and 4. If the goods receipt is posted before the invoice, the GR/IR clearing account is posted at the purchase order value. If the invoice is posted first, the GR/IR clearing account is posted at the invoice value.

The GR/IR clearing account is cleared for a purchase order item if the goods receipt quantity and invoice quantity are the same.

A difference on the GR/IR clearing account means that the process is not yet complete—that is, an invoice or delivery is missing, respectively a credit memo or a return delivery.

If, however, no further invoice or goods receipt is sent, the GR/IR clearing account should be cleared with an extra transaction. A balance on the GR/IR clearing account is an indication that the procurement process was not completed properly.

In the example illustrated in Figure 11.45, 10 pieces of a material are ordered, 9 are delivered, and 10 are invoiced.

Either a further delivery of 1 piece should be made or a corresponding credit memo should be issued by the vendor. However, if this is not the case for any reason, the GR/IR clearing account should be cleared. The offsetting entry debits the stock account, since the 9 pieces delivered have cost as much as 10.

Here's our values:

❶ Purchase order: 10 pieces at €12/piece

❷ Goods receipt: 9 pieces at €12/piece

❸ Invoice receipt: 10 pieces at €12/piece (quantity variance)

❹ GR/IR clearing account maintenance: €12

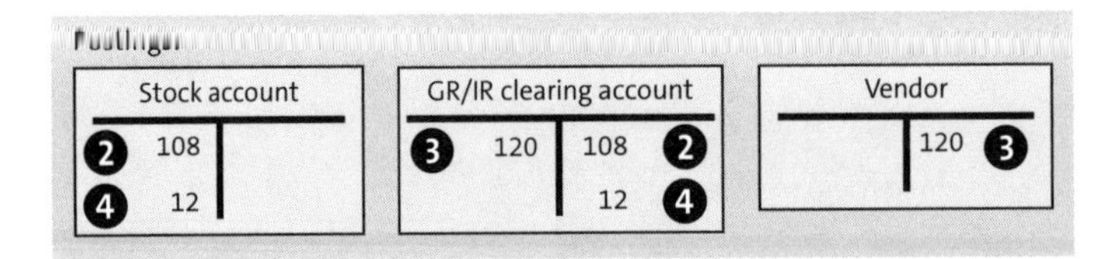

Figure 11.45 Example of GR/IR Clearing Account Maintenance

For a material with a moving average price, the offsetting entry to clear the GR/IR clearing account is posted to the stock account (unless there is no or insufficient stock coverage, in which case a price difference account is posted proportionally to the missing stock).

For a material at a standard price, the system posts the offsetting entry to the price difference account.

For a purchase order item with an account assignment, the system posts the offsetting entry to the consumption account.

Figure 11.46 shows the Clear GR/IR Clearing Account app (Transaction MR11). Here you would first select the items to be cleared ❶ and then specify the processing mode ❷.

Maintain GR/IR Clearing Account
Save as Variant... More
Document Header Data
* Company Code: 1010
* Posting Date: 16.08.2020
Reference:
Doc. Header Txt:
❶ Choose
Supplier: to:
Freight Supplier: to:
Purch. Organization: to:
Purchasing Group: to:
Plant: to:
Purchasing Document: to:
Item: to:
* Purchase Order Date: 01.01.2019 to: 17.06.2020
Purch. Doc. Category:
Order Type: to:
Surplus Types
Delivery Surplus: ☑
Invoice Surplus: ☐
Clear
☑ GR/IR Clearing Account ☐ ERS Purchase Orders
☑ Delivery Cost Accounts ☐ ERS - Delivery Costs
* Last Movement Before Key Date: 16.08.2020
Qty Var. Less Than/Equal To: 100.0 Percentage
Value Variance Less Than/= To:
Origin
☐ Transportation Management
Processing ❷
Automatic Clearance: ○
Prepare List: ◉ Layout:

Figure 11.46 Clear GR/IR Clearing Account App (Transaction MR11)

Important Terminology

The following important terminology was used in this chapter.

- **Logistics Invoice Verification**
  Logistics invoice verification is part of materials management. It is located at the end of the logistics supply chain, which includes purchasing, inventory management, and invoice verification. In logistics invoice verification, invoices are verified in terms of content, price, and calculation. When the invoice is posted, the invoice data is stored in the system. The system updates the data stored in the invoice documents in materials management and financial accounting.
- **Materials management invoice document**
  When you post an invoice using Logistics Invoice Verification, you create an materials management invoice document that records the invoice data, in addition to an accounting document. Both documents usually have different numbers.
- **Posting gross/Posting net**
  This is a method of posting a cash discount and is dependent on the document type.
- **Document parking**
  You can interrupt an invoice entry and simply save the status of the invoice. Depending on its status, you can either hold or park the invoice. Parking can also be used systematically in a company to ensure the four-eyes policy.
- **GR/IR clearing account**
  When entering goods and invoice receipts with purchase order reference, postings are made to the GR/IR clearing account.
- **Planned delivery costs**
  Planned delivery costs are already included in the purchase order item and are posted upon goods receipt. When the invoice is posted, the freight clearing account is simply balanced, provided no differences occur.
- **Unplanned delivery costs**
  Unplanned delivery costs appear for the first time with the invoice and can either be distributed to the invoice items of the reference purchase order or posted to a separate account line.
- **Foreign currency**
  You can post invoices in a currency that differs from the company code currency (so-called foreign currency).
- **Goods-receipt-based invoice verification**
  In goods-receipt-based invoice verification, the system does not check the invoice against the entire order item but against individual deliveries.
- **Blocking reason**
  The blocking reason is an indicator in an invoice item and provides information on why an invoice item contributes to the invoice block.

- **Tolerance key**
  Tolerance limits for blocking invoices are defined in Customizing using tolerance keys per company code.
- **Stochastic blocking**
  Stochastic blocking, if activated, ensures that an invoice will be blocked with a certain probability if it exceeds a certain total value. The threshold value and probability percentage are defined in Customizing.
- **Invoice reduction**
  Invoice reduction is used when you do not want to accept a difference that is disadvantageous for you: in addition to the incoming invoice, the system posts a credit memo for the difference.
- **Evaluated Receipt Settlement (ERS)**
  With this procedure, you agree with the vendor that the vendor will not submit an invoice for a purchase order. The goods receipt is settled directly by posting the invoice document automatically based on the purchase order and the goods receipt.

Practice Questions

These questions will help you evaluate your understanding of the topics covered in this chapter. They are similar in nature to those on the certification examination. Although none of these questions will be found in the exam itself, they will allow you to review your knowledge of the subject. Select the correct answers, and then check the completeness of your answers in the next section. Remember that on the exam, you must select all correct answers—and only correct answers—to receive credit for the question.

1. True or false: When you enter an invoice, the first thing you must do is enter the supplier number.

☐ **A.** True

☐ **B.** False

2. You enter an invoice with reference to a purchase order. Which of the following entries must you make? (There are two correct answers.)

☐ **A.** Gross invoice amount

☐ **B.** Invoicing party

☐ **C.** Item quantity

☐ **D.** Invoice date

☐ **E.** Item amount

3. True or false: When you enter an invoice with reference to a purchase order, the system proposes all purchase order items with the quantities and values ordered.

☐ **A.** True

☐ **B.** False

4. You enter an incoming invoice with Transaction MIRO (Create Supplier Invoice—Advanced app). The invoice items have different tax rates. How can you post these tax amounts? (There are three correct answers.)

☐ **A.** In the **Basic Data** tab, enter the total tax amount manually.

☐ **B.** On the **Tax** tab, enter the tax amount manually.

☐ **C.** On the **Tax** tab, flag the **Calculate Tax** indicator.

☐ **D.** On the **Basic Data** tab, flag the **Calculate Tax** indicator.

☐ **D.** Manually enter a separate G/L account line for each tax rate.

5. Which of the following statements apply to the payment terms? (There are two correct answers.)

☐ **A.** Terms of payment can only be maintained in the master record of the business partner for each company code.

☐ **B.** Terms of payment are inherited in the purchase order from the vendor master record and can be changed there.

☐ **C.** Terms of payment are defined in the purchase order and cannot be changed when the invoice is entered.

☐ **D.** The terms of payment are configured in Customizing for Financial Accounting.

6. True or false: Cash discount posting can depend on the valuation procedure of the material in the invoice item.

☐ **A.** True

☐ **B.** False

7. You are posting an invoice gross. Where is the cash discount amount posted at this time?

☐ **A.** To a cash discount clearing account

☐ **B.** To a cash discount account

☐ **C.** Nowhere. The cash account is not posted at this time.

☐ **D.** To the GR/IR clearing account

8. Which exchange rate can be used to convert an invoice in a foreign currency into the company code currency? (There are three correct answers.)

☐ **A.** The current valid exchange rate

☐ **B.** The rate in the purchase order, if it is fixed

☐ **C.** The rate used at goods receipt

☐ **D.** A manually entered exchange rate

9. You do not want unplanned delivery costs to affect the valuation of your materials. You therefore want to post them to a separate account. What options do you have? (There are two correct answers.)

☐ **A.** Activate direct posting to G/L accounts in Customizing and post the costs directly to the account of your choice.

☐ **B.** Specify in Customizing that unplanned delivery costs are to be posted to a separate G/L account, and the system automatically determines the account using the transaction key UPF.

☐ **C.** Activate direct posting to G/L accounts in Customizing, and the system automatically determines the account using the transaction key UPF.

☐ **D.** Specify in Customizing that unplanned delivery costs are to be posted to a separate G/L account, and post the costs directly to the account of your choice.

10. True or false: A subsequent debit always results in a posting to the GR/IR clearing account.

☐ **A.** True

☐ **B.** False

11. Which of the following effects does parking an invoice have? (There are two correct answers.)

☐ **A.** The purchase order history is updated.

☐ **B.** An MM invoice document is created.

☐ **C.** A financial accounting document is created.

☐ **D.** The GR/IR clearing account is posted to.

12. True or false: In goods-receipt-based invoice verification, no invoices can be entered before a goods receipt is posted.

☐ **A.** True

☐ **B.** False

13. Where do you specify whether a purchase-order-based or a goods-receipt-based invoice verification is to be performed?

☐ **A.** In the purchase order header

☐ **B.** In Customizing per company code

☐ **C.** In the purchase order item

☐ **D.** When entering the invoice

14. True or false: If a vendor invoice contains a price that is greater than planned, you must first pay this higher price. You may receive a credit memo later.

☐ **A.** True

☐ **B.** False

15. With some suppliers, your company would like to pay at most the amount agreed in the purchase order. An invoice that shows a difference of up to 50 euros to your disadvantage should be automatically reduced by the system. How can you achieve this?

☐ **A.** Define tolerance groups per company code and vendor account group in Customizing.

☐ **B.** Enter a tolerance limit in the company code data of the relevant vendor master records.

☐ **C.** Define tolerance groups per company code in Customizing and assign them to vendor master records.

☐ **D.** Post a separate account line with the correction indicator **Supplier error: reduce invoice**.

16. True or false: An invoice is blocked for payment due to a price variance in one item. If the buyer adjusts the price in the relevant purchase order item, the invoice is automatically released without further action.

☐ **A.** True

☐ **B.** False

17. When can a balance arise on the GR/IR clearing account? (There are two correct answers.)

☐ **A.** For a purchase order item, the invoice quantity is greater than the goods receipt quantity.

☐ **B.** For a purchase order item, the invoice price is higher than the order price.

☐ **C.** For a purchase order item, the goods receipt quantity is greater than the invoice quantity.

☐ **D.** For a purchase order item, the order price is higher than the invoice price.

18. Which of the following are prerequisites for applying the Evaluated Receipt Settlement (ERS) to a purchase order item? (There are three correct answers.)

☐ **A.** The GR-based invoice verification indicator must be set in the vendor master record.

☐ **B.** The ERS indicator must be set in the vendor master record.

☐ **C.** The ERS indicator must be set in the purchase order item.

☐ **D.** The No ERS indicator must not be set in the purchasing info record.

☐ **E.** The GR-based invoice verification indicator must be set in the purchase order item.

19. True or false: The system automatically releases blocked invoices only when all items no longer have blocking reasons. Manually, however, you can release an invoice for payment even if not all blocking reasons for the items have been eliminated.

☐ **A.** True

☐ **B.** False

20. How are tolerance limits for blocking invoices defined in Customizing?

☐ **A.** Per plant

☐ **B.** Per company code

☐ **C.** Per supplier group

☐ **D.** Per user group

☐ **E.** Per client

Practice Question Answers and Explanations

1. Correct answer: **B**
   False. If you enter an invoice with reference to a procurement process (with reference to a purchase order, for example), you do not have to enter the supplier number. This is derived from the purchasing document.
2. Correct answers: **A, D**
   When you enter an invoice with reference to a purchase order, you enter the invoice date, the total gross amount, and the reference document. The system derives the other data and the items. If necessary, you can overwrite these default values to match the vendor invoice.

Note

The reference is an optional field in the standard system in which you can enter the original supplier invoice number. This field can be configured as a required entry field to force an entry.

3. Correct answer: **B**

   False. In purchase-order-related invoice verification, all purchase order items are proposed. But the system does not propose the total quantity and value of each purchase order item. Instead, it proposes the quantity already delivered but not yet invoiced with the corresponding value.

4. Correct answers: **B, C, D**

   If an invoice has several tax codes, you can either enter the tax amounts on the **Tax** tab page or let the system calculate the amounts by setting the **Calculate Tax** indicator, either on the **Basic Data** or **Tax** tab page. Enter the total tax amount manually in the **Basic Data** tab only if the invoice has one single tax code.

5. Correct answers: **B, D**

   The terms of payment are configured in Customizing of Financial Accounting. They can be maintained in the master record of the business partner both per company code and per purchasing organization, enabling you to agree and record two different terms of payment for a vendor: those for procurement transactions and those for invoices without reference to a purchasing document. The terms of payment that you have maintained in the purchasing section of the vendor master record are proposed in the purchase orders and can be adjusted there. These are then proposed in the invoices and can also be adjusted there in accordance with the incoming invoice.

6. Correct answer: **A**

   True. Cash discount posting first depends on the document type. If the document type is a net document type, then it depends on whether the invoice item is an account-assigned item or a material item. If it is a material item, the posting does indeed depend on whether the material is valuated at the standard price or at the moving average price.

7. Correct answer: **C**

   When posting gross, the cash account is not considered at the time of posting the invoice. Only at the time of payment.

8. Correct answers: **A, B, D**

   If the exchange rate is fixed in the purchase order, the system uses the exchange rate from the purchase order. If not, the system proposes the exchange rate valid on the day of invoice entry, which you can overwrite.

9. Correct answers: **A, B**

   You can post these unplanned costs manually to a separate account line. You only must make sure that direct posting is activated (possible). If you want to use automatic account determination for this, make the setting in Customizing that unplanned delivery costs are to be posted to a separate G/L account and use the "unplanned delivery costs" field in the invoice header. The system automatically determines the account using the transaction key UPF.

10. Correct answer: **B**

    False. A subsequent debit only leads to an update of the GR/IR clearing account if the subsequent debit occurs before the goods receipt was posted for the corresponding quantity. Otherwise, the system posts to either the stock account, the price difference account, or the cost account.

11. Correct answers: **A, B**

    When an invoice is parked, the purchase order history is updated and an invoice document is created. However, there is no accounting document—only an MM document. Merely an FI document number is assigned, but no accounts are posted to. Only if prepayment is activated are accounting documents created when parking.

12. Correct answer: **A**

    True. In goods-receipt-based invoice verification, the quantity invoiced cannot be greater than the quantity of goods received. Therefore, no invoices can be posted until a goods receipt has been posted.

13. Correct answer: **C**

    You can set the **Goods-receipt-based invoice verification** indicator at the purchase order item level. This indicator can also be set in the supplier master record or in the source of supply (info record or outline agreement). In this case, the system automatically adopts it as a default value in the purchase order item, where it can be changed, if required.

14. Correct answer: **B**

    False. If a vendor invoice shows a price difference or other variance, you can first accept it, in which case you overwrite the default values and post the invoice. Depending on the tolerance values, this invoice may be blocked for payment. You will check the reason for the difference, release the invoice for payment, and you may also receive a credit memo from the vendor. However, you can reject this variance and use the function Reduce Invoice. In this case, the system posts two documents: the invoice as received and a credit memo for the difference to the planned value. You only pay the planned value (the difference between the invoice amount and the credit memo amount).

15. Correct answer: **C**

    You define tolerance groups per company code in Customizing. For each tolerance group, you can set tolerance limits for acceptance posted as small differences and tolerance limits for reductions. You then assign these tolerance groups to vendor master records.

16. Correct answer: **B**

    False. The invoice is not released by itself, even if the blocking reason is no longer valid. You must either remove the blocking reason manually or have it removed automatically by the system using Transaction MRBR.

17. Correct answers: **A, C**

    Only a quantity difference between goods receipt and invoice receipt for a purchase order item results in a balance on the GR/IR clearing account.

18. Correct answer: **B, C, E**

    To use ERS, you must first set the ERS indicator in the supplier master record: it indicates that the agreement with the vendor has been made. Then this ERS indicator must also be selected in the relevant purchase order item. If the **No ERS** indicator is set in the purchasing info record, it is not proposed in the purchase order item but can be set manually. For technical reasons, the indicator for goods-receipt-based invoice verification must be set in the purchase order item. This indicator may already be set in the vendor master record but is not a must.

19. Correct answer: **A**

    True. The system automatically releases blocked invoices only when all items no longer have blocking reasons. Manually, however, you can release an invoice for payment at any time, even if not all blocking reasons for the items have been eliminated.

20. Correct answer: **B**

    Tolerance limits for blocking invoices are defined in Customizing using tolerance keys per company code.

Takeaway

In this chapter, we have reviewed the most important features of Logistics Invoice Verification: You should now know how to enter an invoice with reference, how the system proposes invoice items, what is automatically checked by the system, and how to react in the event of deviations. You should be familiar with the special features for foreign currency, delivery costs, or cash discount. You should also understand why and how an invoice can be blocked for payment, how you can influence this by making settings in Customizing, and what to do to release a blocked invoice. In addition, you should understand the postings that take place when an invoice is posted and be familiar with the requirements for introducing the ERS procedure as a process simplification option.

Summary

The Logistics invoice verification completes the procurement process. Before an invoice is submitted to Financial Accounting for payment, it must be posted correctly and released for payment. By maintaining GR/IR clearing accounts, you ensure that the value flow of the procurement processes is correct. Even though the entry of invoices is becoming increasingly automated, this task does not lose any of its importance and the consultant's task is even more crucial because background processing must run smoothly and is largely dependent on a good understanding of the process and correct configuration.

Chapter 12
Configuration Cross Topics

Techniques You'll Master

- Understand the function of purchasing document types.
- Create purchasing document types.
- Control the field selection in purchasing.
- Understand, maintain, and assign special user parameters for purchasers.
- Configure movement types.
- Control number assignment and field selection for goods movements.
- Configure number assignment and field selection for material master records.

In this last chapter, we will deal with configuration options that we have not covered in the previous chapters. In the earlier chapters, we mainly presented processes, and whenever it was appropriate, we also described and explained configuration steps. We will not go back to these here; instead, we will cover topics that would have disturbed the flow of explanations if we had included them in the process description. We thus decided to create a small extra chapter for them.

Real-World Scenario

As a sourcing and procurement consultant, you have already acquired a broad knowledge of the main processes and functions that SAP S/4HANA offers. You have also learned to what extent you can influence these processes through customizing. We would now like to equip you with general knowledge that you will find useful in any customer project. You will almost always be confronted with the question of how to show or hide certain fields in transaction processing or master data maintenance, how to influence the number assignment of purchase orders, contracts, material documents, and so forth. Then all you need to do is gain some experience to become a great professional!

Objectives of This Portion of the Test

The purpose of this part of the certification exam is to test your general knowledge of configuration topics that are not related to specific processes:

- Role and configuration of document types in purchasing and inventory management
- Field selection control in purchasing, inventory management, and material master
- Number assignment for order documents, material documents, and material master records
- Some important Customizing topics, such as movement types, material types, or material status

Key Concept Refresher

In this section, we discuss important Customizing settings that can be treated separately from the process descriptions. You will learn, for example, about the role of document types in both purchasing and inventory management. We will discuss how you can control the field selection and which influencing factors you can use for this.

Warning

The functions and settings for field selection presented in this chapter mainly refer to the SAP GUI transactions and the corresponding applications on the SAP Fiori launchpad.

Let's start with the role and configuration of document types in purchasing.

Document Types in Purchasing

Document types are defined depending on the following document categories:

- Purchase requisition
- Request for quotation (RFQ)
- Purchase order
- Contract
- Scheduling agreement

Document types in purchasing have an important control function.

In general, the document type influences the following:

- **The document number assignment**
  You can store up to two number ranges per document type: one number range for internal number assignment and one number range for external number assignment.
- **The field selection**
  The field selection key in Customizing for the document type—together with other factors—determines the attributes of the individual fields in the purchasing document. The field attributes can, for example, be *ready for input*, *hidden*, *mandatory*, and so on.
- **The allowed item categories**
  You define which item categories are allowed for each document type. In this way, you can ensure that some document types are only valid for specific procurement processes.
- **The item number interval**
  The item number interval controls the size of the numbering steps between the item numbers.
- **The possible conversion of purchase requisitions into purchasing documents (linkage of documents)**
  Depending on the document types and item categories of the purchase requisition and the purchasing document, you determine which conversions are allowed—for example, whether a standard item of a purchase requisition with document type NB is to be converted into a consignment item of a purchase order with document type NB.

In addition, the document type contains further control indicators, which vary according to the document category.

Let's describe some of the most important specific settings in more detail:

- **Specific settings for purchase orders**
  In the case of document types for purchase orders, you can use the **Control** field to determine whether the document type is for a stock transport order. In this case, you enter a **T** in the field. For stock transport orders, you can furthermore specify whether a vendor master record is required for the supplying plant.

  Figure 12.1 shows an example of the configuration for the purchase order types.

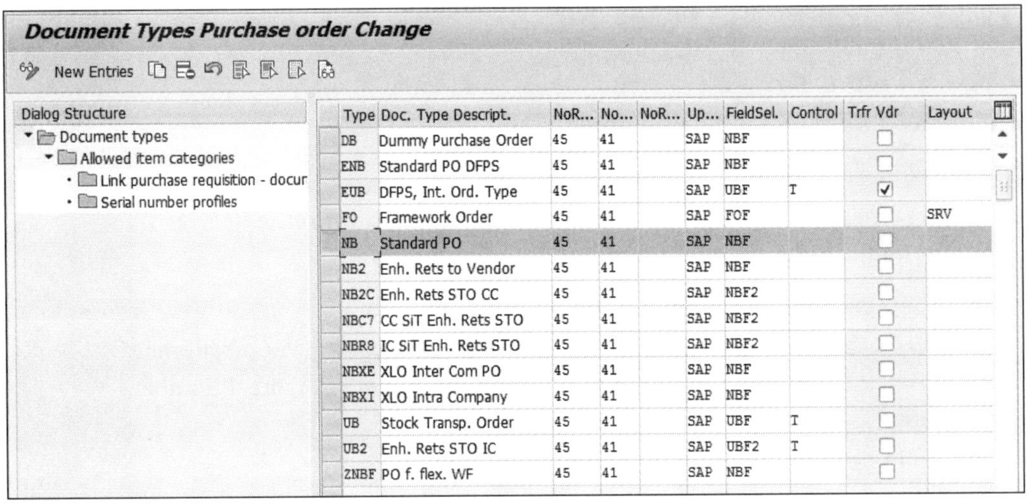

Document Types Purchase order Change

New Entries

Dialog Structure
- Document types
  - Allowed item categories
    - Link purchase requisition - docur
    - Serial number profiles

Type	Doc. Type Descript.	NoR...	No...	NoR...	Up...	FieldSel.	Control	Trfr Vdr	Layout
DB	Dummy Purchase Order	45	41		SAP	NBF		☐	
ENB	Standard PO DFPS	45	41		SAP	NBF		☐	
EUB	DFPS, Int. Ord. Type	45	41		SAP	UBF	T	☑	
FO	Framework Order	45	41		SAP	FOF		☐	SRV
NB	Standard PO	45	41		SAP	NBF		☐	
NB2	Enh. Rets to Vendor	45	41		SAP	NBF		☐	
NB2C	Enh. Rets STO CC	45	41		SAP	NBF2		☐	
NBC7	CC SiT Enh. Rets STO	45	41		SAP	NBF2		☐	
NBR8	IC SiT Enh. Rets STO	45	41		SAP	NBF2		☐	
NBXE	XLO Inter Com PO	45	41		SAP	NBF		☐	
NBXI	XLO Intra Company	45	41		SAP	NBF		☐	
UB	Stock Transp. Order	45	41		SAP	UBF	T	☐	
UB2	Enh. Rets STO IC	45	41		SAP	UBF2	T	☐	
ZNBF	PO f. flex. WF	45	41		SAP	NBF		☐	

Figure 12.1 Customizing: Purchase Order Types

- **Specific settings for scheduling agreements**
  You may use the document type to control whether a scheduling agreement is to be managed with or without release documentation. If you no longer know exactly what this means, refer back to Chapter 5.

  In the same way as for purchase orders, you can use the **Control** field for a document type for scheduling agreements to determine whether it is intended for a stock transport scheduling agreement. In this case, enter a **T** in the field. For a stock transfer scheduling agreement, you can also specify whether a vendor master record is required for the supplying plant.

  Another important field in Customizing for scheduling agreement document types is the indicator in which you specify whether time-dependent conditions are allowed.

- **Specific settings for contracts**
  Set the **Shared Lock Only** indicator to set a shared lock instead of an exclusive lock when a release order is created for a contract of this document type. This allows several users to issue release orders against the contract simultaneously. In doing so, however, the target quantity may be exceeded. If the document

type is intended for an ALE-distributed contract, set the corresponding indicator to track the changes.

- **Specific settings for purchase requisitions**
  By assigning the value **R** in the **Control** field for document types that are intended for purchase requisitions, you indicate that an outline agreement must be created as a follow-on document to the purchase requisition instead of a purchase order. By setting the **Overall Release** indicator, you specify that the purchase requisitions of this document type must be released at the header level and not at the item level.
- **Specific settings for RFQ**
  As for scheduling agreements, you use the document type to control whether time-dependent conditions are allowed for RFQs.

We have already learned that one factor that influences the field selection is the document type. Now we will examine the other factors that influence field selection and how we can change the field selection for purchasing transactions.

Screen Layout for Purchasing Documents

In Customizing, you can change the attributes of some fields depending on the field selection key and account assignment category.

Account Assignment Category

For each account assignment category, you can decide which fields that are relevant to account assignment in the purchase order item must be ready for input, mandatory, or hidden. You may want to refer back to Chapter 4, where we covered Customizing of the account assignment categories in detail.

Field Selection Key

A field selection key either corresponds one-to-one to an influencing factor or allows you to group together other influencing factors for the field selection control. For example, you can assign the same field selection key to document types that should have the same field selection control. You can then use the field selection keys to adjust attributes for fields in the header and in the items of the purchasing documents. Let's go directly to Customizing and look at the field selection keys that represent influencing factors.

Besides general Customizing nodes, the SAP Implementation Guide (IMG) also contains document category-specific Customizing; configuration of the document types and the field selection control are part of this. In Customizing, follow **Materials Management • Purchasing**. Figure 12.2 shows the nodes **Purchase Order** and **Contract** expanded as an example.

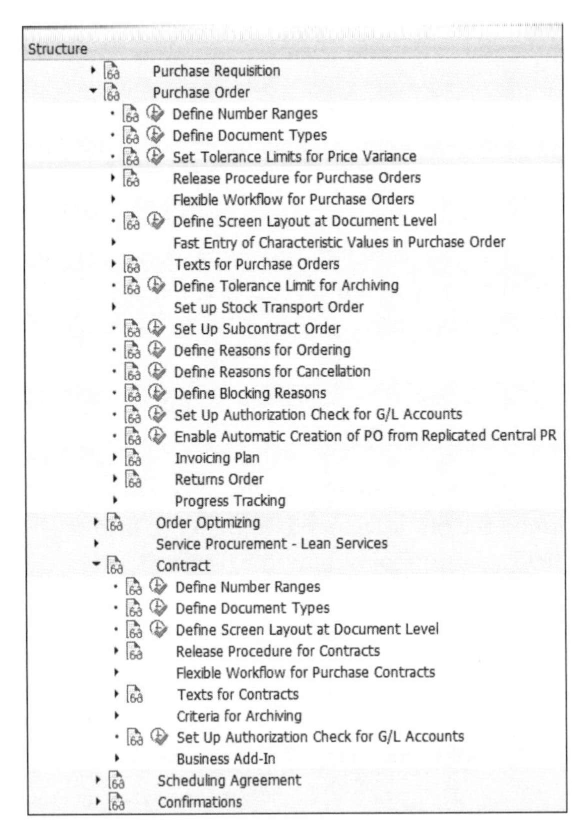

Figure 12.2 Document Category-Specific Customizing

If you choose **Define Screen Layout at Document Level** for a document category, you see the field selection keys available, as shown in Figure 12.3. This figure shows the field selection keys for the following influencing factors:

❶ **The user parameter EFB**
See the note box that follows the figure for more detail.

❷ **The activity category**
The four activity categories (for create, change, display, and extend) are selected internally within the program.

❸ **The transaction**
This field selection key has a fixed and non-changeable assignment to the transaction used.

❹ **The purchasing document type combined with the purchasing document category**
You define a field selection key in Customizing for the document type, which you carry out for each document category. For the standard document types, the field selection key often consists of a document type key and a document category key. The document category key is an internal key (e.g., for purchase orders, the key is F). However, this is not binding; when you create a new document type, you can define a field selection key of your choice for it.

❺ **The item category**

The item category, in combination with the document category, determines this field selection key:

- The first two characters of the field selection key are the letters PT. (This comes from *Positionstyp* in German, which means item category.)
- The third character is the item category (internal).
- The fourth character is the document category.

This character combination is fixed and cannot be changed. The characters assigned to the item categories are as follows:

- A for RFQs
- B for purchase requisitions
- F for purchase orders
- K for contracts
- L for scheduling agreements

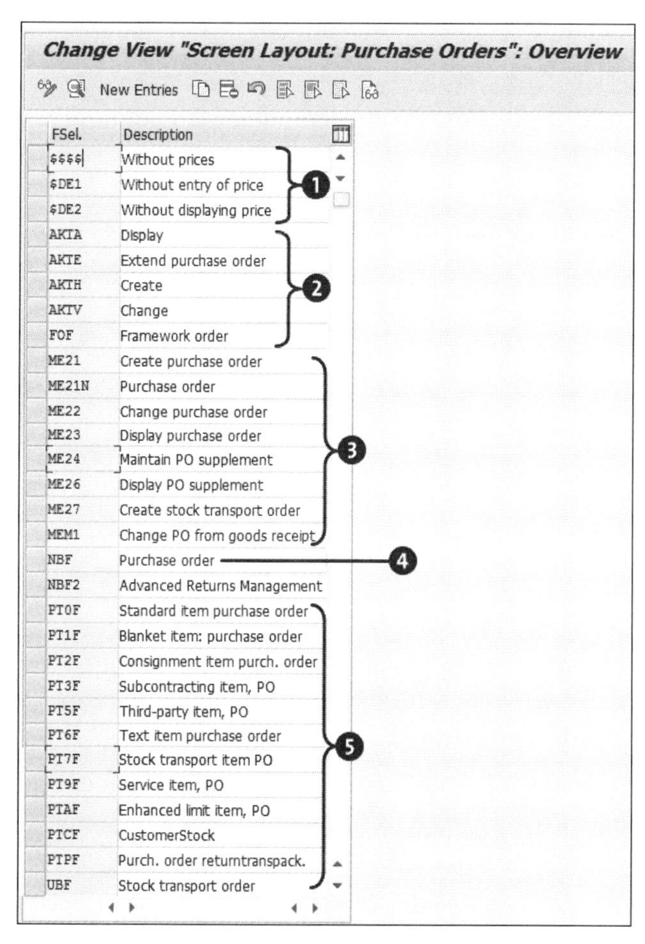

Figure 12.3 Customizing for Purchase Orders; Field Selection Keys

Note

The release status and special procurement transactions can also influence the field attributes.

Note

User authorization can also influence the field selection: in addition to the roles assigned to a user and defining his or her authorizations in the system, there is a user parameter in Purchasing where purchasing-specific authorizations can be assigned or restricted. For example, if a user is not allowed to see prices in a purchase order, according to parameter EFB, the field selection key $$$$ applies. The user parameter EFB is discussed in the next section.

If you select one of the field selection keys, you receive a list of field groups. Figure 12.4 shows the list of field selection groups for the field selection key NBF (which stands for purchase order type NB).

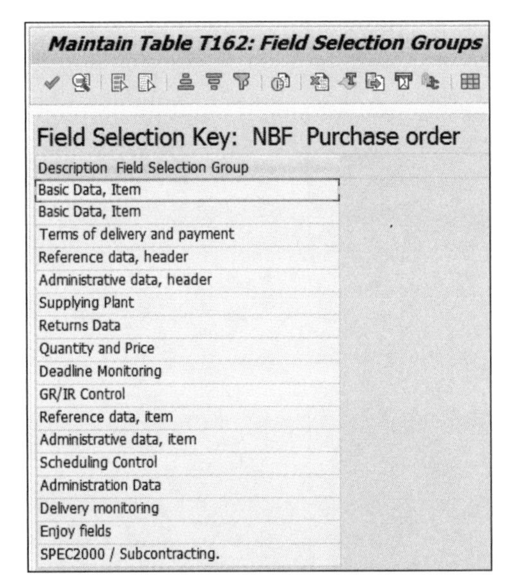

Figure 12.4 Customizing: Field Selection Groups

You then select a line and a list of the corresponding fields appears for which you can maintain the attributes. Figure 12.5 shows the fields corresponding to the basic data of a purchase order item and their attributes.

It may happen that the attribute of a field differs, depending on the influencing factor. In this case, the setting with the higher priority wins:

- Priority 1: Hide
- Priority 2: Display
- Priority 3: Required entry
- Priority 4: Optional entry

Maintain Table T162: Fields for Field Selection Group

Field Selection Key NBF Purchase order
Selection group Basic Data, Item

Fields

Field Label	Reqd.entry	Opt. entry	Display
Plant	☐	☑	☐
Item category	☐	☑	☐
Storage location	☐	☑	☐
Account assignment category	☐	☑	☐
Indicator: "Texts exist"	☐	☑	☐
Short text	☐	☑	☐
Material group	☐	☑	☐
Material description	☐	☑	☐
External service fields	☐	☐	☐
Value limit fields	☐	☐	☐
Manufacturer part number	☐	☑	☐
Advice Code	☐	☑	☐
Requirement Urgency	☐	☑	☐
Product Type Group	☐	☐	☐
Service Performer	☐	☐	☐
Overall Limit	☐	☐	☐

Figure 12.5 Customizing: Field Attributes

For example, if a field is defined as a mandatory field by a field selection key and as an optional field by another selection key, the mandatory attribute applies because it has a higher priority.

User Parameters

Certain Customizing settings can be made user-specific thanks to user parameters. For example, you can utilize the user parameter MSV to output some selected system messages as user-specific warnings or error messages, or even suppress them for some users. You first define both user parameters in Customizing for purchasing and then assign them to the relevant user by entering the keys for the parameter IDs in the user master record.

There are also two purchasing-specific user parameters that control some activities of buyers. These are the parameters EVO (default values for buyers) and EFB (function authorizations EVO). Let's take a look at each, starting with Figure 12.6.

Documentation | Address | Logon Data | SNC | Defaults | Parameters | Roles | Profiles | Groups | Personalization | License

Parameters

SET/GET Parameter ID	Parameter value	Short Description
EVO	01	Default Values for Purchasing
EFB	01	Function Authorization: Purchase Order

Figure 12.6 Parameters EVO and EFP in a User Master Record

Use the parameter EFB, as shown in Figure 12.7, to define whether a user with this parameter in their user master record can see, adopt, and change prices; use certain reference documents; and can carry out a source determination manually. In Customizing, follow **Materials Management • Purchasing • Environment Data • Define Default Values for Buyers.**

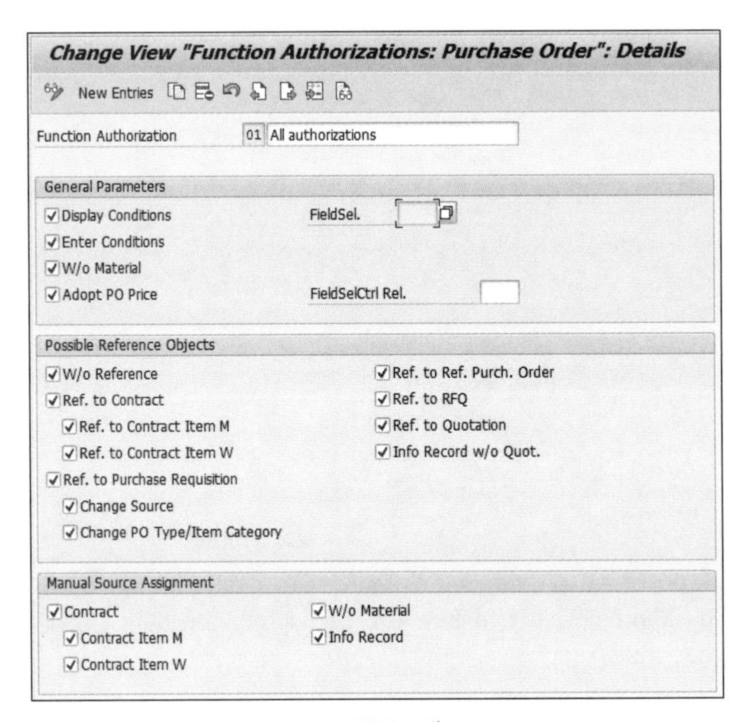

Figure 12.7 Customizing: Parameter EFB Details

The parameter EVO, which is shown in Figure 12.8, allows you to change the standard settings and default values for certain fields on a user-specific basis. Here some examples:

- With the EVO parameter, you can override the general rules for updating the purchasing info record from the purchase order or from the RFQ (see the general rules in Chapter 5).
- You can set the **Order Acknowledgment Requirement** indicator for purchase orders, scheduling agreements, and contracts.
- You can use the parameter EVO to prevent a gross price from being taken as a default value from the last purchase order.

Figure 12.8 shows the available tabs for the configuration of the EVO parameter. In Customizing, follow **Materials Management • Purchasing • Authorization Management • Define Function Authorization for Buyers.**

Now let's move on to some general configuration settings for inventory management.

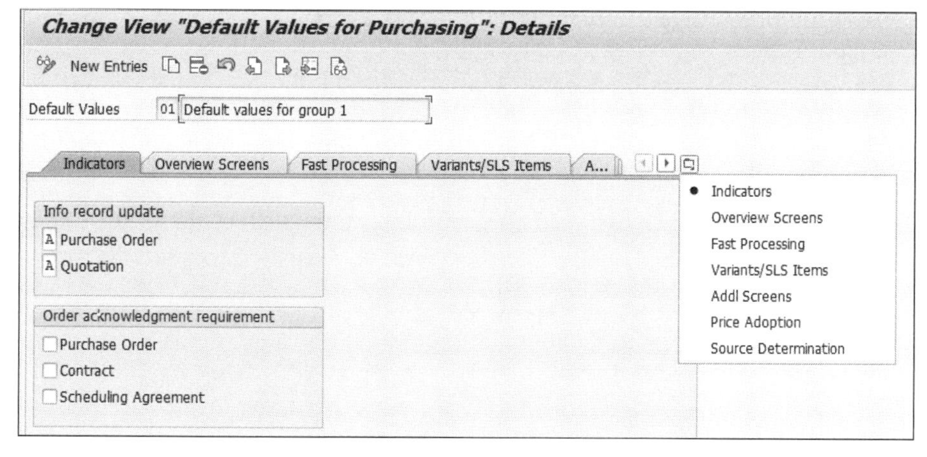

Figure 12.8 Customizing: Parameter EVO Details

Document Types in Inventory Management

We have already learned that a goods movement of a valuated material generates both a material document and an accounting document. Document types for accounting documents including number ranges can be defined in Customizing for financial accounting but also in Customizing for inventory management. In Customizing for inventory management, you also define which accounting document types are determined for which business transactions. Figure 12.9 shows the Customizing options for accounting documents in inventory management.

Document Types
Process the objects in the following sequence:
Set Up Document Types
Financial Accounting Document Types
Financial Accounting Number Ranges
Allocate Doc.Type to Transac.
Goods Issue Document Type
Goods Receipt Document Type
Automatic Movements Document Type
Physical Inventory Document Type

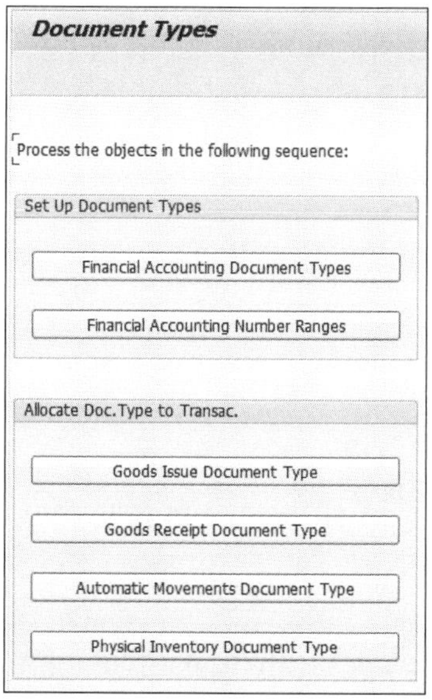

Figure 12.9 Customizing: Accounting Document Types

The number assignment of material and inventory documents is dependent on the transaction type. The transaction types can be grouped together, and you must define a number range interval for each group. Figure 12.10 illustrates this and highlights three key areas on the screen:

❶ The number range for group 01: **Inventory Documents**

❷ The name of group 02: **Material Documents for Goods Movement and Inventory Diffs.**

❸ The transaction types assigned to group 03: **Material Documents for Goods Receipt**

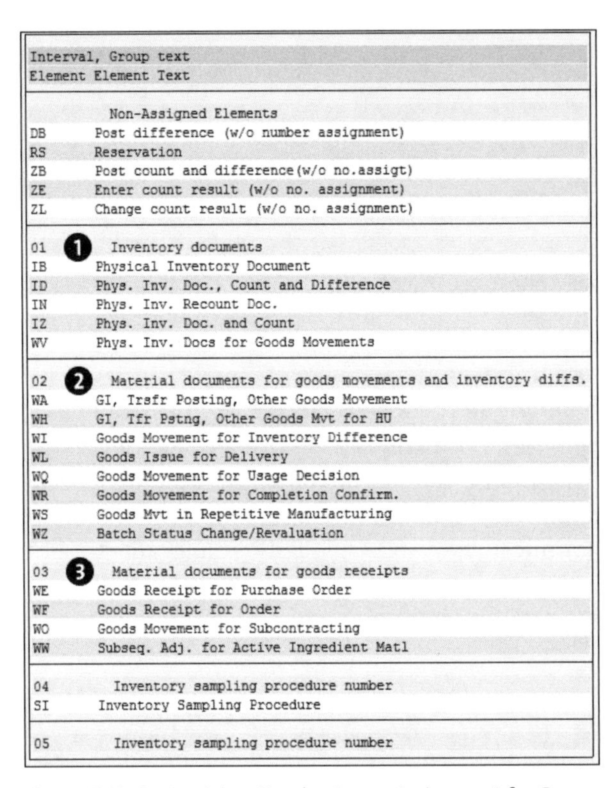

Interval, Group Element	text Element Text
	Non-Assigned Elements
DB	Post difference (w/o number assignment)
RS	Reservation
ZB	Post count and difference(w/o no.assigt)
ZE	Enter count result (w/o no. assignment)
ZL	Change count result (w/o no. assignment)
01	Inventory documents
IB	Physical Inventory Document
ID	Phys. Inv. Doc., Count and Difference
IN	Phys. Inv. Recount Doc.
IZ	Phys. Inv. Doc. and Count
WV	Phys. Inv. Docs for Goods Movements
02	Material documents for goods movements and inventory diffs.
WA	GI, Trsfr Posting, Other Goods Movement
WH	GI, Tfr Pstng, Other Goods Mvt for HU
WI	Goods Movement for Inventory Difference
WL	Goods Issue for Delivery
WQ	Goods Movement for Usage Decision
WR	Goods Movement for Completion Confirm.
WS	Goods Mvt in Repetitive Manufacturing
WZ	Batch Status Change/Revaluation
03	Material documents for goods receipts
WE	Goods Receipt for Purchase Order
WF	Goods Receipt for Order
WO	Goods Movement for Subcontracting
WW	Subseq. Adj. for Active Ingredient Matl
04	Inventory sampling procedure number
SI	Inventory Sampling Procedure
05	Inventory sampling procedure number

Figure 12.10 Customizing: Number Range Assignment for Groups of Transaction Types

You cannot change the transaction types. However, you can do three things:

- Create new groups.
- Change the assignment of transaction types to groups.
- Change the assignment of a number range to a group.

Movement Types

Movement types were covered in detail in Chapter 9. The movement type controls the goods movement. You must enter the movement type when you use the Post Goods Movement app (Transaction MIGO).

SAP Fiori apps are either dedicated to a specific goods movement or one determines the movement using other fields, so that the movement type does not have to be explicitly specified for SAP Fiori apps. However, it is determined in the background and is visible in the material document after posting.

As a refresher, here are the most important control fields of the movement type:

- Quantity and value update
- Account determination for the offsetting posting to the stock posting
- Field selection during goods movement entry
- Output determination

Figure 12.11, Figure 12.12, Figure 12.13, and Figure 12.14 show the customizing of movement type 201 as an example. Let's look at some of the most important fields available:

- **Check SL Expir. Date**: Check the shelf life expiration date.
- **Create SLoc. Automat.**: This automatically creates storage location data for a material master record at the time of the first goods receipt.
- **Automatic PO**: This indicator allows the automatic generation of a purchase order when you post a goods receipt without a reference.
- **Stck Determination Rule**: Stock determination strategies are defined depending on the stock determination rule assigned to a movement type and the stock determination group assigned to a material (see Chapter 9).
- **Consumption Posting**: This field determines whether a movement (usually a goods issue) with this movement type is included in the consumption statistics or not.
- **Control Reason**: Here you specify whether the input of a reason for movement is mandatory, possible, or unforeseen.

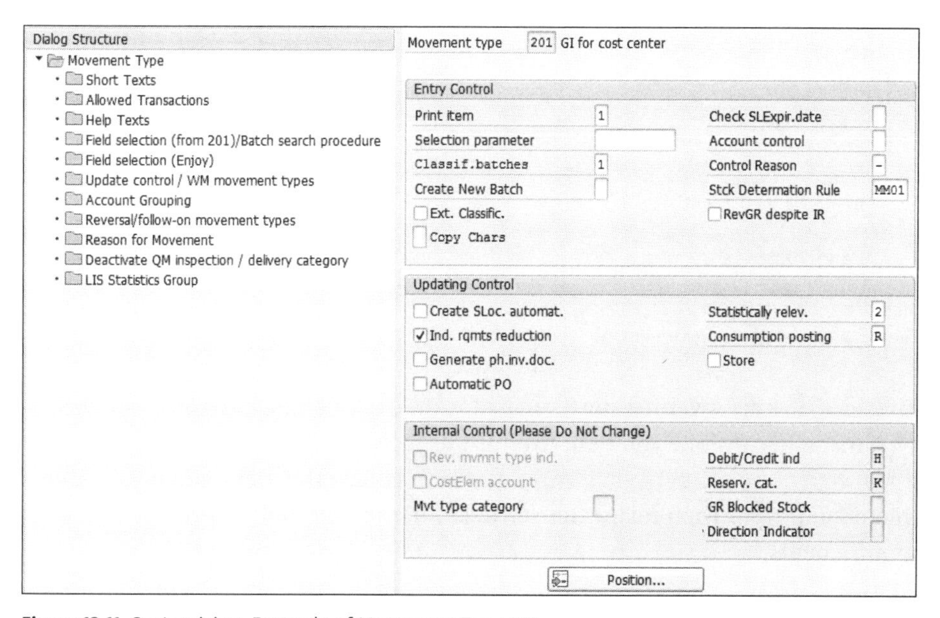

Figure 12.11 Customizing: Example of Movement Type 201

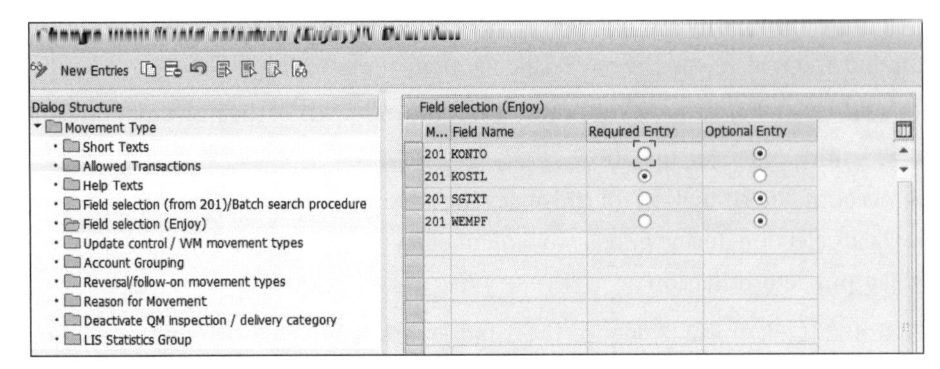

Figure 12.12 Customizing Movement Type: Field Selection

Change View "Account Grouping": Overview

Dialog Structure
- Movement Type
  - Short Texts
  - Allowed Transactions
  - Help Texts
  - Field selection (from 201)/Batch search procedure
  - Field selection (Enjoy)
  - Update control / WM movement types
  - Account Grouping
  - Reversal/follow-on movement types
  - Reason for Movement
  - Deactivate QM inspection / delivery category
  - LIS Statistics Group

M...	S	Val.upd...	Qty up...	Mvt	Cns	Val.strng	Cn	TEKey	Acct Mod.	Check acct.a
201	P	☐	☐			WA03	1	KON	PIP	☐
201	P	☐	☐			WA03	2	GBB	VBR	☑
201	K	☐	☑			WA03	2	GBB	VBR	☑
201	P	☐	☑			WA03	1	KON	PIP	☐
201	P	☐	☑			WA03	2	GBB	VBR	☑
201		☑	☐			WA01	2	GBB	VBR	☑
201		☑	☐			WA01	3	PRD	PRA	☐
201		☑	☑			WA01	2	GBB	VBR	☑
201		☑	☑			WA01	3	PRD	PRA	☐
201	K	☑	☑			WA03	2	GBB	VBR	☑
201	P	☑	☑			WA03	1	KON	PIP	☐
201	P	☑	☑			WA03	2	GBB	VBR	☑

Figure 12.13 Customizing Movement Type: Value String and Account Modification

Change View "Reversal/follow-on movement types": Overview

Dialog Structure
- Movement Type
  - Short Texts
  - Allowed Transactions
  - Help Texts
  - Field selection (from 201)/Batch search procedure
  - Field selection (Enjoy)
  - Update control / WM movement types
  - Account Grouping
  - Reversal/follow-on movement types
  - Reason for Movement
  - Deactivate QM inspection / delivery category
  - LIS Statistics Group

M...	FC	Description	M...	Type of posting
201	RL	Return Delivery	202	1
201	SS	Reversal (Detail)	202	2
201	SSR	Return Del. (Detail)	202	1
201	ST	Reversal	202	2

Figure 12.14 Customizing Movement Type: Reversal

To create a new movement type, first copy an existing one. Choose a movement type that is like the one you want to create, especially regarding quantity and value updating, to use as a reference. After you have copied the reference, save the new movement type. Then adapt the settings for the new movement type to your requirements.

Note that the key of the new movement type must begin with 9, X, Y, or Z. When copying the reference, make sure that you also copy all dependent entries of the reference movement type. It is best to also copy the reversal movement type of the referenced standard movement type and assign it to the new one.

Customizing for Material Master Data

In Chapter 2, we discussed the material master record. In that chapter, we analyzed the structure of the material master record, learned how to create and extend a material master record, and learned about which views are required for the procurement process. We also dealt with some important controlling fields, such as the material type, the industry sector, and the material status. If you need a refresher on these points, please read the "Material Master Data" section in Chapter 2 again (with particular attention paid to the discussion about material types). Remember that the material type controls shown in Figure 12.15 function in various areas, such as the following:

- The procurement type in procurement
- Quantity and value-based inventory management in the respective valuation areas for inventory management
- Account category reference and default values for price control during account determination
- Allowed views in the maintenance and usability of the materials

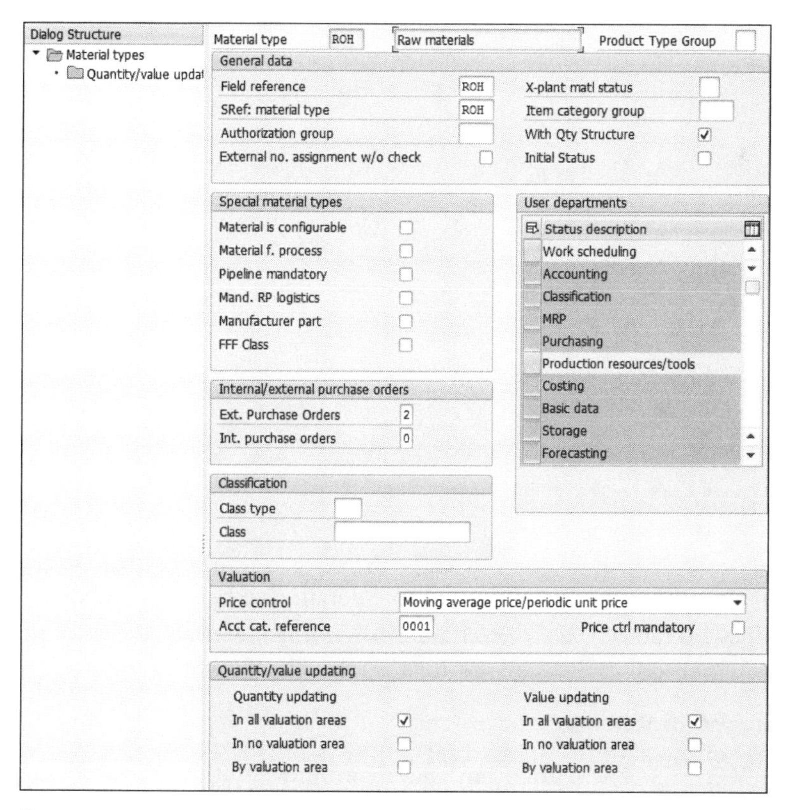

Figure 12.15 Customizing: Material Type

Remember to note the **Material Status** field (also explained in Chapter 2), which can restrict the usability of materials. Figure 12.16 shows an example of the definition of a material status (**BP**), for which the system does not allow any purchasing activities:

- The entry **B** for the **Purchasing** application causes the system to issue an error message in the event of a purchase transaction for a material with the status **BP**. If the system should issue a warning message instead of an error message, for purchase transactions, change the entry and choose **A** for the **Purchasing** application.
- The system allows goods movements and MRP, for example, because the **Inventory Management** and **Material Requirements** statuses are empty.

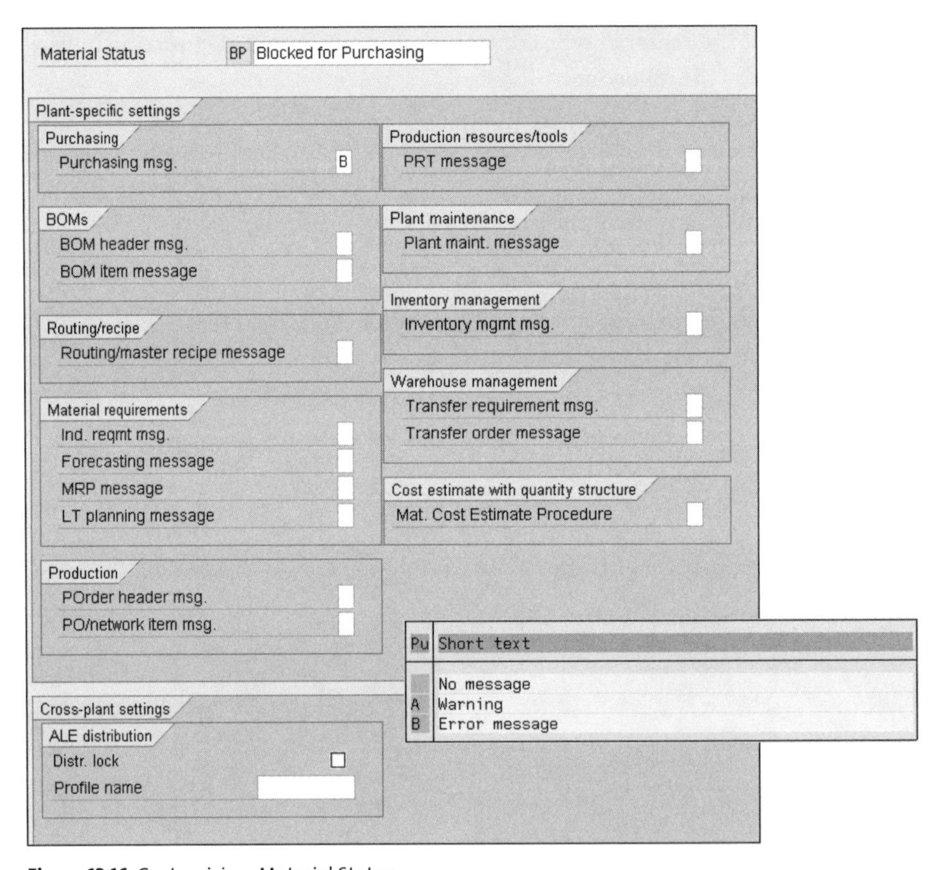

Figure 12.16 Customizing: Material Status

Now we will discuss the following topics in more detail: number assignment for materials and field and screen selection when maintaining materials.

Number Assignment for Materials

A material number in SAP S/4HANA can be a maximum of 40 characters long. Numerical numbers are limited to 18 characters.

The default setting, both after a transition to SAP S/4HANA or for a new installation, is that the length of the material number is 18 characters and the extended material number functionality is not activated. To use a material number with more than 18 characters, you must activate this functionality.

You should first define the internal and external number ranges you need for your materials, as shown in Figure 12.17.

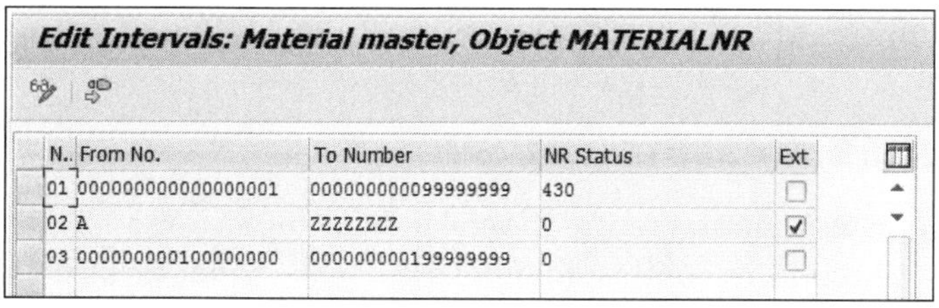

Edit Intervals: Material master, Object MATERIALNR

N..	From No.	To Number	NR Status	Ext
01	000000000000000001	000000000099999999	430	☐
02	A	ZZZZZZZZ	0	☑
03	000000000100000000	000000000199999999	0	☐

Figure 12.17 Materials Number Ranges

You can then define groups of material types and assign one internal and one external number range per group. Finally, assign your material types to the groups.

Figure 12.18 shows the group Group 1 assigned to the internal number range 01 and the external number range 02. The group includes the material types HALB, HIBE, and so forth.

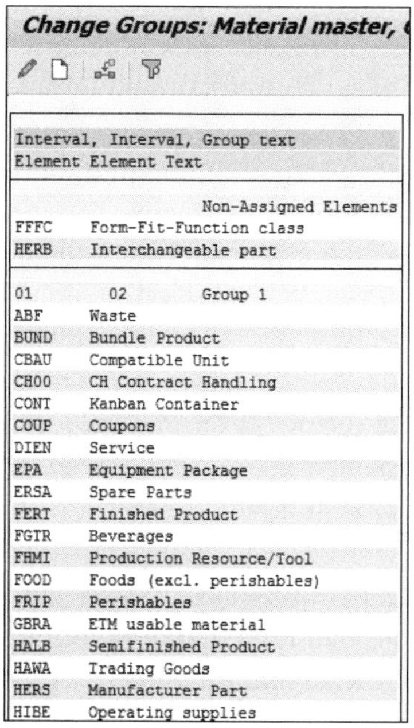

Change Groups: Material master,

Interval, Element	Interval, Element	Group text Text
		Non-Assigned Elements
FFFC	Form-Fit-Function class	
HERB	Interchangeable part	
01	02	Group 1
ABF	Waste	
BUND	Bundle Product	
CBAU	Compatible Unit	
CH00	CH Contract Handling	
CONT	Kanban Container	
COUP	Coupons	
DIEN	Service	
EPA	Equipment Package	
ERSA	Spare Parts	
FERT	Finished Product	
FGTR	Beverages	
FHMI	Production Resource/Tool	
FOOD	Foods (excl. perishables)	
FRIP	Perishables	
GBRA	ETM usable material	
HALB	Semifinished Product	
HAWA	Trading Goods	
HERS	Manufacturer Part	
HIBE	Operating supplies	

Figure 12.18 Customizing: Material Groups for Number Assignment

Figure 12.19 shows the group S4550 assigned to the internal number range 03. The group includes the material type ROH.

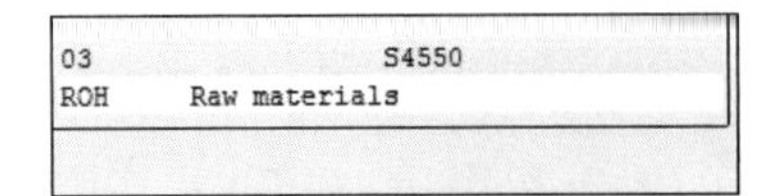

Figure 12.19 Customizing: Material Groups for Number Assignment

In Customizing for a material type (see Figure 12.15), select the checkbox **External No. Assignment w/o Check** to specify that the system should not check whether an external material number is within a number range.

Field and Screen Selection for Materials Master Records

When you maintain a material master record, several influencing factors apply for determining the field selection. In contrast to other areas, you cannot make field selection settings for each individual field of material master records but for groups of fields. In material master record maintenance, the field attributes are controlled by the field selection group to which a field belongs.

For the factors that influence the field selection in the material master, there are so-called field references in the system. You can use these keys to control the field selection, depending on the influencing factor.

The influencing factors for the field selection control in material master maintenance are as follows:

- **The transaction**
  The field selection control logically depends on whether you create, change or display a material. The field references are named identically to the transactions, such as MM01, MM02, and so forth.
- **The material type**
  You can define field references for the material types in Customizing. You use this key to control the field selection, depending on the material type. You can group material types together from the point of view of field selection if you assign the same field reference to several material types.
- **The industry sector**
  You can define field references for the industry sectors in Customizing and group industry sectors together from the point of view of field selection if you assign the same field reference to several industry sectors.
- **The plant**
  You can define field references for the plants in Customizing, as shown in Figure 12.20. Use this key to control the field selection, depending on the plant. You can group plants together from the point of view of field selection if you assign the same field reference to several plants.
- **The procurement indicator**
  The procurement type can be E for inhouse production or F for external procurement.

- **The SAP standard delivery**
  The field selection according to the SAP standard delivery is controlled by the field references starting with SAP and should not be changed.

Change View "Field Selection at Plant Level": Overview

New Entries

Plant	Name 1	Maintenance status	Field reference
0001	Werk 0001	KDEVALBPQSZXCFG	0001
0003	Plant 0003 (is-ht-sw)	KDEVALBPQSZXCFG	0001
1010	Hamburg	KDEVALBPQSZXCFG	0001
1030	Frankfurt	KDEVALBPQSZXCFG	0001
1400	Stuttgart	KDEVALBPQSZXCFG	0001

Figure 12.20 Customizing: Field References for Plants

Figure 12.21 shows field group 52, some fields that belong to it, and then in the lower part, the field selection control per field reference.

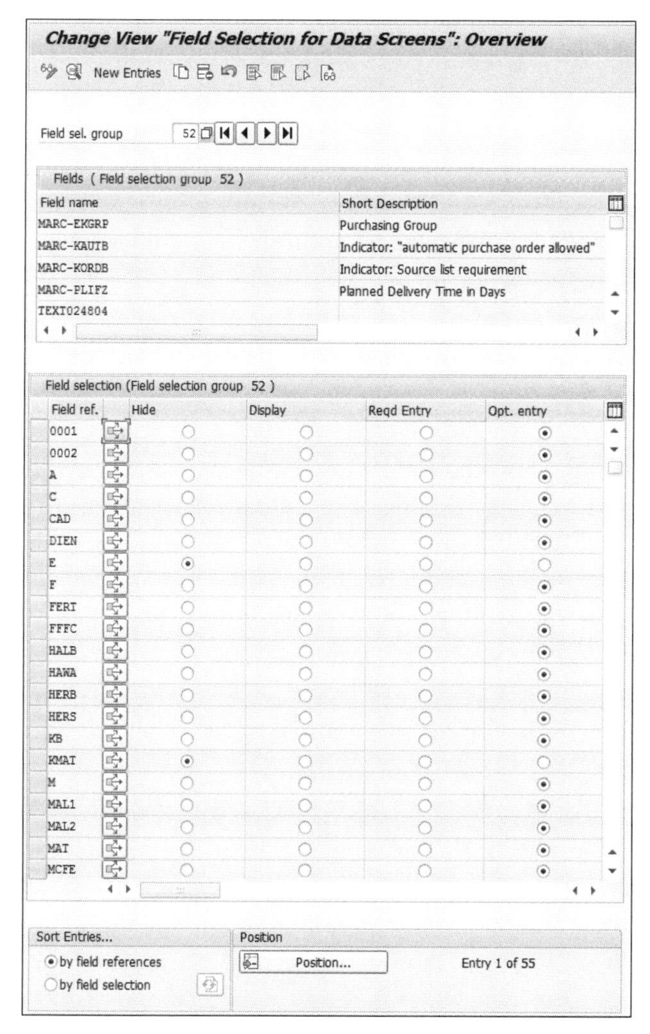

Figure 12.21 Customizing: Example of a Field Group

Important Terminology

The following important terminology was used in this chapter.

- **Document type**
  The document type characterizes the business transaction and controls number assignment and field selection, among other things.
- **Document category**
  The document type in purchasing is assigned internally and determines whether the document is a purchase order, a purchase requisition, an RFQ, a contract, or a scheduling agreement.
- **Field selection key**
  The field selection key in purchasing represents the influencing factors for the field selection control and allows you to group them.
- **User parameter**
  The user parameter EVO controls default values in purchasing. The user parameter EF controls authorizations.
- **Movement type**
  In inventory management, a material movement is defined by the movement type.
- **Transaction type**
  The transaction type in inventory management controls the number assignment of material and inventory documents.
- **Material type**
  Materials that have similar properties are categorized together and assigned to a material type
- **Material status**
  The material status determines how the system deals with the material in different applications, such as purchasing, MRP, or inventory management. The material status enables you to restrict the usability of a material for specific business applications. The material status is defined in Customizing and entered in the material master record for a specific plant or for all plants.
- **Field group**
  For field selection control in material master record maintenance, material fields are grouped together. The field selection control is set per group.
- **Field reference**
  The field reference represents the influencing factors for the field selection control for maintaining material master records.

Practice Questions

These questions will help you evaluate your understanding of the topics covered in this chapter. They are similar in nature to those on the certification examination. Although none of these questions will be found in the exam itself, they will allow you to review your knowledge of the subject.

Select the correct answers, and then check the completeness of your answers in the next section. Remember that on the exam, you must select all correct answers—and only correct answers—to receive credit for the question.

1. True or false: A document type for a purchase requisition and a document type for a purchase order can have the same key.

☐ **A.** True

☐ **B.** False

2. Say you want subcontracting purchase orders to be assigned a separate specific number range. What must you do?

☐ **A.** Create the required number range. Assign the required number range to the item category Subcontracting.

☐ **B.** Create the required number range as an external number range. Assign this external number range to the order type(s) for which the item category Subcontracting is permitted.

☐ **C.** Create the required number range. Create a new purchase order type, assign the required number range, and only allow the item category Subcontracting.

☐ **D.** Create the required number range. Assign this number range to a purchase order type that is linked to a combination of any purchase requisition type and the item category Subcontracting.

3. For which of the following reasons must you create a new document type for purchase orders? (There are three correct answers.)

☐ **A.** Assignment of specific numbers

☐ **B.** Limitation of the permitted account assignments

☐ **C.** Requirement to make certain order header fields mandatory

☐ **D.** Limitation of the permitted procurement processes

☐ **E.** Activation of the time-dependent conditions

4. Which of the following influencing factors can you use to declare a field in the purchase order header that was previously only ready for input as a mandatory field? (There are three correct answers.)

☐ **A.** The activity

☐ **B.** The item category

☐ **C.** The transaction

☐ D The account assignment category

☐ **E.** The document type

5. You want to prevent a material from being procured externally for a certain time. Which field can you use for this?

☐ **A.** Material type

☐ **B.** User parameter EFB

☐ **C.** Material status

☐ **D.** Field selection key

6. Which of the following could be reasons for setting up a custom material type? (There are three correct answers.)

☐ **A.** You want to create configurable materials with a specific external number range.

☐ **B.** You want to temporarily prevent goods movements for certain materials.

☐ **C.** You need materials for which only the Purchasing, MRP, and Accounting views can be created, and only external procurement is allowed.

☐ **D.** You want to make the moving average price mandatory for a category of materials.

☐ **E.** You need differentiated requirements planning within the plant.

7. To which objects are material number ranges assigned?

☐ **A.** Material types

☐ **B.** Groups of material types

☐ **C.** Material groups

☐ **D.** Group of material fields

8. For which purchasing document categories can you define in Customizing of the document type whether time-dependent conditions are allowed? (There are two correct answers.)

☐ **A.** Purchase order

☐ **B.** RFQ

☐ **C.** Purchase requisition

☐ **D.** Contract

☐ **E.** Scheduling agreement

Practice Question Answers and Explanations

1. Correct answer: **A**

   True. Document types are defined per document category. The combination document category document type must be unique, but you can use the same document type key for different categories. This is the case in the standard system: document type NB is used for both purchase requisitions and purchase orders.

2. Correct answer: **C**

   Create the required number range and a new purchase order type; then assign the new number range to the new document type and only allow the item category Subcontracting for this document type. If you create an external number range and assign a number from this number range every time you create a subcontracting purchase order, you can achieve your goal; however, it is cumbersome and there is no guarantee that non-subcontracting purchase orders will not also receive a number from this range.

3. Correct answers: **A, C, D**

   Time-dependent conditions can only be activated for scheduling agreements and requests for quotations. The document type is not relevant for account assignment, but the account assignment category is. The document type influences number assignment, field selection control, and the permitted item categories (procurement processes).

4. Correct answers: **A, C, E**

   For field selection at the header level, you can select the field selection keys that are assigned to the activity, the transaction, or the document type. Using the account assignment category, you only influence item fields that are relevant for account assignment. The field selection key for the item category only allows you to influence item fields. Note that not all fields are available for control. Some fields are permanently controlled by the process itself. For example, you cannot make the price of a consignment purchase order a mandatory entry.

5. Correct answer: **C**

   The material status allows you to limit the usability of a material per application.

6. Correct answers: **A, C, D**

   The material type allows you to control, among other things, whether the material to be created has the following characteristics:

- Is a special material (like configurable or manufacturer part)
- Can be created for certain views
- Should be valuated with the moving average price or with the standard price, and whether this setting is mandatory
- Can only be externally procured, only manufactured in-house, or both

You also control number assignment via the material type. The material status is suitable for limiting the usability of a material for a limited period (no goods movements, for example). If you require differentiated MRP within a plant, you must set up appropriate MRP areas.

7. Correct answer: **B**

   For number assignment, material types are merged into groups and number ranges are assigned to these groups.

8. Correct answers: **B, E**

   Purchase requisitions and purchase orders contain conditions that apply to the business process mapped with them: prices/discounts/surcharges are valid for the quantity and date recorded in the document. These are not time-dependent conditions.

   In contracts, you can always store time-dependent conditions (with validity periods and scales) independently of Customizing. For RFQs and scheduling agreements, this depends on the document type.

Takeaway

This chapter described how to control number assignment for materials, purchasing documents, and material documents. You should be familiar with the special significance of some key fields: the material type in material master record maintenance, the document type in applications in general, and the movement type in goods movements. In addition, you should now understand how to control field selection in these important areas.

Summary

This chapter brings together individual customizing topics that could not be incorporated into the individual process chapters. Here you will find individual topics such as user parameters in purchasing, a summary of the properties and tasks of the movement type, and general topics such as number assignment and field selection. The significance of some key fields, such as document type and material type, has been repeated and, in some cases, deepened.

We have reached the end of this certification study guide. This chapter was the last piece of a complex and diverse puzzle that you must master in order to pass the certification exam for sourcing and procurement in SAP S/4HANA. If you feel comfortable with the presented material and can answer the practice questions with confidence and a real understanding of the solutions presented herein, you should be able to successfully complete your certification exam. We hope that this book will also be useful to you beyond the certification as an easy-to-understand piece of reference material.

The Author

Fabienne Bourdelle is a graduate of the ESCP Europe Business School. She started her career at SAP as a consultant and specialized in the support of SAP implementation projects, her area of expertise being materials management. She is now learning architect at SAP Training and Adoption and owner of the associate certification exams for SAP S/4HANA Sourcing and Procurement.

Index

T

U

V

W

- Configure purchasing, sourcing, invoicing, evaluation, and more
- Run your sourcing and procurement processes in SAP S/4HANA
- Analyze your procurement operations

Justin Ashlock

Sourcing and Procurement with SAP S/4HANA

Your comprehensive guide to SAP S/4HANA sourcing and procurement is here! Get step-by-step instructions to configure sourcing, invoicing, supplier management and evaluation, and centralized procurement. Learn how to integrate SAP S/4HANA with SAP Ariba, SAP Fieldglass, and more. Then, expertly run your system after go-live with predictive analysis and machine learning. See the future of sourcing and procurement!

716 pages, 2nd edition, pub. 02/2020
E-Book: $79.99 | **Print:** $89.95 | **Bundle:** $99.99

www.sap-press.com/5003